Machine Learning, Low-Rank Approximations and Reduced Order Modeling in Computational Mechanics

Machine Learning, Low-Rank Approximations and Reduced Order Modeling in Computational Mechanics

Special Issue Editors

Felix Fritzen
David Ryckelynck

MDPI • Basel • Beijing • Wuhan • Barcelona • Belgrade

Special Issue Editors
Felix Fritzen
University of Stuttgart
Germany

David Ryckelynck
ParisTech
France

Editorial Office
MDPI
St. Alban-Anlage 66
4052 Basel, Switzerland

This is a reprint of articles from the Special Issue published online in the open access journal *Mathematical and Computational Applications* (ISSN 2297-8747) from 2018 to 2019 (available at: https://www.mdpi.com/journal/mca/special_issues/rom_comput_mech)

For citation purposes, cite each article independently as indicated on the article page online and as indicated below:

LastName, A.A.; LastName, B.B.; LastName, C.C. Article Title. *Journal Name* **Year**, *Article Number*, Page Range.

ISBN 978-3-03921-409-9 (Pbk)
ISBN 978-3-03921-410-5 (PDF)

Contents

About the Special Issue Editors vii

Preface to "Machine Learning, Low-Rank Approximations and Reduced Order Modeling in Computational Mechanics" ix

Patrick Buchfink, Ashish Bhatt, Bernard Haasdonk
Symplectic Model Order Reduction with Non-Orthonormal Bases
Reprinted from: *mca* **2019**, *24*, 43, doi:10.3390/mca24020043 1

Felix Selim Göküzüm, Lu Trong Khiem Nguyen, Marc-André Keip
An Artificial Neural Network Based Solution Scheme for Periodic Computational Homogenization of Electrostatic Problems
Reprinted from: *mca* **2019**, *24*, 40, doi:10.3390/mca24020040 27

Julian Lißner and Felix Fritzen
Data-Driven Microstructure Property Relations
Reprinted from: *mca* **2019**, *24*, 57, doi:10.3390/mca24020057 55

Clément Olivier, David Ryckelynck, Julien Cortial
Multiple Tensor Train Approximation of Parametric Constitutive Equations in Elasto-Viscoplasticity
Reprinted from: *mca* **2019**, *24*, 17, doi:10.3390/mca24010017 82

Oliver Kunc, Felix Fritzen
Finite Strain Homogenization Using a Reduced Basis and Efficient Sampling
Reprinted from: *mca* **2019**, *24*, 56, doi:10.3390/mca24020056 99

William Hilth, David Ryckelynck and Claire Menet
Data Pruning of Tomographic Data for the Calibration of Strain Localization Models
Reprinted from: *mca* **2019**, *24*, 18, doi:10.3390/mca24010018 126

Benjamin Brands, Denis Davydov, Julia Mergheim and Paul Steinmann
Reduced-Order Modelling and Homogenisation in Magneto-Mechanics: A Numerical Comparison of Established Hyper-Reduction Methods
Reprinted from: *mca* **2019**, *24*, 20, doi:10.3390/mca24010020 151

Nissrine Akkari, Fabien Casenave, Vincent Moureau
Time Stable Reduced Order Modeling by an Enhanced Reduced Order Basis of the Turbulent and Incompressible 3D Navier–Stokes Equations
Reprinted from: *mca* **2019**, *24*, 45, doi:10.3390/mca24020045 179

Shadi Alameddin, Amélie Fau, David Néron, Pierre Ladevèze, Udo Nackenhorst
Toward Optimality of Proper Generalised Decomposition Bases
Reprinted from: *mca* **2019**, *24*, 30, doi:10.3390/mca24010030 200

Fabien Casenave, Nissrine Akkari
An Error Indicator-Based Adaptive Reduced Order Model for Nonlinear Structural Mechanics—Application to High-Pressure Turbine Blades
Reprinted from: *mca* **2019**, *24*, 41, doi:10.3390/mca24020041 214

About the Special Issue Editors

Felix Fritzen received diploma degrees in Mechanical Engineering as well as in Mathematics in Technology from the Karlsruhe Institute of Technology (KIT). He obtained a PhD degree with distinction at the KIT in 2011. Thereafter he led the KIT Young Investigator Group Computer Aided Material Modeling from 2012-2015. Since 2015 he is the leader of the DFG Emmy Noether Group EMMA—Efficient Methods for Mechanical Analysis. He is about to install a new DFG Heisenberg-professorship Data-Analytics in Engineering at the University of Stuttgart. His research is devoted to computational mechanics of materials with a focus on innovative algorithms. Core technologies involved in his methods are model order reduction, data analysis, FFT-accelerated solvers and innovative kernel regression and interpolation. More recently, machine learning, e.g., in terms of artifical neural networks, has been a significant driver in his research. His publication list comprises 42 articles. His 23 articles recorded in Web of Science have a total of 406 citations excluding self-cites and an h-index of 12 is reported.

David Ryckelynck graduated and received his doctorate in Mechanics form Ecole Normale Supérieur de Cachan in 1993 and 1998 respectively. He is full professor in Mechanics of Mines ParisTech PSL University since 2013. He has been professor in this university during 2008–2013, and assistant professor of Arts et Métiers ParisTech during 1998–2008. He did seminal works on hyper-reduction methods, also termed hyperreduction, in the field of applied mathematics and computational mechanics. Without counting for self citations, hyper-reduction has been cited 544 times in peer reviewed papers, according to Web of Science. Hyper-reduction methods can be seen as an extension of non supervised machine learning for dimension reduction to the field of model order reduction for the approximate solution of partial differential equations. He also did seminal works on "a priori model reduction" with 227 citations in peer reviewed paper without counting self citations. He has been the supervisor of 25 PhD theses in computational mechanics. He was the head of a research team involving 12 permanent researchers at Mines ParisTech, from 2011 to 2015. He is the head of a new lecture on Ingénierie Digitale Des Systèmes Complexes (Data Science for Computational Engineering) at Mines ParisTech PSL University. This lecture has been created by Pr. David Ryckelynck and Pr. Elie Hachem in 2017. In this lecture, a team of 10 lecturers are teaching to students of Mines ParisTech the way that simulation data and experimental data can be used in machine learning. We focus our attention to hybrid approaches that combines machine learning and usual modeling methods in engineering.

Preface to "Machine Learning, Low-Rank Approximations and Reduced Order Modeling in Computational Mechanics"

The use of machine learning in mechanics is booming. Algorithms inspired by developments in the field of artificial intelligence today cover increasingly varied fields of application. This book illustrates recent results on coupling machine learning with computational mechanics, particularly for the construction of surrogate models or reduced order models. The articles contained in this compilation were presented at the EUROMECH Colloquium 597, "Reduced Order Modeling in Mechanics of Materials", held in Bad Herrenalb, Germany, from August 28th to August 31th 2018.

The Colloquium hosted a total of around 40 people including guest speakers for particular slots. The scientific aim of relating machine learning, model order reduction, data-driven modeling and simulation, and error estimation was fully achieved due to the very high quality of the 33 oral presentations and the impressive number of 14 accompanying posters which were extensively discussed during a separate time slot. The ten best papers are presented here in this book.

Felix Fritzen, David Ryckelynck
Special Issue Editors

Mathematical and Computational Applications

MDPI

Article

Symplectic Model Order Reduction with Non-Orthonormal Bases

Patrick Buchfink [1,*], Ashish Bhatt [2] and Bernard Haasdonk [1]

1 Institute of Applied Analysis and Numerical Simulation, University of Stuttgart, 70569 Stuttgart, Germany; bernard.haasdonk@ians.uni-stuttgart.de
2 Indian Institute of Technology (ISM), Dhanbad, Jharkhand 826004, India; ashishbhatt@iitism.ac.in
* Correspondence: patrick.buchfink@ians.uni-stuttgart.de; Tel.: +49-711-685-64778

Received: 26 February 2019; Accepted: 17 April 2019; Published: 21 April 2019

Abstract: Parametric high-fidelity simulations are of interest for a wide range of applications. However, the restriction of computational resources renders such models to be inapplicable in a real-time context or in multi-query scenarios. Model order reduction (MOR) is used to tackle this issue. Recently, MOR is extended to preserve specific structures of the model throughout the reduction, e.g., structure-preserving MOR for Hamiltonian systems. This is referred to as symplectic MOR. It is based on the classical projection-based MOR and uses a symplectic reduced order basis (ROB). Such an ROB can be derived in a data-driven manner with the Proper Symplectic Decomposition (PSD) in the form of a minimization problem. Due to the strong nonlinearity of the minimization problem, it is unclear how to efficiently find a global optimum. In our paper, we show that current solution procedures almost exclusively yield suboptimal solutions by restricting to orthonormal ROBs. As a new methodological contribution, we propose a new method which eliminates this restriction by generating non-orthonormal ROBs. In the numerical experiments, we examine the different techniques for a classical linear elasticity problem and observe that the non-orthonormal technique proposed in this paper shows superior results with respect to the error introduced by the reduction.

Keywords: symplectic model order reduction; proper symplectic decomposition (PSD); structure preservation of symplecticity; Hamiltonian system

1. Introduction

Simulations enable researchers of all fields to run virtual experiments that are too expensive or impossible to be carried out in the real world. In many contexts, high-fidelity models are indispensable to represent the simulated process accurately. These high-fidelity simulations typically come with the burden of a large computational cost such that an application in real time or an evaluation for many different parameters is impossible respecting the given restrictions of computational resources at hand. Model order reduction (MOR) techniques can be used to reduce the computational cost of evaluations of the high-fidelity model by approximating these with a surrogate reduced-order model (ROM) [1].

One class of high-fidelity models are systems of ordinary differential equations (ODEs) with a high order, i.e., a high dimension in the unknown variable. Such models typically arise from fine discretizations of time-dependent partial differential equations (PDEs). Since each point in the discretization requires one or multiple unknowns, fine discretizations with many discretization points yield a system of ODEs with a high order. In some cases, the ODE system takes the form of a finite-dimensional Hamiltonian system. Examples are linear elastic models [2] or gyro systems [3].

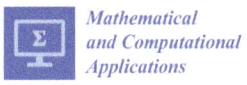

Symplectic MOR [4] allows for deriving a ROM for high-dimensional Hamiltonian systems by lowering the order of the system while maintaining the Hamiltonian structure. Thus, it is also referred

to as structure-preserving MOR for Hamiltonian systems [5]. Technically speaking, a Petrov–Galerkin projection is used in combination with a symplectic reduced-order basis (ROB).

For a data-driven generation of the ROB, the conventional methods, e.g., the Proper Orthogonal Decomposition (POD) [1] is not suited since they do not necessarily compute a symplectic ROB. To this end, the referenced works introduce the Proper Symplectic Decomposition (PSD) which is a data-driven basis generation technique for symplectic ROBs. Due to the high nonlineariy of the optimization problem, an efficient solution strategy is yet unknown for the PSD. The existing PSD methods (Cotangent Lift, Complex Singular Value Decomposition (Complex SVD)), a nonlinear programming approach [4] and a greedy procedure introduced in [5]) each restrict to a specific subset of symplectic ROBs from which they select optimal solutions which might be globally suboptimal.

The present paper classifies the existing symplectic basis generation techniques in two classes of methods which either generate orthonormal or non-orthonormal bases. To this end, we show that the existing basis generation techniques for symplectic bases almost exclusively restrict to orthonormal bases. Furthermore, we prove that Complex SVD is the optimal solution of the PSD on the set of orthonormal, symplectic bases. During the proof, an alternative formulation of the Complex SVD for symplectic matrices is introduced. To leave the class of orthonormal, symplectic bases, we propose a new basis generation technique, namely the PSD SVD-like decomposition. It is based on an SVD-like decomposition of arbitrary matrices $\boldsymbol{B} \in \mathbb{R}^{n \times 2m}$ introduced in [6].

This paper is organized in the following way: Section 2 is devoted to the structure-preserving MOR for autonomous and non-autonomous, parametric Hamiltonian systems and thus introduces symplectic geometry, Hamiltonian systems and symplectic MOR successively. The data-driven generation of a symplectic ROB with PSD is discussed in Section 3. The numerical results are presented and elaborated in Section 4 exemplified by a Lamé–Navier type elasticity model which we introduce at the beginning of that section together with a short comment on the software that is used for the experiments. The paper is summarized and concluded in Section 5.

2. Symplectic Model Reduction

Symplectic MOR for autonomous Hamiltonian systems is introduced in [4]. We repeat the essentials for the sake of completeness and to provide a deeper understanding of the methods used. In the following, $\boldsymbol{\mu} \in \mathcal{P} \subset \mathbb{R}^p$ describes $p \in \mathbb{N}$ parameters of the system from the parameter set $\mathcal{P}$. We might skip the explicit dependence on the parameter vector $\boldsymbol{\mu}$ if it is not relevant in this specific context.

2.1. Symplectic Geometry in Finite Dimensions

Definition 1 (Symplectic form over $\mathbb{R}$). *Let $\mathbb{V}$ be a finite-dimensional vector space over $\mathbb{R}$. We consider a skew-symmetric and non-degenerate bilinear form $\omega : \mathbb{V} \times \mathbb{V} \to \mathbb{R}$, i.e., for all $\boldsymbol{v}_1, \boldsymbol{v}_2 \in \mathbb{V}$, it holds*

$$\omega(\boldsymbol{v}_1, \boldsymbol{v}_2) = -\omega(\boldsymbol{v}_2, \boldsymbol{v}_1) \qquad \text{and} \qquad \omega(\boldsymbol{v}_2, \boldsymbol{v}_3) = 0 \quad \forall \boldsymbol{v}_3 \in \mathbb{V} \implies \boldsymbol{v}_3 = \boldsymbol{0}.$$

The bilinear form ω is called symplectic form on $\mathbb{V}$ and the pair $(\mathbb{V}, \omega)$ is called symplectic vector space.

It can be shown that $\mathbb{V}$ is necessarily of even dimension [7]. Thus, $\mathbb{V}$ is isomorphic to $\mathbb{R}^{2n}$ which is why we are restricted to $\mathbb{V} = \mathbb{R}^{2n}$ and write ω_{2n} instead of ω as follows. In context of the theory of Hamiltonians, $\mathbb{R}^{2n}$ refers to the phase space which consists, in the context of classical mechanics, of position states $\boldsymbol{q} = [q_1, \dots, q_n]^{\mathsf{T}} \in \mathbb{R}^n$ of the configuration space and momentum states $\boldsymbol{p} = [p_1, \dots, p_n]^{\mathsf{T}} \in \mathbb{R}^n$ which form together the state $\boldsymbol{x} = [q_1, \dots, q_n, p_1, \dots, p_n]^{\mathsf{T}} \in \mathbb{R}^{2n}$.

It is guaranteed [7] that there exists a basis $\{\boldsymbol{e}_1, \dots, \boldsymbol{e}_n, \boldsymbol{f}_1, \dots, \boldsymbol{f}_n\} \subset \mathbb{R}^{2n}$ such that the symplectic form takes the canonical structure

$$\omega_{2n}(\boldsymbol{v}_1, \boldsymbol{v}_2) = \boldsymbol{v}_1^{\mathsf{T}} \mathbb{J}_{2n} \boldsymbol{v}_2 \quad \forall \boldsymbol{v}_1, \boldsymbol{v}_2 \in \mathbb{R}^{2n}, \qquad \mathbb{J}_{2n} := \begin{bmatrix} \boldsymbol{0}_n & \boldsymbol{I}_n \\ -\boldsymbol{I}_n & \boldsymbol{0}_n \end{bmatrix}, \tag{1}$$

where $I_n \in \mathbb{R}^{n\times n}$ is the identity matrix, $0_n \in \mathbb{R}^{n\times n}$ is the matrix of all zeros and $\mathbb{J}_{2n}$ is called Poisson matrix. Thus, we restrict to symplectic forms of the canonical structure in the following. For the Poisson matrix, it holds for any $v \in \mathbb{R}^{2n}$

$$\mathbb{J}_{2n}\mathbb{J}_{2n}^{\mathsf{T}} = I_{2n}, \qquad \mathbb{J}_{2n}\mathbb{J}_{2n} = \mathbb{J}_{2n}^{\mathsf{T}}\mathbb{J}_{2n}^{\mathsf{T}} = -I_{2n}, \qquad v^{\mathsf{T}}\mathbb{J}_{2n}v = 0. \tag{2}$$

These properties are intuitively understandable as the Poisson matrix is a $2n$-dimensional, 90° rotation matrix and the matrix $-I_{2n}$ can be interpreted as a rotation by 180° in this context.

Definition 2 (Symplectic map). *Let $A : \mathbb{R}^{2m} \to \mathbb{R}^{2n}$, $y \mapsto Ay$, $A \in \mathbb{R}^{2n\times 2m}$ be a linear mapping for $n, m \in \mathbb{N}$ and $m \leq n$. We call A a linear symplectic map and A a symplectic matrix with respect to ω_{2n} and ω_{2m} if*

$$A^{T}\mathbb{J}_{2n}A = \mathbb{J}_{2m}, \tag{3}$$

where ω_{2m} is the canonical symplectic form on $\mathbb{R}^{2m}$ (and is equal to ω_{2n} if $n = m$).

Let $U \subset \mathbb{R}^{2m}$ be an open set and $g : U \to \mathbb{R}^{2n}$ a differentiable map on U. We call g a symplectic map if the Jacobian matrix $\frac{d}{dy}g(y) \in \mathbb{R}^{2n\times 2m}$ is a symplectic matrix for every $y \in U$.

For a linear map, it is easy to check that condition (3) is equivalent to the preservation of the symplectic form, i.e., for all $v_1, v_2 \in \mathbb{R}^{2m}$

$$\omega_{2n}\left(Av_1,\ Av_2\right) = v_1^{\mathsf{T}}A^{\mathsf{T}}\mathbb{J}_{2n}Av_2 = v_1^{\mathsf{T}}\mathbb{J}_{2m}v_2 = \omega_{2m}\left(v_1,\ v_2\right).$$

Now, we give the definition of the so-called symplectic inverse which will be used in Section 2.3.

Definition 3 (Symplectic inverse). *For each symplectic matrix $A \in \mathbb{R}^{2n\times 2m}$, we define the symplectic inverse*

$$A^{+} = \mathbb{J}_{2m}^{T}A^{T}\mathbb{J}_{2n} \in \mathbb{R}^{2m\times 2n}. \tag{4}$$

The symplectic inverse A^{+} exists for every symplectic matrix and it holds the inverse relation

$$A^{+}A = \mathbb{J}_{2m}^{\mathsf{T}}A^{\mathsf{T}}\mathbb{J}_{2n}A = \mathbb{J}_{2m}^{\mathsf{T}}\mathbb{J}_{2m} = I_{2m}.$$

2.2. Finite-Dimensional, Autonomous Hamiltonian Systems

To begin with, we introduce the Hamiltonian system in a finite-dimensional, autonomous setting. The non-autonomous case is discussed subsequently in Section 2.4.

Definition 4 (Finite-dimensional, autonomous Hamiltonian system). *Let $\mathcal{H} : \mathbb{R}^{2n} \times \mathcal{P} \to \mathbb{R}$ be a scalar-valued function that we require to be continuously differentiable in the first argument and which we call Hamiltonian (function). Hamilton's equation is an initial value problem with the prescribed initial data $t_0 \in \mathbb{R}$, $x_0(\mu) \in \mathbb{R}^{2n}$ which describes the evolution of the solution $x(t, \mu) \in \mathbb{R}^{2n}$ for all $t \in [t_0, t_{\text{end}}]$, $\mu \in \mathcal{P}$ with*

$$\frac{d}{dt}x(t,\mu) = \mathbb{J}_{2n}\nabla_x\mathcal{H}(x(t,\mu),\mu) =: X_{\mathcal{H}}(x(t,\mu),\mu), \qquad x(t_0,\mu) = x_0(\mu), \tag{5}$$

where $X_{\mathcal{H}}(\bullet,\mu)$ is called the Hamiltonian vector field. The triple $(\mathbb{V}, \omega_{2n}, \mathcal{H})$ is referred to as the Hamiltonian system. We denote the flow of a Hamiltonian system as the mapping $\varphi_t : \mathbb{R}^{2n} \times \mathcal{P} \to \mathbb{R}^{2n}$ that evolves the initial state $x_0(\mu) \in \mathbb{R}^{2n}$ to the corresponding solution $x(t, \mu;\ t_0, x_0(\mu))$ of Hamilton's equation

$$\varphi_t(x_0,\mu) := x(t,\mu;\ t_0, x_0(\mu)),$$

where $x(t, \mu;\ t_0, x_0(\mu))$ indicates that it is the solution with the initial data $t_0, x_0(\mu)$.

The two characteristic properties of Hamiltonian systems are (a) the preservation of the Hamiltonian function and (b) the symplecticity of the flow which are presented in the following two propositions.

Proposition 1 (Preservation of the Hamiltonian). *The flow of Hamilton's equation $\boldsymbol{\varphi}_t$ preserves the Hamiltonian function $\mathcal{H}$.*

Proof. We prove the assertion by showing that the evolution over time is constant for any $\boldsymbol{x} \in \mathbb{R}^{2n}$ due to

$$\frac{\mathrm{d}}{\mathrm{d}t}\mathcal{H}(\boldsymbol{\varphi}_t(\boldsymbol{x})) = \left(\nabla_{\boldsymbol{x}}\mathcal{H}(\boldsymbol{\varphi}_t(\boldsymbol{x}))\right)^\mathsf{T} \frac{\mathrm{d}}{\mathrm{d}t}\boldsymbol{\varphi}_t(\boldsymbol{x}) \overset{(5)}{=} \left(\nabla_{\boldsymbol{x}}\mathcal{H}(\boldsymbol{\varphi}_t(\boldsymbol{x}))\right)^\mathsf{T} \mathbb{J}_{2n}\nabla_{\boldsymbol{x}}\mathcal{H}(\boldsymbol{\varphi}_t(\boldsymbol{x})) \overset{(2)}{=} 0.$$

□

Proposition 2 (Symplecticity of the flow). *Let the Hamiltonian function be twice continuously differentiable in the first argument. Then, the flow $\boldsymbol{\varphi}_t(\bullet, \boldsymbol{\mu}) : \mathbb{R}^{2n} \to \mathbb{R}^{2n}$ of a Hamiltonian system is a symplectic map.*

Proof. See ([8], Chapter VI, Theorem 2.4). □

2.3. Symplectic Model Order Reduction for Autonomous Hamiltonian Systems

The goal of MOR [1] is to reduce the order, i.e., the dimension, of high dimensional systems. To this end, we approximate the high-dimensional state $\boldsymbol{x}(t) \in \mathbb{R}^{2n}$ with

$$\boldsymbol{x}(t, \boldsymbol{\mu}) \approx \boldsymbol{x}_{\mathrm{rc}}(t, \boldsymbol{\mu}) = \boldsymbol{V}\boldsymbol{x}_{\mathrm{r}}(t, \boldsymbol{\mu}), \qquad \mathcal{V} = \operatorname{colspan}(\boldsymbol{V}),$$

with the reduced state $\boldsymbol{x}_{\mathrm{r}}(t) \in \mathbb{R}^{2k}$, the reduced-order basis (ROB) $\boldsymbol{V} \in \mathbb{R}^{2n\times 2k}$, the reconstructed state $\boldsymbol{x}_{\mathrm{rc}}(t) \in \mathcal{V}$ and the reduced space $\mathcal{V} \subset \mathbb{R}^{2n}$. The restriction to even-dimensional spaces $\mathbb{R}^{2n}$ and $\mathbb{R}^{2k}$ is not necessary for MOR in general but is required for the symplectic MOR in the following. To achieve a computational advantage with MOR, the approximation should introduce a clear reduction of the order, i.e., $2k \ll 2n$.

For Petrov–Galerkin projection-based MOR techniques, the ROB $\boldsymbol{V}$ is accompanied by a projection matrix $\boldsymbol{W} \in \mathbb{R}^{2n\times 2k}$ which is chosen to be biorthogonal to $\boldsymbol{V}$, i.e., $\boldsymbol{W}^\mathsf{T}\boldsymbol{V} = \boldsymbol{I}_{2k}$. The reduced-order model (ROM) is derived with the requirement that the residual $\boldsymbol{r}(t, \boldsymbol{\mu})$ vanishes in the space spanned by the columns of the projection matrix, i.e., in our case

$$\boldsymbol{r}(t, \boldsymbol{\mu}) = \frac{\mathrm{d}}{\mathrm{d}t}\boldsymbol{x}_{\mathrm{rc}}(t, \boldsymbol{\mu}) - \boldsymbol{X}_{\mathcal{H}}(\boldsymbol{x}_{\mathrm{rc}}(t, \boldsymbol{\mu}), \boldsymbol{\mu}) \in \mathbb{R}^{2n}, \qquad \boldsymbol{W}^\mathsf{T}\boldsymbol{r}(t, \boldsymbol{\mu}) = \boldsymbol{0}_{2k\times 1}, \tag{6}$$

where $\boldsymbol{0}_{2k\times 1} \in \mathbb{R}^{2k}$ is the vector of all zeros. Due to the biorthogonality, this is equivalent to

$$\frac{\mathrm{d}}{\mathrm{d}t}\boldsymbol{x}_{\mathrm{r}}(t, \boldsymbol{\mu}) = \boldsymbol{W}^\mathsf{T}\boldsymbol{X}_{\mathcal{H}}(\boldsymbol{x}_{\mathrm{rc}}(t, \boldsymbol{\mu}), \boldsymbol{\mu}) = \boldsymbol{W}^\mathsf{T}\mathbb{J}_{2n}\nabla_{\boldsymbol{x}}\mathcal{H}(\boldsymbol{x}_{\mathrm{rc}}(t, \boldsymbol{\mu}), \boldsymbol{\mu}), \qquad \boldsymbol{x}_{\mathrm{r}}(t_0, \boldsymbol{\mu}) = \boldsymbol{W}^\mathsf{T}\boldsymbol{x}_0(\boldsymbol{\mu}). \tag{7}$$

In the context of symplectic MOR, the ROB is chosen to be a symplectic matrix (3) which we call a symplectic ROB. Additionally, the transposed projection matrix is the symplectic inverse $\boldsymbol{W}^\mathsf{T} = \boldsymbol{V}^+$ and the projection in (7) is called a symplectic projection or symplectic Galerkin projection [4]. The (possibly oblique) projection reads

$$\boldsymbol{P} = \boldsymbol{V}\left(\boldsymbol{W}^\mathsf{T}\boldsymbol{V}\right)^{-1}\boldsymbol{W}^\mathsf{T} = \boldsymbol{V}(\boldsymbol{V}^+\boldsymbol{V})^{-1}\boldsymbol{V}^+ = \boldsymbol{V}\boldsymbol{V}^+.$$

In combination, this choice of $\boldsymbol{V}$ and $\boldsymbol{W}$ guarantees that the Hamiltonian structure is preserved by the reduction which is shown in the following proposition.

Proposition 3 (Reduced autonomous Hamiltonian system). *Let V be a symplectic ROB with the projection matrix $W^{\mathsf{T}} = V^{+}$. Then, the ROM (7) of a high-dimensional Hamiltonian system $(\mathbb{R}^{2n}, \omega_{2n}, \mathcal{H})$ is a Hamiltonian system $(\mathbb{R}^{2k}, \omega_{2k}, \mathcal{H}_{\mathrm{r}})$ on $\mathbb{R}^{2k}$ with the canonical symplectic form ω_{2k} and the reduced Hamiltonian function $\mathcal{H}_{\mathrm{r}}(x_{\mathrm{r}}, \mu) = \mathcal{H}(Vx_{\mathrm{r}}, \mu)$ for all $x_{\mathrm{r}} \in \mathbb{R}^{2k}$.*

Proof. First, we remark that the symplectic inverse is a valid biorthogonal projection matrix since it fulfils $W^{\mathsf{T}}V = V^{+}V = I_{2k}$. To derive the Hamiltonian form of the ROM in (7), we use the identity

$$W^{\mathsf{T}}\mathbb{J}_{2n} = V^{+}\mathbb{J}_{2n} \overset{(4)}{=} \mathbb{J}_{2k}^{\mathsf{T}}V^{\mathsf{T}}\mathbb{J}_{2n}\mathbb{J}_{2n} = -\mathbb{J}_{2k}^{\mathsf{T}}V^{\mathsf{T}} = \mathbb{J}_{2k}V^{\mathsf{T}}, \tag{8}$$

which makes use of the properties (2) of the Poisson matrix. It follows with (5), (7) and (8)

$$\frac{\mathrm{d}}{\mathrm{d}t}x_{\mathrm{r}}(t) = W^{\mathsf{T}}\mathbb{J}_{2n}\nabla_x\mathcal{H}(x_{\mathrm{rc}}(t)) = \mathbb{J}_{2k}V^{\mathsf{T}}\nabla_x\mathcal{H}(x_{\mathrm{rc}}(t)) = \mathbb{J}_{2k}\nabla_{x_{\mathrm{r}}}\mathcal{H}_{\mathrm{r}}(x_{\mathrm{r}}(t)),$$

where the last step follows from the chain rule. Thus, the evolution of the reduced state takes the form of Hamilton's equation and the resultant ROM is equal to the Hamiltonian system $(\mathbb{R}^{2k}, \omega_{2k}, \mathcal{H}_{\mathrm{r}})$. □

Corollary 1 (Linear Hamiltonian system). *Hamilton's equation is a linear system in the case of a quadratic Hamiltonian $\mathcal{H}(x, \mu) = 1/2\, x^{T}H(\mu)x + x^{T}h(\mu)$ with $H(\mu) \in \mathbb{R}^{2n\times 2n}$ symmetric and $h(\mu) \in \mathbb{R}^{2n}$*

$$\frac{\mathrm{d}}{\mathrm{d}t}x(t,\mu) = A(\mu)x(t,\mu) + b(\mu), \qquad A(\mu) = \mathbb{J}_{2n}H(\mu), \qquad b(\mu) = \mathbb{J}_{2n}h(\mu). \tag{9}$$

The evolution of the reduced Hamiltonian system reads

$$\frac{\mathrm{d}}{\mathrm{d}t}x_{\mathrm{r}}(t,\mu) = A_{\mathrm{r}}(\mu)x_{\mathrm{r}}(t,\mu) + b_{\mathrm{r}}(\mu), \qquad \begin{aligned} A_{\mathrm{r}}(\mu) &= \mathbb{J}_{2k}H_{\mathrm{r}}(\mu) \overset{(8)}{=} W^{T}A(\mu)V, & H_{\mathrm{r}}(\mu) &= V^{T}H(\mu)V,\\ b_{\mathrm{r}}(\mu) &= \mathbb{J}_{2k}h_{\mathrm{r}}(\mu) \overset{(8)}{=} W^{T}b(\mu)V, & h_{\mathrm{r}}(\mu) &= V^{T}h(\mu), \end{aligned}$$

with the reduced Hamiltonian function $\mathcal{H}_{\mathrm{r}}(x_{\mathrm{r}}, \mu) = 1/2\, x_{\mathrm{r}}^{T}H_{\mathrm{r}}(\mu)x_{\mathrm{r}} + x_{\mathrm{r}}^{T}h_{\mathrm{r}}(\mu)$.

Remark 1. *We emphasise that the reduction of linear Hamiltonian systems follows the pattern of the classical projection-based MOR approaches [9] to derive the reduced model with $A_{\mathrm{r}} = W^{T}AV$ and $b_{\mathrm{r}} = W^{T}b$, which allows a straightforward implementation in existing frameworks.*

Since the ROM is a Hamiltonian system, it preserves its Hamiltonian. Thus, it can be shown that the error in the Hamiltonian $e_{\mathcal{H}}(t,\mu) = \mathcal{H}(x(t,\mu),\mu) - \mathcal{H}_{\mathrm{r}}(x_{\mathrm{r}}(t,\mu),\mu)$ is constant [4]. Furthermore, there are a couple of results for the preservation of stability ([5], Theorem 18), ([4], Section 3.4.) under certain assumptions on the Hamiltonian function.

Remark 2 (Offline/online decomposition). *A central concept in the field of MOR for parametric systems is the so-called offline/online decomposition. The idea is to split the procedure in a possibly costly offline phase and a cheap online phase where the terms costly and cheap refer to the computational cost. In the offline phase, the ROM is constructed. The online phase is supposed to evaluate the ROM fast. The ultimate goal is to avoid any computations that depend on the high dimension 2n in the online phase. For a linear system, the offline/online decomposition can be achieved if $A(\mu)$, $b(\mu)$ and $x_0(\mu)$ allow a parameter-separability condition [9].*

Remark 3 (Non-linear Hamiltonian systems). *If the Hamiltonian function is not quadratic, the gradient is nonlinear. Thus, the right-hand side of the ODE system (i.e., the Hamiltonian vector field) is nonlinear. Nevertheless, symplectic MOR, technically, can be applied straightforwardly. The problem is that, without further assumptions, an efficient offline/online decomposition cannot be achieved which results in a low or no* *reduction of the computational costs. Multiple approaches [10,11] exist which introduce an approximation of the nonlinear right-hand side to enable an efficient offline/online decomposition.*

For symplectic MOR, the symplectic discrete empirical interpolation method (SDEIM) was introduced ([4], Section 5.2.) and ([5], Section 4.2.) to preserve the symplectic structure throughout the approximation of the nonlinear terms. The performace of these methods is discussed in in [4,5] for typical examples like the sine-Gordon equation or the nonlinear Schrödinger equation.

Remark 4 (MOR for port-Hamiltonian systems). *Alternatively to symplectic MOR with a snapshot-based basis generation method, MOR for so-called port-Hamiltonian systems can be used [12,13]. The projection scheme of that approach does not require symplecticity of the ROB but instead approximates the gradient of the Hamiltonian with the projection matrix* $\nabla_{\boldsymbol{x}}\mathcal{H}(\boldsymbol{V}\boldsymbol{x}_{\mathrm{r}}) \approx \boldsymbol{W}\nabla_{\boldsymbol{x}_{\mathrm{r}}}\mathcal{H}_{\mathrm{r}}(\boldsymbol{x}_{\mathrm{r}})$*. That alternative approach, like symplectic MOR, preserves the structure of the Hamiltonian equations, but, contrary to symplectic MOR, might result in a non-canonical symplectic structure in the reduced system.*

2.4. Finite-Dimensional, Non-Autonomous Hamiltonian Systems

The implementation of non-autonomous Hamiltonian systems in the symplectic framework is non-trivial as the Hamiltonian might change over time. We discuss the concept of the extended phase space [14] in the following, which redirects the non-autonomous Hamiltonian system to the case of an autonomous Hamiltonian system. The model reduction of these systems is discussed subsequently in Section 2.5.

Definition 5 (Finite-dimensional, non-autonomous Hamiltonian system). *Let* $\mathcal{H} : \mathbb{R} \times \mathbb{R}^{2n} \times \mathcal{P} \to \mathbb{R}$ *be a scalar-valued function function that is continuously differentiable in the second argument. A non-autonomous (or time-dependent) Hamiltonian system* $(\mathbb{R}^{2n}, \omega_{2n}, \mathcal{H})$ *is of the form*

$$\boldsymbol{x}(t,\boldsymbol{\mu}) = \mathbb{J}_{2n}\nabla_{\boldsymbol{x}}\mathcal{H}(t,\boldsymbol{x}(t,\boldsymbol{\mu}),\boldsymbol{\mu}). \tag{10}$$

We therefore call $\mathcal{H}(t,\boldsymbol{x})$ *a time-dependent Hamiltonian function.*

A problem for non-autonomous Hamiltonian systems occurs as the explicit time dependence of the Hamiltonian function introduces an additional variable, the time, and the carrier manifold becomes odd-dimensional. As mentioned in Section 2.1, symplectic vector spaces are always even-dimensional which is why a symplectic description is no longer possible. Different approaches are available to circumvent this issue.

As suggested in ([15], Section 4.3), we use the methodology of the so-called symplectic extended phase space ([14], Chap. VI, Section 10) to redirect the non-autonomous system to an autonomous system. The formulation is based on the extended Hamiltonian function $\mathcal{H}^{\mathrm{e}} : \mathbb{R}^{2n+2} \to \mathbb{R}$ with

$$\mathcal{H}^{\mathrm{e}}(\boldsymbol{x}^{\mathrm{e}}) = \mathcal{H}(q^{\mathrm{e}},\boldsymbol{x}) + p^{\mathrm{e}}, \quad \boldsymbol{x}^{\mathrm{e}} = (q^{\mathrm{e}}\ \boldsymbol{q}\ p^{\mathrm{e}}\ \boldsymbol{p})^{\mathsf{T}} \in \mathbb{R}^{2n+2}, \quad \boldsymbol{x} = (\boldsymbol{q}\ \boldsymbol{p})^{\mathsf{T}} \in \mathbb{R}^{2n}, \quad q^{\mathrm{e}}, p^{\mathrm{e}} \in \mathbb{R}. \tag{11}$$

Technically, the time is added to the extended state $\boldsymbol{x}^{\mathrm{e}}$ with $q^{\mathrm{e}} = t$ and the corresponding momentum $p^{\mathrm{e}} = -\mathcal{H}(t,\boldsymbol{x}(t))$ is chosen such that the extended system is an autonomous Hamiltonian system.

This procedure requires the time-dependent Hamiltonian function to be differentiable in the time variable. Thus, it does for example not allow for the description of loads that are not differentiable in time in the context of mechanical systems. This might, e.g., exclude systems that model mechanical contact since loads that are not differentiable in time are required.

2.5. Symplectic Model Order Reduction of Non-Autonomous Hamiltonian Systems

For the MOR of the, now autonomous, extended system, only the original phase space variable $\boldsymbol{x} \in \mathbb{R}^{2n}$ is reduced. The time and the corresponding conjugate momentum $q^{\mathrm{e}}, p^{\mathrm{e}}$ are not reduced.

To preserve the Hamiltonian structure, a symplectic ROB $V \in \mathbb{R}^{2n \times 2k}$ is used for the reduction of $x \in \mathbb{R}^{2n}$ analogous to the autonomous case. The result is a reduced extended system which again can be written as a non-autonomous Hamiltonian system $(\mathbb{R}^{2k}, \omega_{2k}, \mathcal{H}_r)$ with the time-dependent Hamiltonian $\mathcal{H}_r(t, x_r, \mu) = \mathcal{H}(t, Vx_r, \mu)$ for all $(t, x_r) \in [t_0, t_{end}] \times \mathbb{R}^{2k}$.

An unpleasant side effect of the extended formulation is that the linear dependency on the additional state variable p^e (see (11)) implies that the Hamiltonian cannot have strict extrema. Thus, the stability results listed in [4,5] do not apply if there is a true time-dependence in the Hamiltonian $\mathcal{H}(t, x)$. Nevertheless, symplectic MOR in combination with a non-autonomous Hamiltonian system shows stable results in the numerical experiments.

Furthermore, it is important to note that only the extended Hamiltonian $\mathcal{H}^e$ is preserved throughout the reduction. The time-dependent Hamiltonian $\mathcal{H}(\cdot, t)$ is not necessarily preserved throughout the reduction, i.e.,$\mathcal{H}^e(x^e(t)) = \mathcal{H}^e_r(x^e_r(t))$ but potentially $\mathcal{H}(x(t), t) \neq \mathcal{H}_r(x(t), t)$.

3. Symplectic Basis Generation with the Proper Symplectic Decomposition (PSD)

We introduced the symplectic MOR for finite-dimensional Hamiltonian systems in the previous section. This approach requires a symplectic ROB which is yet not further specified. In the following, we discuss the Proper Symplectic Decomposition (PSD) as a data-driven basis generation approach. To this end, we classify symplectic ROBs as orthogonal and non-orthogonal. The PSD is investigated for these two classes in Sections 3.1 and 3.2 separately. For symplectic, orthogonal ROBs, we prove that an optimal solution can be derived based on an established procedure, the PSD Complex SVD. For symplectic, non-orthogonal ROBs, we provide a new basis generation method, the PSD SVD-like decomposition.

We pursue the approach to generate an ROB from a collection of snapshots of the system [16]. A snapshot is an element of the so-called solution manifold $\mathcal{S}$ which we aim to approximate with a low-dimensional surrogate $\widehat{\mathcal{S}}_{VW}$

$$\mathcal{S} := \{x(t, \mu) \mid t \in [t_0, t_{end}], \mu \in \mathcal{P}\} \subset \mathbb{R}^{2n}, \qquad \widehat{\mathcal{S}}_{VW} := \{Vx_r(t, \mu) \mid t \in [t_0, t_{end}], \mu \in \mathcal{P}\} \approx \mathcal{S},$$

where $x(t, \mu) \in \mathbb{R}^{2n}$ is a solution of the full model (5), $V \in \mathbb{R}^{2n \times 2k}$ is the ROB and $x_r \in \mathbb{R}^{2k}$ is the solution of the reduced system (7). In [4], the Proper Symplectic Decomposition (PSD) is proposed as a snapshot-based basis generation technique for symplectic ROBs. The idea is to derive the ROB from a minimization problem which is suggested in analogy to the very well established Proper Orthogonal Decomposition (POD, also Principal Component Analysis) [1].

Classically, the POD chooses the ROB V_{POD} to minimize the sum over squared norms of all $n_s \in \mathbb{N}$ residuals $(I_{2n} - V_{POD}V_{POD}^T)x_i^s$ of the orthogonal projection $V_{POD}V_{POD}^T x_i^s$ of the $1 \leq i \leq n_s$ single snapshots $x_i^s \in \mathcal{S}$ measured in the 2-norm $\|\bullet\|_2$ with the constraint that the ROB V_{POD} is orthogonal, i.e.,

$$\underset{V_{POD} \in \mathbb{R}^{2n \times 2k}}{\text{minimize}} \sum_{i=1}^{n_s} \left\| \left(I_{2n} - V_{POD}V_{POD}^T \right) x_i^s \right\|_2^2 \qquad \text{subject to} \quad V_{POD}^T V_{POD} = I_{2k}. \tag{12}$$

In contrast, the PSD requires the ROB to be symplectic instead of orthogonal, which is expressed in the reformulated constraint. Furthermore, the orthogonal projection is replaced by the symplectic projection $VV^+x_i^s$ which results in

$$\underset{V \in \mathbb{R}^{2n \times 2k}}{\text{minimize}} \sum_{i=1}^{n_s} \|(I_{2n} - VV^+)x_i^s\|_2^2 \qquad \text{subject to} \quad V^T \mathbb{J}_{2n} V = \mathbb{J}_{2k}. \tag{13}$$

We summarize this in a more compact (matrix-based) formulation in the following definition.

Definition 6 (Proper Symplectic Decomposition (PSD)). *Given n_s snapshots $x_1^s, \dots, x_{n_s}^s \in \mathcal{S}$, we denote the snapshot matrix as $X_s = [x_1^s, \dots, x_{n_s}^s] \in \mathbb{R}^{2n \times n_s}$. Find a symplectic ROB $V \in \mathbb{R}^{2n \times 2k}$ which minimizes*

$$\underset{V \in \mathbb{R}^{2n \times 2k}}{\text{minimize}} \left\| (I_{2n} - VV^+) X_s \right\|_F^2 \qquad \text{subject to} \quad V^T \mathbb{J}_{2n} V = \mathbb{J}_{2k}. \tag{14}$$

We denote the minimization problem (14) in the following as $\texttt{PSD}(X_s)$, where X_s is the given snapshot matrix.

The constraint in (14) ensures that the ROB V is symplectic and thus guarantees the existence of the symplectic inverse V^+. Furthermore, the matrix-based formulation (14) is equivalent to the vector-based formulation presented in (13) due to the properties of the Frobenius norm $\|\bullet\|_F$.

Remark 5 (Interpolation-based ROBs). *Alternatively to the presented snapshot-based basis generation techniques, interpolation-based ROBs might be used. These aim to interpolate the transfer function of linear problems (or the linearized equations of nonlinear problems). For the framework of MOR of port-Hamiltonian systems (see Remark 4), there exists an interpolation-based basis generation technique ([13], Section 2.2.). In the scope of our paper, we focus on symplectic MOR and snapshot-based techniques.*

3.1. Symplectic, Orthonormal Basis Generation

The foremost problem of the PSD is that there is no explicit solution procedure known so far due to the high nonlinearity and possibly multiple local optima. This is an essential difference to the POD as the POD allows to find a global minimum by solving an eigenvalue problem [1].

Current solution procedures for the PSD restrict to a certain subset of symplectic matrices and derive an optimal solution for this subset which might be suboptimal in the class of symplectic matrices. In the following, we show that this subclass almost exclusively restricts to symplectic, orthonormal ROBs.

Definition 7 (Symplectic, orthonormal ROB). *We call an ROB $V \in \mathbb{R}^{2n \times 2k}$ symplectic, orthonormal (also orthosymplectic, e.g., in [5]) if it is symplectic w.r.t. ω_{2n} and ω_{2k} and is orthonormal, i.e., the matrix V has orthonormal columns*

$$V^T \mathbb{J}_{2n} V = \mathbb{J}_{2k} \qquad \text{and} \qquad V^T V = I_{2k}.$$

In the following, we show an alternative characterization of a symplectic and orthonormal ROB. Therefore, we extend the results given, e.g., in [17] for square matrices $Q \in \mathbb{R}^{2n \times 2n}$ in the following Proposition 4 to the case of rectangular matrices $V \in \mathbb{R}^{2n \times 2k}$. This was also partially addressed in ([4], Lemma 4.3.).

Proposition 4 (Characterization of a symplectic matrix with orthonormal columns). *The following statements are equivalent for any matrix $V \in \mathbb{R}^{2n \times 2k}$*

(i) V is symplectic with orthonormal columns,
(ii) V is of the form

$$V = \begin{bmatrix} E & \mathbb{J}_{2n}^T E \end{bmatrix} =: V_E \in \mathbb{R}^{2n \times 2k}, \qquad E \in \mathbb{R}^{2n \times k}, \qquad E^T E = I_k, \qquad E^T \mathbb{J}_{2n} E = 0_k, \tag{15}$$

(iii) V is symplectic and it holds $V^T = V^+$.

We remark that these matrices are characterized in [4] to be elements in $\mathrm{Sp}(2k, \mathbb{R}^{2n}) \cap V_k(\mathbb{R}^{2n})$ where $\mathrm{Sp}(2k, \mathbb{R}^{2n})$ is the symplectic Stiefel manifold and $V_k(\mathbb{R}^{2n})$ is the Stiefel manifold.

Proof. "(i) $\implies$ (ii)": Let $V \in \mathbb{R}^{2n\times 2k}$ be a symplectic matrix with orthonormal columns. We rename the columns to $V = [E \quad F]$ with $E = [e_1, \dots, e_k]$ and $F = [f_1, \dots, f_k]$. The symplecticity of the matrix written in terms of E and F reads

$$V^\mathsf{T}\mathbb{J}_{2n}V = \begin{bmatrix} E^\mathsf{T}\mathbb{J}_{2n}E & E^\mathsf{T}\mathbb{J}_{2n}F \\ F^\mathsf{T}\mathbb{J}_{2n}E & F^\mathsf{T}\mathbb{J}_{2n}F \end{bmatrix} = \begin{bmatrix} 0_k & I_k \\ -I_k & 0_k \end{bmatrix} \iff \begin{array}{c} E^\mathsf{T}\mathbb{J}_{2n}E = F^\mathsf{T}\mathbb{J}_{2n}F = 0_k, \\ -F^\mathsf{T}\mathbb{J}_{2n}E = E^\mathsf{T}\mathbb{J}_{2n}F = I_k. \end{array} \tag{16}$$

Expressed in terms of the columns e_i, f_i of the matrices E, F, this condition reads for any $1 \le i, j \le k$

$$e_i^\mathsf{T}\mathbb{J}_{2n}e_j = 0, \qquad e_i^\mathsf{T}\mathbb{J}_{2n}f_j = \delta_{ij}, \qquad f_i^\mathsf{T}\mathbb{J}_{2n}e_j = -\delta_{ij}, \qquad f_i^\mathsf{T}\mathbb{J}_{2n}f_j = 0,$$

and the orthonormality of the columns of V implies

$$e_i^\mathsf{T}e_j = \delta_{ij}, \qquad f_i^\mathsf{T}f_j = \delta_{ij}.$$

For a fixed $i \in \{1, \dots, k\}$, it is easy to show with $\mathbb{J}_{2n}^\mathsf{T}\mathbb{J}_{2n} = I_{2n}$ that $\mathbb{J}_{2n}f_i$ is of unit length

$$1 = \delta_{ii} = f_i^\mathsf{T}f_i = f_i^\mathsf{T}\mathbb{J}_{2n}^\mathsf{T}\mathbb{J}_{2n}f_i = \|\mathbb{J}_{2n}f_i\|_2^2.$$

Thus, e_i and $\mathbb{J}_{2n}f_i$ are both unit vectors which fulfill $e_i^\mathsf{T}\mathbb{J}_{2n}f_i = \langle e_i, \mathbb{J}_{2n}f_i\rangle_{\mathbb{R}^{2n}} = 1$. By the Cauchy–Schwarz inequality, it holds $\langle e_i, \mathbb{J}_{2n}f_i\rangle = \|e_i\| \, \|\mathbb{J}_{2n}f_i\|$ if and only if the vectors are parallel. Thus, we infer $e_i = \mathbb{J}_{2n}f_i$, which is equivalent to $f_i = \mathbb{J}_{2n}^\mathsf{T}e_i$. Since this holds for all $i \in \{1, \dots, k\}$, we conclude that V is of the form proposed in (15).

"(ii) $\implies$ (iii)": Let V be of the form (15). Direct calculation yields

$$V^\mathsf{T}\mathbb{J}_{2n}V = \begin{bmatrix} E^\mathsf{T} \\ E^\mathsf{T}\mathbb{J}_{2n} \end{bmatrix} \mathbb{J}_{2n} \begin{bmatrix} E & \mathbb{J}_{2n}^\mathsf{T}E \end{bmatrix} = \begin{bmatrix} E^\mathsf{T}\mathbb{J}_{2n}E & E^\mathsf{T}E \\ -E^\mathsf{T}E & E^\mathsf{T}\mathbb{J}_{2n}E \end{bmatrix} \overset{(15)}{=} \begin{bmatrix} 0_k & I_k \\ -I_k & 0_k \end{bmatrix} = \mathbb{J}_{2k},$$

which shows that V is symplectic. Thus, the symplectic inverse V^+ exists. The following calculation shows that it equals the transposed V^T

$$V^+ = \mathbb{J}_{2k}^\mathsf{T}V^\mathsf{T}\mathbb{J}_{2n} = \mathbb{J}_{2k}^\mathsf{T}\begin{bmatrix} E^\mathsf{T} \\ E^\mathsf{T}\mathbb{J}_{2n} \end{bmatrix}\mathbb{J}_{2n} = \begin{bmatrix} -E^\mathsf{T}\mathbb{J}_{2n} \\ E^\mathsf{T} \end{bmatrix}\mathbb{J}_{2n} = \begin{bmatrix} -E^\mathsf{T}\mathbb{J}_{2n}\mathbb{J}_{2n} \\ E^\mathsf{T}\mathbb{J}_{2n} \end{bmatrix} = \begin{bmatrix} E^\mathsf{T} \\ E^\mathsf{T}\mathbb{J}_{2n} \end{bmatrix} = V^\mathsf{T}.$$

"(iii) $\implies$ (i)": Let V be symplectic with $V^\mathsf{T} = V^+$. Then, we know that V has orthonormal columns since

$$I_k = V^+V = V^\mathsf{T}V.$$

□

Proposition 4 essentially limits the symplectic, orthonormal ROB V to be of the form (15). Later in the current section, we see how to solve the PSD for ROBs of this type. In Section 3.2, we are interested in ridding the ROB V of this requirement to explore further solution methods of the PSD.

As mentioned before, the current solution procedures for the PSD almost exclusively restrict to the class of symplectic, orthonormal ROBs introduced in Proposition 4. This includes the Cotangent Lift [4], the Complex SVD [4], partly the nonlinear programming algorithm from [4] and the greedy procedure presented in [5]. We briefly review these approaches in the following proposition.

Proposition 5 (Symplectic, orthonormal basis generation)**.** *The Cotangent Lift (CT), Complex SVD (cSVD) and the greedy procedure for symplectic basis generation all derive a symplectic and orthonormal ROB. The nonlinear programming (NLP) admits a symplectic, orthonormal ROB if the coefficient matrix* C *in ([4],*

Algorithm 3) is symplectic and has orthonormal columns, i.e., it is of the form $C_G = [G \quad \mathbb{J}_{2k}^T G]$. *The methods can be rewritten with* $V_E = [E \quad \mathbb{J}_{2n}^T E]$, *where the different formulations of* E *read*

$$E_{CT} = \begin{bmatrix} \Phi_{CT} \\ 0_{n\times k} \end{bmatrix} \qquad E_{cSVD} = \begin{bmatrix} \Phi_{cSVD} \\ \Psi_{cSVD} \end{bmatrix}, \qquad E_{greedy} = [e_1, \dots, e_k], \qquad E_{NLP} = \widetilde{V}_E G,$$

where

(i) $\Phi_{CT}, \Phi_{cSVD}, \Psi_{cSVD} \in \mathbb{R}^{n\times k}$ *are matrices that fulfil*

$$\Phi_{CT}^T \Phi_{CT} = I_k, \qquad \Phi_{cSVD}^T \Phi_{cSVD} + \Psi_{cSVD}^T \Psi_{cSVD} = I_k, \qquad \Phi_{cSVD}^T \Psi_{cSVD} = \Psi_{cSVD}^T \Phi_{cSVD},$$

which is technically equivalent to $E^T E = I_k$ *and* $E^T \mathbb{J}_{2n} E = 0_k$ *(see (15)) for* E_{CT} *and* E_{cSVD},

(ii) $e_1, \dots, e_k \in \mathbb{R}^{2n}$ *are the basis vectors selected by the greedy algorithm,*

(iii) $\widetilde{V}_E \in \mathbb{R}^{2n\times 2k}$ *is an ROB computed from CT or cSVD and* $G \in \mathbb{R}^{2k\times r}$, $r \le k$, *stems from the coefficient matrix* $C_G = [G \quad \mathbb{J}_{2k}^T G]$ *computed by the NLP algorithm.*

Proof. All of the listed methods derive a symplectic ROB of the form $V_E = [E \quad \mathbb{J}_{2n}^T E]$ which satisfies (15). By Proposition 4, these ROBs are each a symplectic, orthonormal ROB. □

In the following, we show that PSD Complex SVD is the solution of the PSD in the subset of symplectic, orthonormal ROBs. This was partly shown in [4] which yet lacked the final step that, restricting to orthonormal, symplectic ROBs, a solution of $\mathrm{PSD}([X_s \quad -\mathbb{J}_{2n} X_s])$ solves $\mathrm{PSD}(X_s)$ and vice versa. This proves that the PSD Complex SVD is not only near optimal in this set but indeed optimal. Furthermore, the proof we show is alternative to the original and naturally motivates an alternative formulation of the PSD Complex SVD which we call the POD of Y_s in the following. To begin with, we reproduce the definition of PSD Complex SVD from [4].

Definition 8 (PSD Complex SVD). *We define the complex snapshot matrix*

$$C_s = [q_1^s + \mathrm{i} p_1^s, \dots, q_{n_s}^s + \mathrm{i} p_{n_s}^s] \in \mathbb{C}^{n\times n_s}, \qquad x_j^s = \begin{bmatrix} q_j^s \\ p_j^s \end{bmatrix} \text{ for all } 1 \le j \le n_s, \tag{17}$$

which is derived with the imaginary unit i. *The PSD Complex SVD is a basis generation technique that requires the auxiliary complex matrix* $U_{C_s} \in \mathbb{C}^{n\times k}$ *to fulfil*

$$\underset{U_{C_s} \in \mathbb{C}^{n\times k}}{\text{minimize}} \left\| C_s - U_{C_s} (U_{C_s})^* C_s \right\|_F^2 \qquad \text{subject to} \quad (U_{C_s})^* U_{C_s} = I_k \tag{18}$$

and builds the actual ROB $V_E \in \mathbb{R}^{2n\times 2k}$ *with*

$$V_E = [E \quad \mathbb{J}_{2n}^T E], \qquad E = \begin{bmatrix} \mathrm{Re}(U_{C_s}) \\ \mathrm{Im}(U_{C_s}) \end{bmatrix}.$$

The solution of (18) is known to be based on the left-singular vectors of C_s *which can be explicitly computed with a complex version of the SVD.*

We emphasize that we denote this basis generation procedure as PSD Complex SVD in the following to avoid confusions with the usual complex SVD algorithm.

Proposition 6 (Minimizing PSD in the set of symplectic, orthonormal ROBs). *Given the snapshot matrix* $X_s \in \mathbb{R}^{2n\times n_s}$, *we augment this with "rotated" snapshots to* $Y_s = [X_s \quad \mathbb{J}_{2n} X_s]$. *We assume that 2k is such that* *we obtain a gap in the singular values of Y_s, i.e., $\sigma_{2k}(Y_s) > \sigma_{2k+1}(Y_s)$. Then, minimizing the PSD in the set of symplectic, orthonormal ROBs is equivalent to the following minimization problem*

$$\underset{V\in\mathbb{R}^{2n\times 2k}}{\text{minimize}} \left\| (I_{2n} - VV^T) \begin{bmatrix} X_s & \mathbb{J}_{2n}X_s \end{bmatrix} \right\|_F^2 \qquad \text{subject to} \quad V^T V = I_{2k}. \tag{19}$$

Clearly, this is equivalent to the POD (12) applied to the snapshot matrix Y_s. We, thus, call this procedure the POD of Y_s in the following. A minimizer can be derived with the SVD as it is common for POD [1].

Proof. The proof proceeds in three steps: we show

(i) that (u, v) is a pair of left- and right-singular vectors of Y_s to the singular value σ if and only if $(\mathbb{J}_{2n}^T u, \mathbb{J}_{2n_s}^T v)$ also is a pair of left- and right-singular vectors of Y_s to the same singular value σ,
(ii) that a solution of the POD of Y_s is a symplectic, orthonormal ROB, i.e., $V = V_E = [E \;\; \mathbb{J}_{2n}^T E]$,
(iii) that the POD of Y_s is equivalent to the PSD for symplectic, orthonormal ROBs.

We start with the first step (i). Let (u, v) be a pair of left- and right-singular vectors of Y_s to the singular value σ. We use that the left-singular (or right-singular) vectors of Y_s are a set of orthonormal eigenvectors of $Y_s Y_s^T$ (or $Y_s^T Y_s$). To begin with, we compute

$$\begin{aligned} \mathbb{J}_{2n}^T Y_s Y_s^T \mathbb{J}_{2n} &= \mathbb{J}_{2n}^T (X_s X_s^T + \mathbb{J}_{2n} X_s X_s^T \mathbb{J}_{2n}^T) \mathbb{J}_{2n} = \mathbb{J}_{2n}^T X_s X_s^T \mathbb{J}_{2n} + X_s X_s^T = Y_s Y_s^T, \\ \mathbb{J}_{2n_s}^T Y_s^T Y_s \mathbb{J}_{2n_s} &= \mathbb{J}_{2n_s}^T \begin{bmatrix} X_s^T X_s & X_s^T \mathbb{J}_{2n} X_s \\ X_s^T \mathbb{J}_{2n}^T X_s & X_s^T X_s \end{bmatrix} \mathbb{J}_{2n_s} = \begin{bmatrix} X_s^T X_s & -X_s^T \mathbb{J}_{2n}^T X_s \\ -X_s^T \mathbb{J}_{2n} X_s & X_s^T X_s \end{bmatrix} = Y_s^T Y_s, \end{aligned} \tag{20}$$

where we use $\mathbb{J}_{2n_s}^T = -\mathbb{J}_{2n_s}$. Thus, we can reformulate the eigenvalue problems of $Y_s Y_s^T$ and, respectively, $Y_s^T Y_s$ as

$$\begin{aligned} \sigma u = Y_s Y_s^T u = \mathbb{J}_{2n} \mathbb{J}_{2n}^T Y_s Y_s^T \mathbb{J}_{2n} \mathbb{J}_{2n}^T u \quad &\overset{\mathbb{J}_{2n}^T \cdot |}{\Longleftrightarrow} \quad \sigma \mathbb{J}_{2n}^T u = \mathbb{J}_{2n}^T Y_s Y_s^T \mathbb{J}_{2n} \mathbb{J}_{2n}^T u \overset{(20)}{=} Y_s Y_s^T \mathbb{J}_{2n}^T u, \\ \sigma v = Y_s^T Y_s v = \mathbb{J}_{2n_s} \mathbb{J}_{2n_s}^T Y_s^T Y_s \mathbb{J}_{2n_s} \mathbb{J}_{2n_s}^T v \quad &\overset{\mathbb{J}_{2n_s}^T \cdot |}{\Longleftrightarrow} \quad \sigma \mathbb{J}_{2n_s}^T v = \mathbb{J}_{2n_s}^T Y_s^T Y_s \mathbb{J}_{2n_s} \mathbb{J}_{2n_s}^T v \overset{(20)}{=} Y_s^T Y_s \mathbb{J}_{2n_s}^T v. \end{aligned}$$

Thus, $(\mathbb{J}_{2n}^T u, \mathbb{J}_{2n_s}^T v)$ is necessarily another pair of left- and right-singular vectors of Y_s with the same singular value σ. We infer that the left-singular vectors u_i, $1 \le i \le 2n$, ordered by the magnitude of the singular values in a descending order can be written as

$$U = [u_1 \;\; \mathbb{J}_{2n}^T u_1 \;\; u_2 \;\; \mathbb{J}_{2n}^T u_2 \;\; \dots u_n \;\; \mathbb{J}_{2n}^T u_n] \in \mathbb{R}^{2n\times 2n}. \tag{21}$$

For the second step (ii), we remark that the solution of the POD is explicitly known to be any matrix which stacks in its columns $2k$ left-singular vectors of the snapshot matrix Y_s with the highest singular value [1]. Due to the special structure (21) of the singular vectors for the snapshot matrix Y_s, a minimizer of the POD of Y_s necessarily adopts this structure. We are allowed to rearrange the order of the columns in this matrix and thus the result of the POD of Y_s can always be rearranged to the form

$$V_E = [E \;\; \mathbb{J}_{2n}^T E], \qquad E = [u_1 \;\; u_2 \;\; \dots \;\; u_k], \qquad \mathbb{J}_{2n}^T E = [\mathbb{J}_{2n}^T u_1 \;\; \mathbb{J}_{2n}^T u_2 \;\; \dots \;\; \mathbb{J}_{2n} u_k].$$

Note that it automatically holds that $E^T E = I_k$ and $E^T(\mathbb{J}_{2n} E) = 0_k$ since, in both products, we use the left-singular vectors from the columns of the matrix U from (21) which is known to be orthogonal from properties of the SVD. Thus, (15) holds and we infer from Proposition 4 that the POD of Y_s indeed is solved by a symplectic, orthonormal ROB.

For the final step (iii), we define the orthogonal projection operators

$$P_{V_E} = V_E (V_E)^T = EE^T + \mathbb{J}_{2n}^T E E^T \mathbb{J}_{2n}, \qquad P_{V_E}^{\perp} = I_{2n} - P_{V_E}.$$

Both are idempotent and symmetric, thus $\left(P_{V_E}^{\perp}\right)^{\mathsf{T}} P_{V_E}^{\perp} = P_{V_E}^{\perp} P_{V_E}^{\perp} = P_{V_E}^{\perp}$. Due to $\mathbb{J}_{2n}\mathbb{J}_{2n}^{\mathsf{T}} = I_{2n}$, it further holds

$$\mathbb{J}_{2n} \left(P_{V_E}^{\perp}\right)^{\mathsf{T}} P_{V_E}^{\perp} \mathbb{J}_{2n}^{\mathsf{T}} = \mathbb{J}_{2n} P_{V_E}^{\perp} \mathbb{J}_{2n}^{\mathsf{T}} = \mathbb{J}_{2n}\mathbb{J}_{2n}^{\mathsf{T}} - \mathbb{J}_{2n} E E^{\mathsf{T}} \mathbb{J}_{2n}^{\mathsf{T}} - \mathbb{J}_{2n}\mathbb{J}_{2n}^{\mathsf{T}} E E^{\mathsf{T}} \mathbb{J}_{2n}\mathbb{J}_{2n}^{\mathsf{T}} = P_{V_E}^{\perp} = \left(P_{V_E}^{\perp}\right)^{\mathsf{T}} P_{V_E}^{\perp}.$$

Thus, it follows

$$\left\| P_{V_E}^{\perp} X_{\mathrm{s}} \right\|_{\mathrm{F}}^2 = \operatorname{trace}\left(X_{\mathrm{s}}^{\mathsf{T}} \left(P_{V_E}^{\perp}\right)^{\mathsf{T}} P_{V_E}^{\perp} X_{\mathrm{s}} \right) = \operatorname{trace}\left(X_{\mathrm{s}}^{\mathsf{T}} \mathbb{J}_{2n} \left(P_{V_E}^{\perp}\right)^{\mathsf{T}} P_{V_E}^{\perp} \mathbb{J}_{2n}^{\mathsf{T}} X_{\mathrm{s}} \right) = \left\| P_{V_E}^{\perp} \mathbb{J}_{2n}^{\mathsf{T}} X_{\mathrm{s}} \right\|_{\mathrm{F}}^2$$

and with $Y_{\mathrm{s}} = [X_{\mathrm{s}} \quad \mathbb{J}_{2n}^{\mathsf{T}} X_{\mathrm{s}}]$

$$2 \left\| P_{V_E}^{\perp} X_{\mathrm{s}} \right\|_{\mathrm{F}}^2 = \left\| P_{V_E}^{\perp} X_{\mathrm{s}} \right\|_{\mathrm{F}}^2 + \left\| P_{V_E}^{\perp} \mathbb{J}_{2n}^{\mathsf{T}} X_{\mathrm{s}} \right\|_{\mathrm{F}}^2 = \left\| P_{V_E}^{\perp} [X_{\mathrm{s}} \quad \mathbb{J}_{2n}^{\mathsf{T}} X_{\mathrm{s}}] \right\|_{\mathrm{F}}^2 = \left\| P_{V_E}^{\perp} Y_{\mathrm{s}} \right\|_{\mathrm{F}}^2,$$

where we use in the last step that for two matrices $A \in \mathbb{R}^{2n \times u}$, $B \in \mathbb{R}^{2n \times v}$ for $u, v \in \mathbb{N}$, it holds $\|A\|_{\mathrm{F}}^2 + \|B\|_{\mathrm{F}}^2 = \|[A \quad B]\|_{\mathrm{F}}^2$ for the Frobenius norm $\|\bullet\|_{\mathrm{F}}$.

Since it is equivalent to minimize a function $f : \mathbb{R}^{2n \times 2k} \to \mathbb{R}$ or a multiple cf of it for any positive constant $c \in \mathbb{R}_{>0}$, minimizing $\left\| P_{V_E}^{\perp} X_{\mathrm{s}} \right\|_{\mathrm{F}}^2$ is equivalent to minimizing $2 \left\| P_{V_E}^{\perp} X_{\mathrm{s}} \right\|_{\mathrm{F}}^2 = \left\| P_{V_E}^{\perp} Y_{\mathrm{s}} \right\|_{\mathrm{F}}^2$. Additionally, for an ROB of the form $V_E = [E \quad \mathbb{J}_{2n}^{\mathsf{T}} E]$, the constraint of orthonormal columns is equivalent to the requirements in (15). Thus, to minimize the PSD in the class of symplectic, orthonormal ROBs is equivalent to the POD of Y_{s} (19). □

Remark 6. *We remark that, in the same fashion as the proof of step (iii) in Proposition 6, it can be shown that, restricting to symplectic, orthonormal ROBs, a solution of* $\mathtt{PSD}([X_{\mathrm{s}} \quad \mathbb{J}_{2n} X_{\mathrm{s}}])$ *is a solution of* $\mathtt{PSD}(X_{\mathrm{s}})$ *and vice versa, which is one detail that was missing in [4] to show the optimality of PSD Complex SVD in the set of symplectic, orthonormal ROBs.*

We next prove that PSD Complex SVD is equivalent to POD of Y_{s} from (19) and thus also minimizes the PSD in the set of symplectic, orthonormal bases. To this end, we repeat the optimality result from [4] and extend it with the results of the present paper.

Proposition 7 (Optimality of PSD Complex SVD). *Let* $\mathbb{M}_2 \subset \mathbb{R}^{2n \times 2k}$ *denote the set of symplectic bases with the structure* $V_E = [E \quad \mathbb{J}_{2n}^{T} E]$. *The PSD Complex SVD solves* $\mathtt{PSD}([X_{\mathrm{s}} \quad -\mathbb{J}_{2n} X_{\mathrm{s}}])$ *in* $\mathbb{M}_2$.

Proof. See ([4], Theorem 4.5). □

Proposition 8 (Equivalence of POD of Y_{s} and PSD Complex SVD). *PSD Complex SVD is equivalent to the POD of* Y_{s}. *Thus, PSD Complex SVD yields a minimizer of the PSD for symplectic, orthonormal ROBs.*

Proof. By Proposition 7, PSD Complex SVD minimizes (19) in the set $\mathbb{M}_2$ of symplectic bases with the structure $V_E = [E \quad \mathbb{J}_{2n}^{\mathsf{T}} E]$. Thus, (16) holds with $F = \mathbb{J}_{2n}^{\mathsf{T}} E$ which is equivalent to the conditions on E required in (15). By Proposition 4, we infer that $\mathbb{M}_2$ equals the set of symplectic, orthonormal bases.

Furthermore, we can show that, in the set $\mathbb{M}_2$, a solution of $\mathtt{PSD}([X_{\mathrm{s}} \quad -\mathbb{J}_{2n} X_{\mathrm{s}}])$ is a solution of $\mathtt{PSD}(X_{\mathrm{s}})$ and vice versa (see Remark 6). Thus, PSD Complex SVD minimizes the PSD for the snapshot matrix X_{s} in the set of orthonormal, symplectic matrices and PSD Complex SVD and the POD of Y_{s} solve the same minimization problem. □

We emphasize that the computation of a minimizer of (19) via PSD Complex SVD requires less memory storage than the computation via POD of Y_{s}. The reason is that the complex formulation uses the complex snapshot matrix $C_{\mathrm{s}} \in \mathbb{C}^{n \times n_{\mathrm{s}}}$ which equals $2 \cdot n \cdot n_{\mathrm{s}}$ floating point numbers while the solution with the POD of Y_{s} method artificially enlarges the snapshot matrix to $Y_{\mathrm{s}} \in \mathbb{R}^{2n \times 2n_{\mathrm{s}}}$ which are

$4 \cdot n \cdot n_s$ floating point numbers. Still, the POD of $\boldsymbol{Y}_s$ might be computationally more efficient since it is a purely real formulation and thereby does not require complex arithmetic operations.

3.2. Symplectic, Non-Orthonormal Basis Generation

In the next step, we want to give an idea how to leave the class of symplectic, orthonormal ROBs. We call a basis generation technique symplectic, non-orthonormal if it is able to compute a symplectic, non-orthonormal basis.

In Proposition 5, we briefly showed that most existing symplectic basis generation techniques generate a symplectic, orthonormal ROB. The only exception is the NLP algorithm suggested in [4]. It is able to compute a non-orthonormal, symplectic ROB. The algorithm is based on a given initial guess $\boldsymbol{V}_0 \in \mathbb{R}^{2n \times 2k}$ which is a symplectic ROB, e.g., computed with PSD Cotangent Lift or PSD Complex SVD. Nonlinear programming is used to leave the class of symplectic, orthonormal ROBs and derive an optimized symplectic ROB $\boldsymbol{V} = \boldsymbol{V}_0\boldsymbol{C}$ with the symplectic coefficient matrix $\boldsymbol{C} \in \mathbb{R}^{2k \times 2r}$ for some $r \leq k$. Since this procedure searches a solution spanned by the columns of $\boldsymbol{V}_0$, it is not suited to compute a global optimum of the PSD which we are interested in the scope of this paper.

In the following, we present a new, non-orthonormal basis generation technique that is based on an SVD-like decomposition for matrices $\boldsymbol{B} \in \mathbb{R}^{2n \times m}$ presented in [6]. To this end, we introduce this decomposition in the following. Subsequently, we present first theoretical results which link the symplectic projection error with the "singular values" of the SVD-like decomposition which we call symplectic singular values. Nevertheless, the optimality with respect to the PSD functional (14) of this new method is yet an open question.

Proposition 9 (SVD-like decomposition [6]). *Any real matrix $\boldsymbol{B} \in \mathbb{R}^{2n \times m}$ can be decomposed as the product of a symplectic matrix $\boldsymbol{S} \in \mathbb{R}^{2n \times 2n}$, a sparse and potentially non-diagonal matrix $\boldsymbol{D} \in \mathbb{R}^{2n \times m}$ and an orthogonal matrix $\boldsymbol{Q} \in \mathbb{R}^{m \times m}$ with*

$$\boldsymbol{B} = \boldsymbol{S}\boldsymbol{D}\boldsymbol{Q}, \qquad \boldsymbol{D} = \begin{array}{c} \begin{matrix} \;\; p & \;\; q & \;\; p & m-2p-q \end{matrix} \\ \left(\begin{array}{cccc} \boldsymbol{\Sigma}_s & \boldsymbol{0} & \boldsymbol{0} & \boldsymbol{0} \\ \boldsymbol{0} & \boldsymbol{I} & \boldsymbol{0} & \boldsymbol{0} \\ \boldsymbol{0} & \boldsymbol{0} & \boldsymbol{0} & \boldsymbol{0} \\ \boldsymbol{0} & \boldsymbol{0} & \boldsymbol{\Sigma}_s & \boldsymbol{0} \\ \boldsymbol{0} & \boldsymbol{0} & \boldsymbol{0} & \boldsymbol{0} \\ \boldsymbol{0} & \boldsymbol{0} & \boldsymbol{0} & \boldsymbol{0} \end{array}\right) \end{array} \begin{array}{l} p \\ q \\ n-p-q \\ p \\ q \\ n-p-q \end{array}, \qquad \begin{array}{c} \boldsymbol{\Sigma}_s = \operatorname{diag}(\sigma_1^s, \dots, \sigma_p^s) \in \mathbb{R}^{p \times p}, \\ \sigma_i^s > 0 \quad \text{for } 1 \leq i \leq p, \end{array} \tag{22}$$

with $p, q \in \mathbb{N}$ and $\operatorname{rank}(\boldsymbol{B}) = 2p + q$ and where we indicate the block row and column dimensions in $\boldsymbol{D}$ by small letters. The diagonal entries σ_i^s, $1 \leq i \leq p$, of the matrix $\boldsymbol{\Sigma}_s$ are related to the pairs of purely imaginary eigenvalues $\lambda_j(\boldsymbol{M}), \lambda_{p+j}(\boldsymbol{M}) \in \mathbb{C}$ of $\boldsymbol{M} = \boldsymbol{B}^T \mathbb{J}_{2n} \boldsymbol{B} \in \mathbb{R}^{m \times m}$ with

$$\lambda_j(\boldsymbol{M}) = -(\sigma_j^s)^2 \mathrm{i}, \qquad \lambda_{p+j}(\boldsymbol{M}) = (\sigma_j^s)^2 \mathrm{i}, \qquad 1 \leq j \leq p.$$

Remark 7 (Singular values). *We call the diagonal entries σ_i^s, $1 \leq i \leq p$, of the matrix $\boldsymbol{\Sigma}_s$ from Proposition 9 in the following the symplectic singular values. The reason is the following analogy to the classical SVD.*

The classical SVD decomposes $\boldsymbol{B} \in \mathbb{R}^{2n \times m}$ as $\boldsymbol{B} = \boldsymbol{U}\boldsymbol{\Sigma}\boldsymbol{V}^T$ where $\boldsymbol{U} \in \mathbb{R}^{2n \times 2n}$, $\boldsymbol{V} \in \mathbb{R}^{m \times m}$ are each orthogonal matrices and $\boldsymbol{\Sigma} \in \mathbb{R}^{2n \times m}$ is a diagonal matrix with the singular values σ_i on its diagonal $\operatorname{diag}(\boldsymbol{\Sigma}) = [\sigma_1, \dots, \sigma_r, 0, \dots, 0] \in \mathbb{R}^{\min(2n,m)}$, $r = \operatorname{rank}(\boldsymbol{B})$. The singular values are linked to the real eigenvalues of $\boldsymbol{N} = \boldsymbol{B}^T\boldsymbol{B}$ with $\lambda_i(\boldsymbol{N}) = \sigma_i^2$. Furthermore, for the SVD, it holds due to the orthogonality of $\boldsymbol{U}$ and $\boldsymbol{V}$, respectively, $\boldsymbol{B}^T\boldsymbol{B} = \boldsymbol{V}^T\boldsymbol{\Sigma}^2\boldsymbol{V}$ and $\boldsymbol{B}\boldsymbol{B}^T = \boldsymbol{U}^T\boldsymbol{\Sigma}^2\boldsymbol{U}$.

A similar relation can be derived for an SVD-like decomposition from Proposition 9. Due to the structure of the decomposition (22) and the symplecticity of $\mathbf{S}$*, it holds*

$$\begin{aligned} \boldsymbol{B}^T\mathbb{J}_{2n}\boldsymbol{B} &= \boldsymbol{Q}^T\boldsymbol{D}^T\overbrace{\boldsymbol{S}^T\mathbb{J}_{2n}\boldsymbol{S}}^{=\mathbb{J}_{2n}}\boldsymbol{D}\boldsymbol{Q} \\ &= \boldsymbol{Q}^T\boldsymbol{D}^T\mathbb{J}_{2n}\boldsymbol{D}\boldsymbol{Q}, \end{aligned} \qquad \boldsymbol{D}^T\mathbb{J}_{2n}\boldsymbol{D} = \begin{array}{c} \begin{array}{cccc} p & q & p & m-2p-q \end{array} \\ \left(\begin{array}{cccc} \mathbf{0} & \mathbf{0} & \boldsymbol{\Sigma}_{\mathrm{s}}^2 & \mathbf{0} \\ \mathbf{0} & \mathbf{0} & \mathbf{0} & \mathbf{0} \\ -\boldsymbol{\Sigma}_{\mathrm{s}}^2 & \mathbf{0} & \mathbf{0} & \mathbf{0} \\ \mathbf{0} & \mathbf{0} & \mathbf{0} & \mathbf{0} \end{array}\right) \end{array} \begin{array}{l} p \\ q \\ p \\ m-2p-q \end{array}. \tag{23}$$

This analogy is why we call the diagonal entries σ_i^{s}*,* $1 \leq i \leq p$*, of the matrix* $\boldsymbol{\Sigma}_{\mathrm{s}}$ *symplectic singular values.*

The idea for the basis generation now is to select $k \in \mathbb{N}$ pairs of columns of $\boldsymbol{S}$ in order to compute a symplectic ROB. The selection should be based on the importance of these pairs which we characterize by the following proposition by linking the Frobenius norm of a matrix with the symplectic singular values.

Proposition 10. *Let* $\boldsymbol{B} \in \mathbb{R}^{2n \times m}$ *with an SVD-like decomposition* $\boldsymbol{B} = \boldsymbol{S}\boldsymbol{D}\boldsymbol{Q}$ *with* $p, q \in \mathbb{N}$ *from Proposition 9. The Frobenius norm of* $\boldsymbol{B}$ *can be rewritten as*

$$\|\boldsymbol{B}\|_{\mathrm{F}}^2 = \sum_{i=1}^{p+q} (w_i^{\mathrm{s}})^2, \qquad w_i^{\mathrm{s}} = \begin{cases} \sigma_i^{\mathrm{s}}\sqrt{\|\boldsymbol{s}_i\|_2^2 + \|\boldsymbol{s}_{n+i}\|_2^2}, & 1 \leq i \leq p, \\ \|\boldsymbol{s}_i\|_2, & p+1 \leq i \leq p+q, \end{cases} \tag{24}$$

where $\boldsymbol{s}_i \in \mathbb{R}^{2n}$ *is the i-th column of* $\boldsymbol{S}$ *for* $1 \leq i \leq 2n$*. In the following, we refer to each* w_i^{s} *as the weighted symplectic singular value.*

Proof. We insert the SVD-like decomposition $\boldsymbol{B} = \boldsymbol{S}\boldsymbol{D}\boldsymbol{Q}$ and use the orthogonality of $\boldsymbol{Q}$ to reformulate

$$\begin{aligned} \|\boldsymbol{B}\|_{\mathrm{F}}^2 = \|\boldsymbol{S}\boldsymbol{D}\boldsymbol{Q}\|_{\mathrm{F}}^2 = \|\boldsymbol{S}\boldsymbol{D}\|_{\mathrm{F}}^2 = \mathrm{trace}(\boldsymbol{D}^{\mathsf{T}}\boldsymbol{S}^{\mathsf{T}}\boldsymbol{S}\boldsymbol{D}) &= \sum_{i=1}^{p}(\sigma_i^{\mathrm{s}})^2\boldsymbol{s}_i^{\mathsf{T}}\boldsymbol{s}_i + \sum_{i=1}^{p}(\sigma_i^{\mathrm{s}})^2\boldsymbol{s}_{n+i}^{\mathsf{T}}\boldsymbol{s}_{n+i} + \sum_{i=1}^{q}\boldsymbol{s}_{p+i}^{\mathsf{T}}\boldsymbol{s}_{p+i} \\ &= \sum_{i=1}^{p}(\sigma_i^{\mathrm{s}})^2\left(\|\boldsymbol{s}_i\|_2^2 + \|\boldsymbol{s}_{n+i}\|_2^2\right) + \sum_{i=1}^{q}\|\boldsymbol{s}_{p+i}\|_2^2, \end{aligned}$$

which is equivalent to (24). □

It proves true in the following Proposition Proposition 11 that we can delete single addends w_i^{s} in (24) with the symplectic projection used in the PSD if we include the corresponding pair of columns in the ROB. This will be our selection criterion in the new basis generation technique that we denote PSD SVD-like decomposition.

Definition 9 (PSD SVD-like decomposition). *We compute an SVD-like decomposition (22) as* $\boldsymbol{X}_{\mathrm{s}} = \boldsymbol{S}\boldsymbol{D}\boldsymbol{Q}$ *of the snapshot matrix* $\boldsymbol{X}_{\mathrm{s}} \in \mathbb{R}^{2n \times n_{\mathrm{s}}}$ *and define* $p, q \in \mathbb{N}$ *as in Proposition 9. In order to compute an ROB* $\boldsymbol{V}$ *with 2k columns, find the k indices* $i \in \mathcal{I}_{PSD} = \{i_1, \dots, i_k\} \subset \{1, \dots, p+q\}$ *which have large contributions* w_i^{s} *in (24) with*

$$\mathcal{I}_{PSD} = \underset{\substack{\mathcal{I} \subset \{1,\dots,p+q\} \\ |\mathcal{I}|=k}}{\mathrm{argmax}} \left(\sum_{i \in \mathcal{I}} (w_i^{\mathrm{s}})^2\right). \tag{25}$$

To construct the ROB, we choose the k pairs of columns $s_i \in \mathbb{R}^{2n}$ *from* S *corresponding to the selected indices* $\mathcal{I}_{\mathrm{PSD}}$ *such that*

$$V = [s_{i_1}, \ldots, s_{i_k}, s_{n+i_1}, \ldots, s_{n+i_k}] \in \mathbb{R}^{2n \times 2k}.$$

The special choice of the ROB is motivated by the following theoretical result which is very analogous to the results known for the classical POD in the framework of orthogonal projections.

Proposition 11 (Projection error by neglegted weighted symplectic singular values). *Let* $V \in \mathbb{R}^{2n\times 2k}$ *be an ROB constructed with the procedure described in Definition 9 to the index set* $\mathcal{I}_{\mathrm{PSD}} \subset \{1, \ldots, p+q\}$ *with* $p, q \in \mathbb{N}$ *from Proposition 9. The PSD functional can be calculated by*

$$\left\|(I_{2n} - VV^+)X_s\right\|_F^2 = \sum_{i\in\{1,\ldots,p+q\}\setminus\mathcal{I}_{\mathrm{PSD}}} (w_i^s)^2, \tag{26}$$

which is the cumulative sum of the squares of the neglected weighted symplectic singular values.

Proof. Let $V \in \mathbb{R}^{2n\times 2k}$ be an ROB constructed from an SVD-like decomposition $X_s = SDQ$ of the snapshot matrix $X_s \in \mathbb{R}^{2n\times 2k}$ with the procedure described in Definition 9. Let $p, q \in \mathbb{N}$ be defined as in Proposition 9 and $\mathcal{I}_{\mathrm{PSD}} = \{i_1, \ldots, i_k\} \subset \{1, \ldots, p+q\}$ be the set of indices selected with (25).

For the proof, we introduce a slightly different notation of the ROB V. The selection of the columns s_i of S is denoted with the selection matrix $I_{\mathcal{I}^{2k}_{\mathrm{PSD}}} \in \mathbb{R}^{2n\times 2k}$ based on

$$\left(I_{\mathcal{I}_{\mathrm{PSD}}}\right)_{\alpha,\beta} = \begin{cases} 1, & \alpha = i_\beta \in \mathcal{I}_{\mathrm{PSD}}, \\ 0, & \text{else}, \end{cases} \qquad \text{for} \quad \begin{matrix} 1 \le \alpha \le 2n, \\ 1 \le \beta \le k, \end{matrix} \qquad I_{\mathcal{I}^{2k}_{\mathrm{PSD}}} = [I_{\mathcal{I}_{\mathrm{PSD}}}, \mathbb{J}_{2n}^{\mathsf{T}} I_{\mathcal{I}_{\mathrm{PSD}}}],$$

which allows us to write the ROB as the matrix–matrix product $V = S I_{\mathcal{I}^{2k}_{\mathrm{PSD}}}$. Furthermore, we can select the neglected entries with $I_{2n} - I_{\mathcal{I}^{2k}_{\mathrm{PSD}}} \left(I_{\mathcal{I}^{2k}_{\mathrm{PSD}}}\right)^{\mathsf{T}}$.

We insert the SVD-like decomposition and the representation of the ROB introduced in the previous paragraph in the PSD which reads

$$\left\|(I_{2n} - VV^+)X_s\right\|_F^2 = \left\|(I_{2n} - S I_{\mathcal{I}^{2k}_{\mathrm{PSD}}} \mathbb{J}_{2k}^{\mathsf{T}} I^{\mathsf{T}}_{\mathcal{I}^{2k}_{\mathrm{PSD}}} S^{\mathsf{T}} \mathbb{J}_{2n}) SDQ\right\|_F^2 = \left\|S(I_{2n} - I_{\mathcal{I}^{2k}_{\mathrm{PSD}}} \mathbb{J}_{2k}^{\mathsf{T}} I^{\mathsf{T}}_{\mathcal{I}^{2k}_{\mathrm{PSD}}} \overbrace{S^{\mathsf{T}} \mathbb{J}_{2n} S}^{=\mathbb{J}_{2n}}) D\right\|_F^2,$$

where we use the orthogonality of Q and the symplecticity of S in the last step. We can reformulate the product of Poisson matrices and the selection matrix as

$$\mathbb{J}_{2k}^{\mathsf{T}} I^{\mathsf{T}}_{\mathcal{I}^{2k}_{\mathrm{PSD}}} \mathbb{J}_{2n} = \mathbb{J}_{2k}^{\mathsf{T}} \begin{bmatrix} I^{\mathsf{T}}_{\mathcal{I}_{\mathrm{PSD}}} \\ I^{\mathsf{T}}_{\mathcal{I}_{\mathrm{PSD}}} \mathbb{J}_{2n} \end{bmatrix} \mathbb{J}_{2n} = \begin{bmatrix} 0_k & -I_k \\ I_k & 0_k \end{bmatrix} \begin{bmatrix} I^{\mathsf{T}}_{\mathcal{I}_{\mathrm{PSD}}} \mathbb{J}_{2n} \\ -I^{\mathsf{T}}_{\mathcal{I}_{\mathrm{PSD}}} \end{bmatrix} = I^{\mathsf{T}}_{\mathcal{I}^{2k}_{\mathrm{PSD}}}.$$

Thus, we can further reformulate the PSD as

$$\left\|(I_{2n} - VV^+)X_s\right\|_F^2 = \left\|S\left(I_{2n} - I_{\mathcal{I}^{2k}_{\mathrm{PSD}}} \left(I_{\mathcal{I}^{2k}_{\mathrm{PSD}}}\right)^{\mathsf{T}}\right) D\right\|_F^2 = \sum_{i\in\{1,\ldots,p+q\}\setminus\mathcal{I}_{\mathrm{PSD}}} (w_i^s)^2,$$

where w_i^s are the weighted symplectic singular values from (24). In the last step, we use that the resultant diagonal matrix in the braces sets all rows of D with indices $i, n+i$ to zero for $i \in \mathcal{I}_{\mathrm{PSD}}$. Thus, the last step can be concluded analogously to the proof of Proposition 10. □

A direct consequence of Proposition 11 is that the decay of the PSD functional is proportional to the decay of the sum over the neglected weighted symplectic singular values w_i^s from (24). In the numerical example Section 4.2.1, we observe an exponential decrease of this quantities which induces an exponential decay of the PSD functional.

Remark 8 (Computation of the SVD-like decomposition). *To compute an SVD-like decompostion (22) of $\boldsymbol{B}$, several approaches exist. The original paper [6] derives a decomposition based on the product $\boldsymbol{B}^T\mathbb{J}_{2n}\boldsymbol{B}$ which is not good for a numerical computation since errors can arise from cancellation. In [3], an implicit version is presented that does not require the computation of the full product $\boldsymbol{B}^T\mathbb{J}_{2n}\boldsymbol{B}$ but derives the decomposition implicitly by transforming $\boldsymbol{B}$. Furthermore, Ref. [18] introduces an iterative approach to compute an SVD-like decomposition which computes parts of an SVD-like decomposition with a block-power iterative method. In the present case, we use the implicit approach [3].*

To conclude the new method, we display the computational steps in Algorithm 1. All methods in this algorithm are standard MATLAB® functions except for $[\boldsymbol{S}, \boldsymbol{D}, \boldsymbol{Q}, p, q] = \texttt{SVD\_like\_decomp}(\boldsymbol{X}_s)$ which is supposed to return the matrices $\boldsymbol{S}$, $\boldsymbol{D}$, $\boldsymbol{Q}$ and integers p, q of the SVD-like decomposition (22). The matrix $\boldsymbol{Q}$ is not required and thus, replaced with $\sim$ as usual in MATLAB® notation.

Algorithm 1: PSD SVD-like decomposition in MATLAB® notation.

Input: snapshot matrix $\boldsymbol{X}_s \in \mathbb{R}^{2n \times n_s}$, size $2k$ of the ROB
Output: symplectic ROB $\boldsymbol{V} \in \mathbb{R}^{2n \times 2k}$

$[\boldsymbol{S}, \boldsymbol{D}, \sim, p, q] \leftarrow \texttt{SVD\_like\_decomp}(\boldsymbol{X}_s)$ // compute SVD-like decomposition, Q is not required
$\boldsymbol{\sigma}^s \leftarrow \texttt{diag}(\boldsymbol{D}(1:p, 1:p))$ // extract symplectic singular values
$\boldsymbol{r} \leftarrow \texttt{sum}(\texttt{power}(\boldsymbol{S}, 2), 1)$ // compute squares of the 2-norm of each column of S
$\boldsymbol{w}^s \leftarrow \texttt{times}(\boldsymbol{\sigma}^s, \texttt{sqrt}(\boldsymbol{r}(1:p) + \boldsymbol{r}(n + (1:p))))$ // weighted sympl. singular values $w_1^s, \dots, w_p^s$
$\boldsymbol{w}^s \leftarrow [\boldsymbol{w}^s, \boldsymbol{r}(p + (1:q))]$ // append weighted symplectic singular values $w_{p+1}^s, \dots, w_{p+q}^s$
$[\sim, \mathcal{I}_{\text{PSD}}] \leftarrow \texttt{maxk}(\boldsymbol{w}^s, k)$ // find indices of k highest weighted symplectic singular values
$\boldsymbol{V} \leftarrow \boldsymbol{S}(:, [\mathcal{I}_{\text{PSD}}, n + \mathcal{I}_{\text{PSD}}])$ // select columns with indices $\mathcal{I}_{\text{PSD}}$ and $n + \mathcal{I}_{\text{PSD}}$

3.3. Interplay of Non-Orthonormal and Orthonormal ROBs

We give further results on the interplay of non-orthonormal and orthonormal ROBs. The fundamental statement in the current section is the Orthogonal SR decomposition [6,19].

Proposition 12 (Orthogonal SR decomposition). *For each matrix $\boldsymbol{B} \in \mathbb{R}^{2n \times m}$ with $m \leq n$, there exists a symplectic, orthogonal matrix $\boldsymbol{S} \in \mathbb{R}^{2n \times 2n}$, an upper triangular matrix $\boldsymbol{R}_{11} \in \mathbb{R}^{m \times m}$ and a strictly upper triangular matrix $\boldsymbol{R}_{21} \in \mathbb{R}^{m \times m}$ such that*

$$\boldsymbol{B} = \boldsymbol{S} \begin{bmatrix} \boldsymbol{R}_{11} \\ \boldsymbol{0}_{(n-m) \times m} \\ \boldsymbol{R}_{21} \\ \boldsymbol{0}_{(n-m) \times m} \end{bmatrix} = \begin{bmatrix} \boldsymbol{S}_m & \mathbb{J}_{2n}^T \boldsymbol{S}_m \end{bmatrix} \begin{bmatrix} \boldsymbol{R}_{11} \\ \boldsymbol{R}_{21} \end{bmatrix}, \qquad \begin{aligned} \boldsymbol{S}_i &= [\boldsymbol{s}_1, \dots, \boldsymbol{s}_i], \quad 1 \leq i \leq n, \\ \boldsymbol{S} &= [\boldsymbol{s}_1, \dots, \boldsymbol{s}_n, \mathbb{J}_{2n}^T \boldsymbol{s}_1, \dots, \mathbb{J}_{2n}^T \boldsymbol{s}_n]. \end{aligned}$$

We remark that a similar result can be derived for the case $m > n$ [6], but it is not introduced since we do not need it in the following.

Proof. Let $\boldsymbol{B} \in \mathbb{R}^{2n \times m}$ with $m \leq n$. We consider the QR decomposition

$$\boldsymbol{B} = \boldsymbol{Q} \begin{bmatrix} \boldsymbol{R} \\ \boldsymbol{0}_{(2n-m) \times m} \end{bmatrix},$$

where $\boldsymbol{Q} \in \mathbb{R}^{2n \times 2n}$ is an orthogonal matrix and $\boldsymbol{R} \in \mathbb{R}^{2n \times m}$ is upper triangular. The original Orthogonal SR decomposition ([19], Corollary 4.5.) for the square matrix states that we can decompose $\boldsymbol{Q} \in \mathbb{R}^{2n \times 2n}$

as a symplectic, orthogonal matrix $S \in \mathbb{R}^{2n \times 2n}$, an upper triangular matrix $\widetilde{R}_{11} \in \mathbb{R}^{n \times n}$, a strictly upper triangular matrix $\widetilde{R}_{21} \in \mathbb{R}^{n \times n}$ and two (possibly) full matrices $\widetilde{R}_{12}, \widetilde{R}_{22} \in \mathbb{R}^{n \times n}$

$$Q = S \begin{bmatrix} \widetilde{R}_{11} & \widetilde{R}_{12} \\ \widetilde{R}_{21} & \widetilde{R}_{22} \end{bmatrix} \qquad \text{and thus} \qquad B = S \begin{bmatrix} \widetilde{R}_{11} & \widetilde{R}_{12} \\ \widetilde{R}_{21} & \widetilde{R}_{22} \end{bmatrix} \begin{bmatrix} R \\ 0_{(2n-m) \times m} \end{bmatrix} = S \begin{bmatrix} \widetilde{R}_{11} \\ \widetilde{R}_{21} \end{bmatrix} \begin{bmatrix} R \\ 0_{(n-m) \times m} \end{bmatrix}.$$

Since R is upper triangular, it does preserve the (strictly) upper triangular pattern in $\widetilde{R}_{11}$ and $\widetilde{R}_{21}$ and we obtain the (strictly) upper triangular matrices $R_{11}, R_{21} \in \mathbb{R}^{m \times m}$ from

$$\begin{bmatrix} R_{11} \\ 0_{(n-m) \times m} \\ R_{21} \\ 0_{(n-m) \times m} \end{bmatrix} = \begin{bmatrix} \widetilde{R}_{11} \\ \widetilde{R}_{21} \end{bmatrix} \begin{bmatrix} R \\ 0_{(n-m) \times m} \end{bmatrix}.$$

□

Based on the Orthogonal SR decomposition, the following two propositions prove bounds for the projection errors of PSD which allows an estimate for the quality of the respective method. In both cases, we require the basis size to satisfy $k \leq n$ or $2k \leq n$, respectively. This restriction is not limiting in the context of symplectic MOR as in all application cases $k \ll n$.

Similar results have been presented in ([20], Proposition 3.11) for PSD Cotangent Lift. In comparison to these results, we are able to extend the bound to the case of PSD Complex SVD and thereby improve the bound for the projection error by a factor of $\frac{1}{2}$.

Proposition 13. *Let $V \in \mathbb{R}^{2n \times k}$ be a minimizer of POD with $k \leq n$ basis vectors and $V_E \in \mathbb{R}^{2n \times 2k}$ be a minimizer of the PSD in the class of orthonormal, symplectic matrices with $2k$ basis vectors. Then, the orthogonal projection errors of V_E and V satisfy*

$$\left\| (I_{2n} - V_E V_E^T) X_s \right\|_F^2 \leq \left\| \left(I_{2n} - V V^T \right) X_s \right\|_F^2.$$

Proof. The Orthogonal SR decomposition (see Proposition 12) guarantees that a symplectic, orthogonal matrix $S \in \mathbb{R}^{2n \times 2k}$ and $R \in \mathbb{R}^{2k \times k}$ exist with $V = SR$. Since both matrices V and S are orthogonal and $\operatorname{img}(V) \subset \operatorname{img}(S)$, we can show that S yields a lower projection error than V with

$$\begin{aligned} \left\| \left(I_{2n} - SS^{\mathsf{T}} \right) X_s \right\|_F^2 &= \left\| \left(I_{2n} - SS^{\mathsf{T}} \right) \left(I_{2n} - VV^{\mathsf{T}} \right) X_s \right\|_F^2 = \sum_{i=1}^{n_s} \left\| \left(I_{2n} - SS^{\mathsf{T}} \right) \left(I_{2n} - VV^{\mathsf{T}} \right) x_i^s \right\|_2^2 \\ &\leq \underbrace{\left\| I_{2n} - SS^{\mathsf{T}} \right\|_2^2}_{\leq 1} \sum_{i=1}^{n_s} \left\| \left(I_{2n} - VV^{\mathsf{T}} \right) x_i^s \right\|_2^2 \leq \left\| \left(I_{2n} - VV^{\mathsf{T}} \right) X_s \right\|_F^2. \end{aligned}$$

Let $V_E \in \mathbb{R}^{2n \times 2k}$ be a minimizer of the PSD in the class of symplectic, orthonormal ROBs. By definition of V_E, it yields a lower projection error than S. Since both ROBs are symplectic and orthonormal, we can exchange the symplectic inverse with the transposition (see Proposition 4, (iii)). This proves the assertion with

$$\left\| \left(I_{2n} - VV^{\mathsf{T}} \right) X_s \right\|_F^2 \geq \left\| \left(I_{2n} - SS^{\mathsf{T}} \right) X_s \right\|_F^2 \geq \left\| \left(I_{2n} - V_E V_E^{\mathsf{T}} \right) X_s \right\|_F^2.$$

□

Proposition 13 proves that we require at most twice the number of basis vectors to generate a symplectic, orthonormal basis with an orthogonal projection error at least as small as the one of the classical POD. An analogous result can be derived in the framework of a symplectic projection which is proven in the following proposition.

Proposition 14. *Assume that there exists a minimizer* $V \in \mathbb{R}^{2n\times 2k}$ *of the general PSD for a basis size* $2k \leq n$ *with potentially non-orthonormal columns. Let* $V_E \in \mathbb{R}^{2n\times 4k}$ *be a minimizer of the PSD in the class of symplectic, orthogonal bases of size* $4k$*. Then, we know that the symplectic projection error of* V_E *is less than or equal to the one of* V*, i.e.,*

$$\left\|(I_{2n} - V_E V_E^+)X_s\right\|_F^2 \leq \left\|(I_{2n} - VV^+)X_s\right\|_F^2 .$$

Proof. Let $V \in \mathbb{R}^{2n\times 2k}$ be a minimizer of PSD with $2k \leq n$. By Proposition 12, we can determine a symplectic, orthogonal matrix $S \in \mathbb{R}^{2n\times 4k}$ and $R \in \mathbb{R}^{4k\times 2k}$ with $V = SR$. Similar to the proof of Proposition 13, we can bound the projection errors. We require the identity

$$(I_{2n} - SS^+)(I_{2n} - VV^+) = I_{2n} - SS^+ - VV^+ + \underbrace{S\overbrace{S^+S}^{=I_{4k}}R}_{=V}\underbrace{\mathbb{J}_{2k}^{\mathsf{T}}R^{\mathsf{T}}S^{\mathsf{T}}\mathbb{J}_{2n}}_{=V^+} = I_{2n} - SS^+.$$

With this identity, we proceed analogously to the proof of Proposition 13 and derive for a minimizer $V_E \in \mathbb{R}^{2n\times 4k}$ of PSD in the class of symplectic, orthonormal ROBs

$$\begin{aligned}\left\|(I_{2n} - V_E V_E^+)X_s\right\|_F^2 &\leq \left\|(I_{2n} - SS^+)X_s\right\|_F^2 = \left\|(I_{2n} - SS^+)(I_{2n} - VV^+)X_s\right\|_F^2 \\ &\leq \underbrace{\left\|(I_{2n} - SS^+)\right\|_2^2}_{\leq 1}\left\|(I_{2n} - VV^+)X_s\right\|_F^2 \leq \left\|(I_{2n} - VV^+)X_s\right\|_F^2 .\end{aligned}$$

□

Proposition 14 proves that we require at most twice the number of basis vectors to generate a symplectic, orthonormal basis with a symplectic projection error at least as small as the one of a (potentially non-orthonormal) minimizer of PSD.

4. Numerical Results

The numerical experiments in the present paper are based on a two-dimensional plane strain linear elasticity model which is described by a Lamé–Navier equation

$$\rho_0 \frac{\partial^2}{\partial^2 t^2} u(\xi, t, \mu) - \mu_L \Delta_\xi u(\xi, t, \mu) + (\lambda_L + \mu_L)\nabla_\xi \left(\mathrm{div}_\xi \left(u(\xi, t, \mu)\right)\right) = \rho_0\, g(\xi, t)$$

for $\xi \in \Omega \subset \mathbb{R}^2$ and $t \in [t_0, t_{\mathrm{end}}]$ with the density $\rho_0 \in \mathbb{R}_{>0}$, the Lamé constants $\mu = (\lambda_L, \mu_L) \in \mathbb{R}^2_{>0}$, the external force $g : \Omega \times [t_0, t_{\mathrm{end}}] \to \mathbb{R}^2$ and Dirichlet boundary conditions on $\Gamma_u \subset \Gamma := \partial\Omega$ and Neumann boundary conditions on $\Gamma_t \subset \Gamma$. We apply non-dimensionalization (e.g., ([21], Chapter 4.1)), apply the Finite Element Method (FEM) with piecewise linear Lagrangian ansatz functions on a triangular mesh (e.g., [22]) and rewrite the system as a first-order system to derive a quadratic Hamiltonian system (see Corollary 1) with

$$x(t, \mu) = \begin{bmatrix} q(t,\mu) \\ p(t,\mu) \end{bmatrix}, \qquad H(\mu) = \begin{bmatrix} K(\mu) & 0_n \\ 0_n & M^{-1} \end{bmatrix}, \qquad h(t) = \begin{bmatrix} -f(t) \\ 0_{n\times 1} \end{bmatrix}, \tag{27}$$

where $q(t, \mu) \in \mathbb{R}^n$ is the vector of displacement DOFs, $p(t, \mu) \in \mathbb{R}^n$ is the vector of linear momentum DOFs, $K(\mu) \in \mathbb{R}^{n\times n}$ is the stiffness matrix, $M^{-1} \in \mathbb{R}^{n\times n}$ is the inverse of the mass matrix and $f(t, \mu)$ is the vector of external forces.

We remark that a Hamiltonian formulation with the velocity DOFs $\boldsymbol{v}(t) = \frac{\mathrm{d}}{\mathrm{d}t}\boldsymbol{x}(t) \in \mathbb{R}^n$ instead of the linear momentum DOFs $\boldsymbol{p}(t)$ is possible if a non-canonical symplectic structure is used. Nevertheless, in ([4], Remark 3.8.), it is suggested to switch to a formulation with a canonical symplectic structure for the MOR of Hamiltonian systems.

In order to solve the system (27) numerically with a time-discrete approximation $\boldsymbol{x}_i(\boldsymbol{\mu}) \approx \boldsymbol{x}(t_i, \boldsymbol{\mu})$ for each of $n_\mathrm{t} \in \mathbb{N}$ time steps $t_i \in [t_0, t_\mathrm{end}]$, $1 \leq i \leq n_\mathrm{t}$, a numerical integrator is required. The preservation of the symplectic structure in the time-discrete system requires a so-called symplectic integrator [8,23]. In the context of our work, the implicit midpoint scheme is used in all cases for the sake of simplicity. Higher-order symplectic integrators exist and could as well be applied.

Remark 9 (Modified Hamiltonian). *We remark that, even though the symplectic structure is preserved by symplectic integrators, the Hamiltonian may be modified in the time-discrete system compared to the original Hamiltonian. In the case of a quadratic Hamiltonian (see Corollary 1) and a symplectic Runge–Kutta integrator, the modified Hamiltonian equals the original Hamiltonian since these integrators preserve quadratic first integrals. For further details, we refer to ([8], Chapter IX.) or ([24], Sections 5.1.2 and 5.2).*

The model parameters are the first and second Lamé constants with $\boldsymbol{\mu} = (\lambda_\mathrm{L}, \mu_\mathrm{L}) \in \mathcal{P} = [35 \times 10^9, 125 \times 10^9]\ \mathrm{N/m^2} \times [35 \times 10^9, 83 \times 10^9]\ \mathrm{N/m^2}$ which varies between cast iron and steel with approx. 12% chromium ([25], App. E 1 Table 1). The density is set to $\rho_0 = 7856\ \mathrm{kg/m^3}$. The non-dimensionalization constants are set to $\lambda_\mathrm{L}^\mathrm{c} = \mu_\mathrm{L}^\mathrm{c} = 81 \times 10^9\ \mathrm{N/m^2}$, $\xi^\mathrm{c} = 1\ \mathrm{m}$, $g^\mathrm{c} = 9.81\ \mathrm{m/s^2}$. The geometry is a simple cantilever beam clamped on the left side with a force applied to the right boundary (see Figure 1). The time interval is chosen to be $t \in [t_0, t_\mathrm{end}]$ with $t_0 = 0\,\mathrm{s}$ and $t_\mathrm{end} = 7.2 \times 10^{-2}\,\mathrm{s}$ which is one oscillation of the beam. For the numerical integration, $n_\mathrm{t} = 151$ time steps are used.

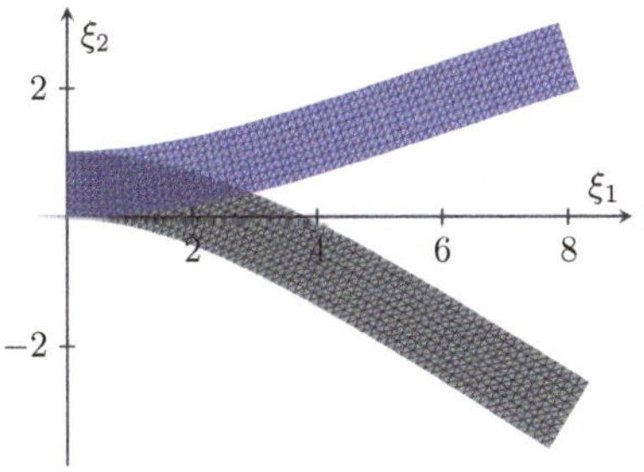

Figure 1. An exaggerated illustration of the displacements $\boldsymbol{q}(t, \boldsymbol{\mu})$ of the non-autonomous beam model (a) at the time with the maximum displacement (gray) and (b) at the final time (blue).

The symplectic MOR techniques examined are PSD Complex SVD (Definition 8), the greedy procedure [5] and the newly introduced PSD SVD-like decomposition (Definition 9). The MOR techniques that do not necessarily derive a symplectic ROB are called non-symplectic MOR techniques in the following. The non-symplectic MOR techniques investigated in the scope of our numerical results are the POD applied to the full state $\boldsymbol{x}(t, \boldsymbol{\mu})$ (POD full state) and a POD applied to the displacement $\boldsymbol{q}(t, \boldsymbol{\mu})$ and linear momentum states $\boldsymbol{p}(t, \boldsymbol{\mu})$ separately (POD separate states). To summarize the basis generation methods, let us enlist them in Table 1 where `SVD(•)` and `cSVD(•)` denote the SVD and the complex SVD, respectively.

Table 1. Basis generation methods used in the numerical experiments in summary, where we use the MATLAB® notation to denote the selection of the first k columns of a matrix, e.g., in $\boldsymbol{U}(:,1:k)$.

method	solution	solution procedure	ortho-norm.	sympl.
POD full	$V_k = \boldsymbol{U}(:,1:k)$	$\boldsymbol{U} = \mathtt{SVD}(\boldsymbol{X}_s)$	✓	✗
POD separate	$V_k = \begin{bmatrix} \boldsymbol{U}_p(:,1:k) \\ \boldsymbol{U}_q(:,1:k) \end{bmatrix}$	$\boldsymbol{U}_p = \mathtt{SVD}\left([p_1,\dots,p_{n_s}]\right)$ $\boldsymbol{U}_q = \mathtt{SVD}\left([q_1,\dots,q_{n_s}]\right)$	✓	✗
PSD cSVD	$V_{2k} = [\boldsymbol{E}(:,1:k)\quad \mathbb{J}_{2n}^{\mathrm{T}}\boldsymbol{E}(:,1:k)]$	$\boldsymbol{E} = \begin{bmatrix} \boldsymbol{\Phi} \\ \boldsymbol{\Psi} \end{bmatrix}, \boldsymbol{\Phi} + \mathrm{i}\boldsymbol{\Psi} = \mathtt{cSVD}\left(\boldsymbol{C}_s\right)$ $\boldsymbol{C}_s = [p_1 + \mathrm{i}q_1,\dots,p_{n_s} + \mathrm{i}q_{n_s}]$	✓	✓
PSD greedy	$V_{2k} = [\boldsymbol{E}(:,1:k)\quad \mathbb{J}_{2n}^{\mathrm{T}}\boldsymbol{E}(:,1:k)]$	$\boldsymbol{E}$ from greedy algorithm	✓	✓
PSD SVD-like	$V_{2k} = [s_{i_1},\dots,s_{i_k},s_{n+i_1},\dots,s_{n+i_k}]$	$\boldsymbol{S} = [s_1,\dots,s_{2n}]$ from (22), $\mathcal{I}_{\mathrm{PSD}} = \{i_1,\dots,i_k\}$ from (25)	✗	✓

All presented experiments are generalization experiments, i.e., we choose nine different training parameter vectors $\boldsymbol{\mu} \in \mathcal{P}$ on a regular grid to generate the snapshots and evaluate the reduced models for 16 random parameter vectors that are distinct from the nine training parameter vectors. Thus, the number of snapshots is $n_s = 9 \cdot 151 = 1359$. The size $2k$ of the ROB $\boldsymbol{V}$ is varied in steps of 20 with $2k \in \{20, 40, \dots, 280, 300\}$.

Furthermore, all experiments consider the performance of the reduced models based on the error introduced by the reduction. We do not compare the computational cost of the different basis generation techniques in the offline-phase since the current (non-optimized) MATLAB® implementation of the SVD-like decomposition does not allow a meaningful numerical comparisons of offline-runtimes as the methods using a MATLAB®-internal, optimized SVD implementation will be faster.

The software used for the numerical experiments is RBmatlab (https://www.morepas.org/software/rbmatlab/) which is an open-source library based on the proprietary software package MATLAB® and contains several reduced simulation approaches. An add-on to RBmatlab is provided in the Supplementary Materials of the current paper which includes all the additional code to reproduce the results of the present paper. The versions used in the present paper are RBmatlab 1.16.09 and MATLAB® 2017a.

4.1. Autonomous Beam Model

In the first model, we load the beam on the free end (far right) with a constant force which induces an oscillation. Due to the constant force, the discretized system can be formulated as an autonomous Hamiltonian system. Thus, the Hamiltonian is constant and its preservation in the reduced models can be analysed. All other reduction results are very similar to the non-autonomous case and thus are exclusively presented for the non-autonomous case in the following Section 4.2.

Preservation over Time of the Modified Hamiltonian in the Reduced Model

In the following, we investigate the preservation of the Hamiltonian of our reduced models. With respect to Remark 9, we mean the preservation over time of the modified Hamiltonian. Since the Hamiltonian is quadratic in our example and the implicit midpoint is a symplectic Runge-Kutta integrator, the modified Hamiltonian equals the original which is why we speak of "the Hamiltonian" in the following.

We present in Figure 2 the count of the total 240 simulations which show a preservation (over time) of the reduced Hamiltonian in the reduced model. The solution x_r of a reduced simulation preserves the reduced Hamiltonian over time if $(\mathcal{H}_r(x_r(t_i), \mu) - \mathcal{H}_r(x_r(t_0), \mu))/\mathcal{H}_{rel}(\mu) < 10^{-10}$ for all discrete times $t_i \in [t_0, t_{end}]$, $1 \leq i \leq n_t$ where $\mathcal{H}_{rel}(\mu) > 0$ is a parameter-dependent normalization factor. The heat map shows that no simulation in the non-symplectic case preserves the Hamiltonian, whereas the symplectic methods all preserve the Hamiltonian which is what was expected from theory.

In Figure 3, we exemplify the non-constant evolution of the reduced Hamiltonian for three non-symplectic bases generated by POD separate states with different basis sizes and one selected test parameter $(\lambda, \mu) \in \mathcal{P}$. It shows that, in all three cases, the Hamiltonian starts to grow exponentially.

POD full state	0/240
POD separate states	0/240
PSD complex SVD	240/240
PSD greedy	240/240
PSD SVD-like	240/240

Figure 2. Heat map which shows the preservation of the reduced Hamiltonian in the reduced model in x of y cases (x/y).

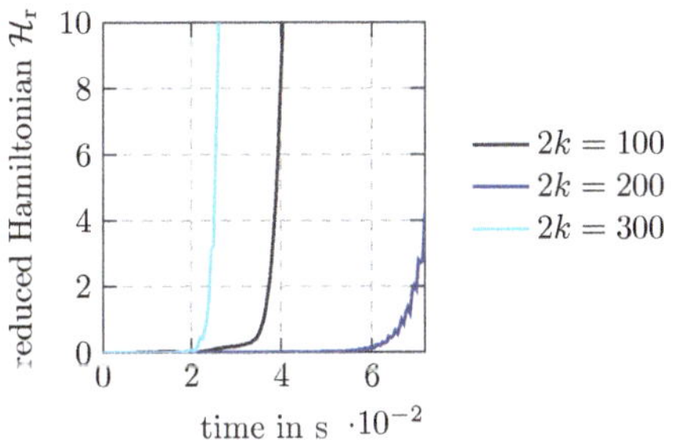

Figure 3. Evolution of the reduced Hamiltonian for POD separate states for a selected parameter $(\lambda, \mu) \in \mathcal{P}$.

4.2. Non-Autonomous Beam Model

The second model is similar to the first one. The only difference is that the force on the free (right) end of the beam is loaded with a time-varying force. The force is chosen to act in phase with the beam. The time dependence of the force necessarily requires a non-autonomous formulation which requires in the framework of the Hamiltonian formulation a time-dependent Hamiltonian function which we introduced in Section 2.4.

We use the model to investigate the quality of the reduction for the considered MOR techniques. To this end, we investigate the projection error, i.e., the error on the training data, the orthogonality and symplecticity of the ROB and the error in the reduced model for the test parameters.

4.2.1. Projection Error of the Snapshots and Singular Values

The projection error is the error on the training data collected in the snapshot matrix X_s, i.e.,

$$e_{l_2}(2k) = \left\| (I_{2n} - VW^\mathsf{T})X_s \right\|_F^2, \quad \begin{array}{l} \text{POD}: W^\mathsf{T} = V^\mathsf{T}, \\ \text{PSD}: W^\mathsf{T} = V^+ (= V^\mathsf{T} \text{for orthosymplectic ROBs, Proposition 4}). \end{array}$$

It is a measure for the approximation qualities of the ROB based on the training data. Figure 4 (left) shows this quantity for the considered MOR techniques and different ROB sizes $2k$. All basis generation techniques show an exponential decay. As expected from theory, POD full state minimizes the projection error for the orthonormal basis generation techniques (see Table 1). PSD SVD-like decomposition shows a lower projection error than the other PSD methods for $2k \geq 80$ and yields a similar projection error for $k \leq 60$. Concluding this experiment, one might expect the full-state POD to yield decent results or even the best results. The following experiments prove this expectation to be wrong.

The decay of (a) the classical singular values σ_i, (b) the symplectic singular values σ_i^{s} (see Remark 7) and (c) the weighted symplectic singular values w_i^{s} (see (24)) sorted by the magnitude of the symplectic singular values is displayed in Figure 4 (right). All show an exponential decrease. The weighting introduced in (24) for w_i^{s} does not influence the exponential decay rate of σ_i^{s}. The decrease in the classical singular values is directly linked to the exponential decrease of the projection error of POD full state due to properties of the Frobenius norm (see [1]). A similar result was deduced in the scope of the present paper for PSD SVD-like decomposition and the PSD functional (see Proposition 11).

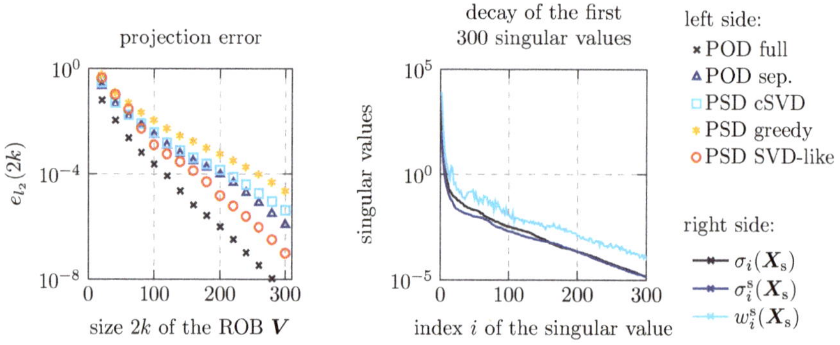

Figure 4. Projection error (left) and decay of the singular values from Remark 7 and (24) (right).

4.2.2. Orthonormality and Symplecticity of the Bases

To verify the orthonormality and the symplecticity numerically, we consider the two functions

$$o_{\boldsymbol{V}}(2k) = \left\| \boldsymbol{V}^{\mathsf{T}} \boldsymbol{V} - \boldsymbol{I}_{2k} \right\|_{\mathrm{F}}, \qquad s_{\boldsymbol{V}}(2k) = \left\| \mathbb{J}_{2k}^{\mathsf{T}} \boldsymbol{V}^{\mathsf{T}} \mathbb{J}_{2n} \boldsymbol{V} - \boldsymbol{I}_{2k} \right\|_{\mathrm{F}}, \tag{28}$$

which are zero/numerically zero if and only if the basis is orthonormal or symplectic, respectively. In Figure 5, we show both values for the considered basis generation techniques and RB sizes.

The orthonormality of the bases is in accordance with the theory. All procedures compute symplectic bases except for PSD SVD like-decomposition. PSD greedy shows minor loss in the orthonormality which is a known issue for the $\mathbb{J}_{2n}$-orthogonalization method used (modified symplectic Gram–Schmidt procedure with re-orthogonalization [26]). However, no major impact on the reduction results could be attributed to this deficiency in the scope of this paper.

In addition, the symplecticity (or $\mathbb{J}_{2n}$-orthogonality) of the bases behaves as expected. All PSD methods generate symplectic bases, whereas the POD methods do not. A minor loss of symplecticity is recorded for PSD SVD-like decomposition which is objected to the computational method that is used to compute an SVD-like decomposition. Further research on algorithms for the computation of an SVD-like decomposition should improve this result. Nevertheless, no major impact on the reduction results could be attributed to this deficiency in the scope of this paper.

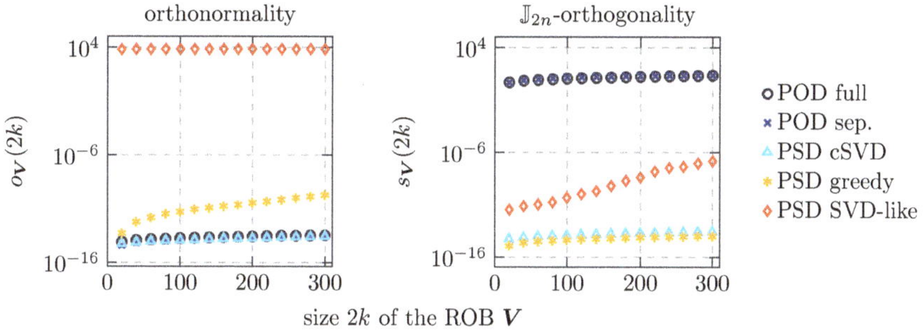

Figure 5. The orthonormality (left) and the $\mathbb{J}_{2n}$-orthogonality (right) from (28).

4.2.3. Relative Error in the Reduced Model

We investigate the error introduced by MOR in the reduced model based on 16 test parameters distinct from the training parameters. The error is measured in the relative ∞-norm $\|\bullet\|_\infty$ in time and space

$$\bar{e}(2k, \boldsymbol{\mu}) := \frac{\max_{i \in \{1,\dots,n_t\}} \|\boldsymbol{x}(t_i, \boldsymbol{\mu}) - \boldsymbol{V}\boldsymbol{x}_r(t_i, \boldsymbol{\mu})\|_\infty}{\max_{i \in \{1,\dots,n_t\}} \|\boldsymbol{x}(t_i, \boldsymbol{\mu})\|_\infty}, \tag{29}$$

where $2k$ indicates the size of the ROB $\boldsymbol{V} \in \mathbb{R}^{2n \times 2k}$, $\boldsymbol{\mu} \in \mathcal{P}$ is one of the test parameters, $\boldsymbol{x}(t, \boldsymbol{\mu}) \in \mathbb{R}^{2n}$ is the solution of the full model (5) and $\boldsymbol{x}_r(t, \boldsymbol{\mu}) \in \mathbb{R}^{2k}$ is the solution of the reduced model (7). This error is used for testing purposes only since it requires the computation of the full solution $\boldsymbol{x}(t, \boldsymbol{\mu})$. It may be instead estimated in the online phase with an a posteriori error estimator as, e.g., in [9,27].

To display the results for all 16 test parameters at once, we use box plots in Figure 6. The box represents the 25%-quartile, the median and the 75%-quartile. The whiskers indicate the range of data points which lay within 1.5 times the interquartile range (IQR). The crosses show outliers. For the sake of a better overview, we truncated relative errors above $10^0 = 100\%$.

The experiments show that the non-symplectic MOR techniques show a strongly non-monotonic behaviour for an increasing basis size. For many of the basis sizes, there exists a parameter which shows crude approximation results which lay above 100% relative error. The POD full state is unable to produce results with a relative error below 2%.

On the other hand, the symplectic MOR techniques show an exponentially decreasing relative error. Furthermore, the IQRs are much lower than for the non-symplectic methods. We stress that the logarithmic scale of the y-axis distorts the comparison of the IQRs—but only in favour of the non-symplectic methods. The low IQRs for the symplectic methods show that the symplectic MOR techniques derive a reliable reduced model that yields good results for any of the 16 randomly chosen test parameters. Furthermore, none of the systems shows an error above 0.19%—for PSD SVD-like decomposition, this bound is 0.018%, i.e., one magnitude lower.

In the set of the considered symplectic, orthogonal MOR techniques, PSD greedy shows the best result for most of the considered ROB sizes. This superior behaviour of PSD greedy in comparison to PSD complex SVD is unexpected since PSD greedy showed inferior results for the projection error in Section 4.2.1. This was also observed in [5].

Within the set of investigated symplectic MOR techniques, PSD SVD-like decomposition shows the best results followed by PSD greedy and PSD complex SVD. While the two orthonormal procedures show comparable results, PSD SVD like-decomposition shows an improvement in the relative error. Comparing the best results of either PSD greedy or PSD complex SVD with the worst result of PSD SVD-like decomposition considering the 16 different test parameters for a fixed basis size—which is pretty much in favour of the orthonormal basis generation techniques—, the improvement of PSD SVD-like decomposition ranges from factor 3.3 to 11.3 with a mean of 6.7.

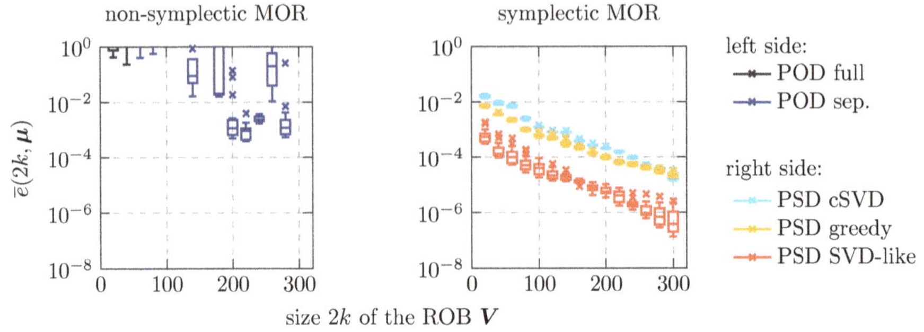

Figure 6. Relative error in the reduced model.

5. Conclusions

We gave an overview of autonomous and non-autonomous Hamiltonian systems and the structure-preserving model order reduction (MOR) techniques for these kinds of systems [4,5,15]. Furthermore, we classified the techniques in orthonormal and non-orthonormal procedures based on the capability to compute a symplectic, (non-)orthonormal reduced order basis (ROB). To this end, we introduced a characterization of rectangular, symplectic matrices with orthonormal columns. Based thereon, an alternative formulation of the PSD Complex SVD [4] was derived which we used to prove the optimality with respect to the PSD functional in the set of orthonormal, symplectic ROBs. As a new method, we presented a symplectic, non-orthonormal basis generation procedure that is based on an SVD-like decomposition [6]. First theoretical results show that the quality of approximation can be linked to a quantity we referred to as weighted symplectic singular values.

The numerical examples show advantages for the considered linear elasticity model for symplectic MOR if a symplectic integrator is used. We were able to reduce the error introduced by the reduction with the newly introduced non-orthonormal method.

We conclude that non-orthonormal methods are able to derive bases with a lower error for both, the training and the test data. However, it is still unclear if the newly introduced method computes the global optimum of the PSD functional. Further work should investigate if a global optimum of the PSD functional can be computed with an SVD-like decomposition.

Furthermore, the application of symplectic MOR techniques in real-time scenarios and multi-query context should be further extended. This includes inverse problems or uncertainty quantification which often require solutions of the model for many different parameters. A suitable framework for uncertainty quantification in combination with symplectic MOR is, e.g., the approach discussed in [28].

Supplementary Materials: The following are available at http://www.mdpi.com/2297-8747/24/2/43/s1.

Author Contributions: Conceptualization, P.B., A.B., and B.H.; methodology, P.B. and A.B.; software, P.B.; validation, P.B.; formal analysis, P.B.; investigation, P.B.; resources, B.H.; data curation, P.B.; writing–original draft preparation, P.B.; writing–review and editing, A.B. and B.H.; visualization, P.B.; supervision, A.B. and B.H.; project administration, B.H.; and funding acquisition, B.H.

Funding: This research was partly funded by the German Research Foundation (DFG) Grant No. HA5821/5-1 and within the GRK 2198/1.

Acknowledgments: We thank the German Research Foundation (DFG) for funding this work. The authors thank Dominik Wittwar for inspiring discussions. The valuable input of the anonymous reviewers is acknowledged.

Conflicts of Interest: The authors declare no conflict of interest.

References

1. Benner, P.; Ohlberger, M.; Cohen, A.; Willcox, K. *Model Reduction and Approximation*; Society for Industrial and Applied Mathematics: Philadelphia, PA, USA, 2017.
2. Buchfink, P. Structure-Preserving Model Order Reduction of Hamiltonian Systems for Linear Elasticity. *ARGESIM Rep.* **2018**, *55*, 35–36. [CrossRef]
3. Xu, H. A Numerical Method for Computing an SVD-like Decomposition. *SIAM J. Matrix Anal. Appl.* **2005**, *26*, 1058–1082. [CrossRef]
4. Peng, L.; Mohseni, K. Symplectic Model Reduction of Hamiltonian Systems. *SIAM J. Sci. Comput.* **2016**, *38*, A1–A27. [CrossRef]
5. Maboudi Afkham, B.; Hesthaven, J.S. Structure Preserving Model Reduction of Parametric Hamiltonian Systems. *SIAM J. Sci. Comput.* **2017**, *39*, A2616–A2644. [CrossRef]
6. Xu, H. An SVD-like matrix decomposition and its applications. *Linear Algebra Its Appl.* **2003**, *368*, 1–24. [CrossRef]
7. Cannas da Silva, A. *Lectures on Symplectic Geometry*; Springer: Berlin/Heidelberg, Germany, 2008.
8. Hairer, E.; Wanner, G.; Lubich, C. *Geometric Numerical Integration: Structure-Preserving Algorithms for Ordinary Differential Equations*; Springer: Berlin/Heidelberg, Germany, 2006.
9. Haasdonk, B.; Ohlberger, M. Efficient Reduced Models and A-Posteriori Error Estimation for Parametrized Dynamical Systems by Offline/Online Decomposition. *Math. Comput. Model. Dyn. Syst.* **2011**, *17*, 145–161. [CrossRef]
10. Barrault, M.; Maday, Y.; Nguyen, N.C.; Patera, A.T. An 'empirical interpolation' method: Application to efficient reduced-basis discretization of partial differential equations. *C. R. Math.* **2004**, *339*, 667–672. [CrossRef]
11. Chaturantabut, S.; Sorensen, D.C. Discrete Empirical Interpolation for nonlinear model reduction. In Proceedings of the 48th IEEE Conference on Decision and Control (CDC) Held Jointly with the 2009 28th Chinese Control Conference, Shanghai, China, 15–18 December 2009; pp. 4316–4321. [CrossRef]
12. Beattie, C.; Gugercin, S. Structure-preserving model reduction for nonlinear port-Hamiltonian systems. In Proceedings of the 2011 50th IEEE Conference on Decision and Control and European Control Conference, Orlando, FL, USA, 12–15 December 2011; pp. 6564–6569. [CrossRef]
13. Chaturantabut, S.; Beattie, C.; Gugercin, S. Structure-Preserving Model Reduction for Nonlinear Port-Hamiltonian Systems. *SIAM J. Sci. Comput.* **2016**, *38*, B837–B865. [CrossRef]
14. Lanczos, C. *The Variational Principles of Mechanics*; Dover Books on Physics; Dover Publications: Mineola, NY, USA, 1970.
15. Maboudi Afkham, B.; Hesthaven, J.S. Structure-Preserving Model-Reduction of Dissipative Hamiltonian Systems. *J. Sci. Comput.* **2018**, [CrossRef]
16. Sirovich, L. Turbulence the dynamics of coherent structures. Part I: coherent structures. *Q. Appl. Math.* **1987**, *45*, 561–571. [CrossRef]
17. Paige, C.; Loan, C.V. A Schur decomposition for Hamiltonian matrices. *Linear Algebra Appl.* **1981**, *41*, 11–32. [CrossRef]
18. Agoujil, S.; Bentbib, A.H.; Kanber, A. An iterative method for computing a symplectic SVD-like decomposition. *Comput. Appl. Math.* **2018**, *37*, 349–363. [CrossRef]
19. Bunse-Gerstner, A. Matrix factorizations for symplectic QR-like methods. *Linear Algebra Appl.* **1986**, *83*, 49–77. [CrossRef]
20. Peng, L.; Mohseni, K. Structure-Preserving Model Reduction of Forced Hamiltonian Systems. *arXiv* **2016**, arXiv:1603.03514.
21. Langtangen, H.P.; Pedersen, G.K. *Scaling of Differential Equations*, 1st ed.; Springer International Publishing: New York, NY, USA, 2016.
22. Ern, A.; Guermond, J.L. Finite Element Interpolation. In *Theory and Practice of Finite Elements*; Springer: New York, NY, USA, 2004; pp. 3–80.
23. Bhatt, A.; Moore, B. Structure-preserving Exponential Runge–Kutta Methods. *SIAM J. Sci. Comput.* **2017**, *39*, A593–A612. [CrossRef]
24. Leimkuhler, B.; Reich, S. *Simulating Hamiltonian Dynamics*; Cambridge University Press: Cambridge, UK, 2005. [CrossRef]

25. Oechsner, M.; Kloos, K.H.; Pyttel, B.; Berger, C.; Kübler, M.; Müller, A.K.; Habig, K.H.; Woydt, M. Anhang E: Diagramme und Tabellen. In *Dubbel: Taschenbuch für den Maschinenbau*; Grote, K.H., Feldhusen, J., Eds.; Springer: Berlin/Heidelberg, Germany, 2014; pp. 324–357. [CrossRef]
26. Al-Aidarous, E. Symplectic Gram–Schmidt Algorithm with Re-Orthogonalization. *J. King Abdulaziz Univ. Sci.* **2011**, *23*, 11–20. [CrossRef]
27. Ruiner, T.; Fehr, J.; Haasdonk, B.; Eberhard, P. A-posteriori error estimation for second order mechanical systems. *Acta Mechanica Sinica* **2012**, *28*, 854–862. [CrossRef]
28. Attia, A.; Ştefănescu, R.; Sandu, A. The reduced-order hybrid Monte Carlo sampling smoother. *Int. J. Numer. Methods Fluids* **2016**, *83*, 28–51. [CrossRef]

Mathematical and Computational Applications

Article

An Artificial Neural Network Based Solution Scheme for Periodic Computational Homogenization of Electrostatic Problems

Felix Selim Göküzüm *, Lu Trong Khiem Nguyen and Marc-André Keip

Institute of Applied Mechanics, Chair of Materials Theory, University of Stuttgart, Pfaffenwaldring 7, 70569 Stuttgart, Germany; nguyen@mechbau.uni-stuttgart.de (L.T.K.N.); keip@mechbau.uni-stuttgart.de (M.-A.K.)

* Correspondence: goekuezuem@mechbau.uni-stuttgart.de

Received: 22 March 2019; Accepted: 9 April 2019; Published: 17 April 2019

Abstract: The present work addresses a solution algorithm for homogenization problems based on an artificial neural network (ANN) discretization. The core idea is the construction of trial functions through ANNs that fulfill a priori the periodic boundary conditions of the microscopic problem. A global potential serves as an objective function, which by construction of the trial function can be optimized without constraints. The aim of the new approach is to reduce the number of unknowns as ANNs are able to fit complicated functions with a relatively small number of internal parameters. We investigate the viability of the scheme on the basis of one-, two- and three-dimensional microstructure problems. Further, global and piecewise-defined approaches for constructing the trial function are discussed and compared to finite element (FE) and fast Fourier transform (FFT) based simulations.

Keywords: machine learning; artificial neural networks; computational homogenization

1. Introduction

Artificial neural networks (ANNs) have attracted a lot of attention in the last few years due to their excellent universal approximation properties. Originally developed to model nervous activity in living brains [1,2], they nowadays grow popular in data-driven approaches. Tasks such as image and speech recognition [3,4] or the prediction of users' behavior on social and commercial websites are characterized by a large amount of accessible data compared to a difficult analytic mathematical description. The use of ANNs or other *machine learning algorithms* such as *anomaly detection* [5] and *support vector machines* [6] is suited for such problems as it enables the fitting of even highly complex data with high accuracy. Recent trends in machine learning concern the physical constraining of data driven methods for even higher convergence rate and accuracy, as done in [7].

Due to the aforementioned advantages and improvements, machine learning algorithms gained entry into the field of continuum mechanics and material modeling as well. Successful implementations were performed for the prediction of material response based on the fitting of experimental data through ANNs [8–10]. Another interesting application is the reduction of microstructure data of a given material through pattern recognition in order to reduce computational demands (see, e.g., [11,12]).

In the present work, we employ ANNs to seek the solution of *homogenization problems*. Homogenization aims for the prediction of the macroscopic response of materials that have microstructures described on length scales far lower than the macroscopic dimension. In terms of *analytical* homogenization, Voigt [13] and Reuss [14] were the first to provide *bounds* of effective properties [15]. Hashin and Shtrikman [16] and Willis [17] improved the theory in terms of tighter bounds. Further estimates were developed using the self-consistent method [18,19] and the Mori–Tanaka method [20] afterwards.

In the case of a rather fine microstructures with complex topology and non-linear material behavior, those bounds are only able to make rough predictions on the effective properties. To describe microscopic fields and effective properties in a more detailed and accurate fashion, several *computational* approaches have been developed in the last decades. Two of the most commonly used discretization techniques are finite element (FE) methods [21,22] and fast Fourier transform (FFT) based solvers [23,24]. However, for materials with a microstructure that need fine discretization, the memory cost and solution time of the solvers increases vastly, making multiscale simulations uneconomical. Promising approaches to mitigate these problems are model order reduction methods (see, e.g., [25,26]).

In the present work, a memory efficient solution scheme based on the discretization through ANNs is presented. We therefore follow the idea of Lagaris et al. [27], who introduced a method for solving differential equations through ANN-based trial functions. The functions are constructed in a way that they a priori fulfill the given boundary conditions and the squared error residual of the differential equation is used as an objective that needs to be optimized with respect to the ANNs' weights. The construction of the trial function might be a difficult task for complicated boundary value problems. However, in conventional homogenization problems, we usually deal with rather simple boundary geometries. In the present work, we consider square representative volume elements (RVE) under periodic boundary conditions, as described in Section 2. In Section 3.1, the concept of the ANN-based trial function according to Lagaris et al. [27] is explained. In contrast to Lagaris et al. [27], the optimization objective in our problem is not the squared error residual of a differential equation but emerges from a global energy potential. Sections 3.2 and 3.3 give the ANNs' structure used in the present work as well as the derivatives necessary to optimize the objective function. Finally, in Section 4, the robustness of the presented method is validated for one-, two- and three-dimensional problems and is compared to FE- and FFT-based computations. Further, we compare a global with a piecewise-defined approach for constructing the trial function. In the global approach, the solution is represented by only one global function using several ANNs and the topology of the microstructure must be learned by the neural networks itself. On the other hand, in a piecewise-defined approach, the solution is represented by many neural networks that are piecewise defined on different sub-domains of the RVE. A conclusion is provided in Section 5.

2. Homogenization Framework

In the present work, we consider first-order homogenization of electrostatic problems. The fundamental laws of electrostatics and the corresponding variational formulation are given. In terms of the homogenization framework, we assume separation of length scales between macro- and microscale in the sense that the length scale at which the material properties vary quickly, namely the microscale, is much smaller than the length scale at the same order of the body's dimensions. A connection for the micro- and macrofields is given by the celebrated Hill–Mandel condition [28], which allows for the derivation of consistent boundary conditions for the microscopic boundary value problem.

2.1. Energy Formulation for Electrostatic Problems

We now recall the fundamental equations of electrostatics in the presence of matter [29]. The focus lies on a variational formulation as the energetic potential is later needed in the optimization principle. Assuming that there are neither free currents nor charges, the fundamental physics of electric fields in a body $\mathcal{B}$ are governed by the reduced Maxwell equations

$$\operatorname{curl} \boldsymbol{E} = 0 \quad \text{and} \quad \operatorname{div} \boldsymbol{D} = 0 \quad \text{in} \quad \mathcal{B}, \tag{1}$$

where $\boldsymbol{E}$ denotes the electric field and $\boldsymbol{D}$ the electric displacement. In vacuum, the electric field and displacement are connected through the vacuum permittivity $\kappa_0 \approx 8.854 \cdot 10^{-12} \frac{\text{As}}{\text{Vm}}$ as $\boldsymbol{D} = \kappa_0 \boldsymbol{E}$. In the presence of matter, the permittivity $\kappa = \kappa_0 \cdot \kappa_r$ must be adapted accordingly. To solve Equation (1), we choose the electric field as our primary variable and construct it as the negative gradient of some scalar electric potential ϕ according to

$$E := -\operatorname{grad}\phi \quad \Rightarrow \quad \operatorname{curl} E \equiv 0, \tag{2}$$

and thus Equation (1)$_1$ is automatically fulfilled. Next, we want to solve Equation (1)$_2$ under some boundary conditions. We therefore introduce an energy potential

$$\Pi(\phi) = \int_{\mathcal{B}} \Psi(E)\, dV + \int_{\partial\mathcal{B}_q} q\,\phi\, dA \quad \text{and} \quad \phi = \phi^* \ \text{ on } \partial\mathcal{B}_\phi, \tag{3}$$

where $\Psi(E)$ is a constitutive energy density, $\partial\mathcal{B}_q$ and $\partial\mathcal{B}_\phi$ denote boundaries along with electric charges q and electric potential ϕ^* as prescribed. From the latter potential, it can be shown that, for equilibrium, i.e. $\delta\Pi = 0$, Equation (1)$_2$ is solved in a weak sense

$$\delta\Pi(\phi) = -\int_{\mathcal{B}} (\operatorname{div} D)\delta\phi\, dV + \int_{\partial\mathcal{B}_q} (q + D\cdot N)\,\delta\phi\, dA = 0 \quad \text{with} \quad D = -\frac{\partial\Psi}{\partial E}, \tag{4}$$

for all $\delta\phi$. Here, N denotes the unit normal vector pointing outwards of $\partial\mathcal{B}$. By choosing $\Psi = -1/2\kappa\, E\cdot E$, the static Maxwell equations are recovered.

2.2. Microscopic Boundary Value Problem

In the present work, we consider homogenization problems governed by the existence of so-called representative volume elements (RVEs). They are chosen in a way that they are statistically representative for the overall microstructure of the material. Fields emerging on the microscale are driven by macroscopic fields, which are assumed to be constant in the RVE due to the separation of length scales. The separation of length scales leads to a degeneration of the RVEs to points on the macroscale. Scale transition rules can be derived from the Hill–Mandel condition [28] by postulating energy conservation between the microscopic RVE and a macroscopic observation in the form

$$\overline{\Pi} = \sup_{E} \frac{1}{|\mathcal{B}|}\Pi(\phi), \tag{5}$$

where the macroscopic energy potential $\overline{\Pi}$ is obtained at equilibrium of the microscopic energy potential $\Pi(\phi)$. According to Equation (3), the internal energy density appearing in the potential is now a function of the electric field. In line with the assumption of first-order homogenization and separation of length scales, the electric field vector

$$E = \overline{E} - \nabla\widetilde{\phi} \tag{6}$$

is decomposed into a macroscopic contribution

$$\overline{E} = \frac{1}{|\mathcal{B}|}\int_{\mathcal{B}} E\, \mathrm{d}V, \tag{7}$$

which is constant on the microscale, and the gradient of the fluctuative scalar electric potential $\widetilde{\phi}$ that acts as primary variable. The system is closed by applying appropriate boundary conditions on the RVE. There are several boundary conditions that fulfill the Hill–Mandel condition (5). In the present work, we focus on periodic boundary conditions of the form

$$\widetilde{\phi}(x^+) = \widetilde{\phi}(x^-) \quad \text{and} \quad D\cdot N(x^+) = -D\cdot N(x^-) \quad \text{on} \quad \partial\mathcal{B}, \tag{8}$$

where $x^{\pm}$ indicate opposite coordinates at the boundary of the RVE [30–32]. Note that we only need to prescribe one of the two boundary conditions given in the latter equation as the other one will emerge naturally in equilibrium.

3. Artificial Neural Network Based Solution Scheme

In this section, a solution procedure based on *artificial neural networks* (ANNs) for finding the equilibrium state of the potential in Equation (3) is presented. Core idea is the construction of trial functions, which consist of ANNs and multiplicative factors fulfilling the given set of boundary conditions in Equation (8). We then recapitulate the two main ANN structures used in the present work, namely the single layer perceptron net and the multilayer perceptron. Additionally, the derivatives of those nets with respect to its inputs as well as with respect to its weights are given as they are needed when using gradient descent methods.

3.1. Optimization Principle

Consider the previously introduced RVE occupying the space $\mathcal{B}$. Our goal is to find the microscopic equilibrium state of a given global potential for this body. Under prescribed periodic Dirichlet boundary conditions in Equation $(8)_1$, the potential in Equation (3) takes the form

$$\Pi(\phi) = \int_{\mathcal{B}} \Psi(\boldsymbol{x}, \boldsymbol{E})\, dV. \tag{9}$$

Having the physical problem covered, the question arises how to approximate the solution fields. While finite element approaches employ local shape functions and Fourier transform based methods use global trigonometric basis functions, in the present work, we want to investigate an approximation method based on artificial neural networks. Following the idea of Lagaris et al. [27], we construct a trial function ϕ_t using arbitrary artificial neural networks $N_i(\boldsymbol{x}, \boldsymbol{p}_i)$ as

$$\widetilde{\phi}_t(\boldsymbol{x}, \boldsymbol{p}) = A_0(\boldsymbol{x}) + A_1(\boldsymbol{x})N_1(\boldsymbol{x}, \boldsymbol{p}_1) + A_2(\boldsymbol{x})N_2(\boldsymbol{x}, \boldsymbol{p}_2) + ... + A_n(\boldsymbol{x})N_n(\boldsymbol{x}, \boldsymbol{p}_n), \tag{10}$$

where $A_i(\boldsymbol{x})$ are functions ensuring that the boundary conditions are fulfilled a priori. As a generic one-dimensional example, one could think of a scalar electric potential that should fulfill the boundary conditions $\widetilde{\phi}(0) = \widetilde{\phi}(1) = 1$. In this case, we would have $A_0 = 1$ and $A_1 = x(1-x)$, yielding the trial function according to Equation (10) as $\widetilde{\phi}_t(x, \boldsymbol{p}) = 1 + x(1-x)N_1(x, \boldsymbol{p}_1)$. The corresponding electric field $\boldsymbol{E}_t(\boldsymbol{x}, \boldsymbol{p})$ in line with Equation (6) can be calculated analytically according to the neural network derivatives given in Sections 3.2 and 3.3. Using the gradient field along with Equation (9) gives the global potential in terms of the neural network's parameters as follows

$$\Pi(\boldsymbol{p}) = \int_{\mathcal{B}} \Psi(\boldsymbol{x}, \boldsymbol{E}_t(\boldsymbol{x}, \boldsymbol{p}))\, dV. \tag{11}$$

Finally, the objective function in our machine learning problem is obtained from the Hill–Mandel condition in Equation (5) in combination with the global potential in Equation (11) approximated by neural networks. It appears as

$$\boxed{\overline{\Pi} = \sup_{\boldsymbol{p}} \frac{1}{|\mathcal{B}|}\Pi(\boldsymbol{p}),} \tag{12}$$

where the optimization is carried out with respect to the neural network's parameters. By having the periodic boundary conditions fulfilled by construction, the optimization can be carried out without any constraints on the parameters $\boldsymbol{p}$.

3.2. Single Layer Perceptron (SLP)

The *single layer perceptron* (SLP) is one of the most basic versions of artificial neural networks [33]. It is a reduced version of the general structure depicted in Figure 1. An SLP only consists of one hidden layer that connects the input and the output. Assuming an input vector $\boldsymbol{x}$ of dimension d, the response N of an SLP is calculated as

$$N(\boldsymbol{x},\boldsymbol{p}) = \sum_{i=1}^{H} v_i \sigma(z_i) + \bar{b} \quad \text{with} \quad z_i = \sum_{j=1}^{d} w_{ij} x_j + u_i, \tag{13}$$

where v_i, w_{ij}, u_i and $\bar{b}$ are the weights and biases of the hidden unit and the output bias, respectively, and H is the overall number of neurons in the hidden layer. Those weights and biases are assembled in the neural network parameter vector $\boldsymbol{p}$. Here, σ denotes an activation function of a neuron. The activation functions may be chosen problem-dependent and can have a large impact on the training behavior of the artificial neural network. Figure 2 shows the three activation functions used in the present work, namely the logistic sigmoid, the hyperbolic tangent and the softplus function. Despite its popularity in machine learning tasks, we are not using the rectifier linear unit (ReLu) activation function in the present work. First tests using the ReLu function resulted in poor convergence rates. We suspect that this stems from errors in the numerical integration when using the ReLu function and its discontinuous derivative. The derivatives of the SLP net with respect to its input then appear as

$$N_{,j} = \frac{\partial N}{\partial x_j} = \sum_{i=1}^{H} v_i w_{ij} \sigma'(z_i), \tag{14}$$

where $\sigma'(z_i)$ denotes the derivative of the sigmoid function with respect to its argument. The spatial derivative can be perceived as an SLP with modified weights and activation function but it is now a gradient field. To use efficient gradient based solvers when optimizing the weights of the neural network, it is convenient to have the explicit derivatives of the ANNs with respect to the weights. These can be obtained as

$$\boxed{\begin{aligned} \frac{\partial N}{\partial u_i} &= v_i \sigma'(z_i), & \frac{\partial N_{,j}}{\partial u_i} &= v_i w_{ij} \sigma''(z_i), \\ \frac{\partial N}{\partial w_{ij}} &= v_i x_j \sigma'(z_i), & \frac{\partial N_{,j}}{\partial w_{im}} &= x_m v_i w_{ij} \sigma''(z_i) + v_i \sigma'(z_i) \delta_{jm}, \\ \frac{\partial N}{\partial \bar{b}} &= 1, & \frac{\partial N_{,j}}{\partial \bar{b}} &= 0, \\ \frac{\partial N}{\partial v_i} &= \sigma(z_i), & \frac{\partial N_{,j}}{\partial v_i} &= w_{ij} \sigma'(z_i). \end{aligned}} \tag{15}$$

Note that, in the derivatives above, the indices are *not* treated by Einstein's summation convention. Indices appearing twice are rather multiplied pointwise in MATLAB [34] convention.

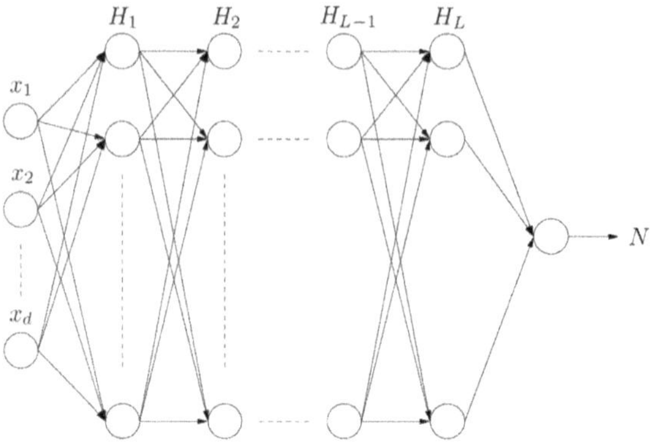

Figure 1. General structure of a multilayer perceptron with input $\boldsymbol{x}$, L hidden layers and output N. Each neuron evaluates its inputs through predefined activation functions.

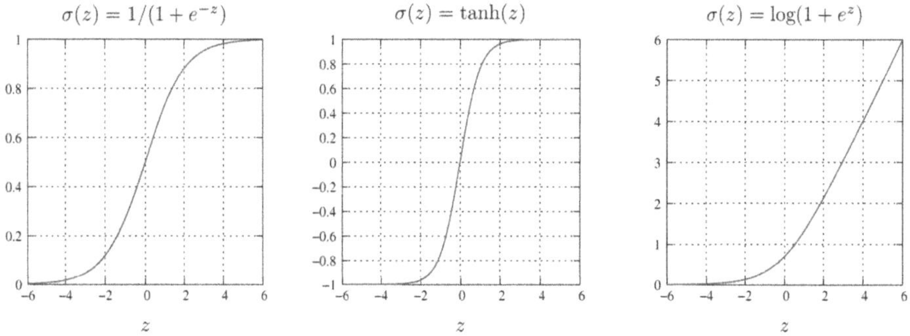

Figure 2. Different types of popular activation functions: logistic sigmoid, hyperbolic tangent and softplus function.

3.3. Multilayer Perceptron (MLP)

The multilayer perceptron (MLP) works similarly to the single layer perceptron. However, it is constructed by a higher number L of hidden layers. It can be shown that this deep structure enables a more general approximation property of the neural network and might lead to better training behavior [35]. In the present work, we focus on MLPs with only two hidden layers. The output is then computed as

$$N(\boldsymbol{x}, \boldsymbol{p}) = \sum_{k=1}^{H_2} v_k \sigma(r_k) + \bar{b}, \quad r_k = \sum_{i=1}^{H_1} \theta_{ki} \sigma(z_i) + c_k \quad \text{with} \quad z_i = \sum_{j=1}^{d} w_{ij} x_j + u_i, \tag{16}$$

where we have now additional weights θ_{ki} and biases c_k associated with the second hidden layer. The spatial derivative of the MLP appears as

$$N_{,j} = \frac{\partial N}{\partial x_j} = \sum_{k=1}^{H_2} \sum_{i=1}^{H_1} v_k \sigma'(r_k) \theta_{ki} \sigma'(z_i) w_{ij}. \tag{17}$$

The derivatives of the MLP with respect to the weights are computed as

$$
\begin{aligned}
\frac{\partial N}{\partial u_i} &= \sum_{k=1}^{H_2} v_k \sigma'(r_k)\theta_{ki}\sigma'(z_i), & \frac{\partial N_{,j}}{\partial u_i} &= \sum_{k=1}^{H_2} \Big[v_k \sigma''(r_k)\theta_{ki}\sigma'(z_i) \sum_{i=1}^{H_1}[w_{ij}\theta_{ki}\sigma'(z_i)] \\
& & &\quad + v_k\sigma'(r_k)\theta_{ki}\sigma''(z_i)w_{ij} \Big], \\
\frac{\partial N}{\partial w_{ij}} &= \sum_{k=1}^{H_2} v_k \sigma'(r_k)\theta_{ki}\sigma'(z_i)x_j, & \frac{\partial N_{,j}}{\partial w_{im}} &= \sum_{k=1}^{H_2} \Big[v_k \sigma''(r_k)\theta_{ki}\sigma'(z_i)x_m \sum_{i=1}^{H_1}[w_{ij}\theta_{ki}\sigma'(z_i)] \\
& & &\quad + v_k\sigma'(r_k)\theta_{ki}\sigma''(z_i)w_{ij}x_m \\
& & &\quad + v_k\sigma'(r_k)\theta_{ki}\sigma'(z_i)\delta_{im} \Big], \\
\frac{\partial N}{\partial c_k} &= v_k\sigma'(r_k), & \frac{\partial N_{,j}}{\partial c_k} &= v_k\sigma''(r_k) \sum_{i=1}^{H_1}[w_{ij}\theta_{ki}\sigma'(z_i)], \\
\frac{\partial N}{\partial \theta_{ki}} &= v_k\sigma'(r_k)\sigma(z_i), & \frac{\partial N_{,j}}{\partial \theta_{ki}} &= v_k\sigma''(r_k)\sigma'(z_i) \sum_{i=1}^{H_1}[w_{ij}\theta_{ki}\sigma(z_i)] \\
& & &\quad + v_k\sigma'(r_k)\sigma'(z_i)w_{ij}, \\
\frac{\partial N}{\partial \bar{b}} &= 1, & \frac{\partial N_{,j}}{\partial \bar{b}} &= 0, \\
\frac{\partial N}{\partial v_k} &= \sigma(r_k), & \frac{\partial N_{,j}}{\partial v_k} &= \sigma'(r_k) \sum_{i=1}^{H_1}[w_{ij}\theta_{ki}\sigma'(z_i)].
\end{aligned}
\tag{18}
$$

4. Numerical Examples

In this section, we test the robustness and reliability of the proposed method on a set of one-, two- and three-dimensional problems. The influence of global versus piecewise-defined constructions of the trial functions as well as the impact of the neuron count on the simulation results are explored. The one-dimensional example is implemented in MATLAB [34] while the two- and three-dimensional examples are carried out by a Fortran 77 code that utilizes the L-BFGS-B optimization algorithms [36,37]. For the sake of simplicity, all the following simulations are performed with normalized units.

4.1. One-Dimensional Example

We first consider a one-dimensional problem to demonstrate the features of the proposed method. The RVE is a simple two-phase laminate of unit length $l = 1$ and unit cross section, which is loaded with a macroscopic electric field $\overline{E} = 0.01$. The global potential takes the form

$$
\Pi(\phi) = \int_{\mathcal{B}} \Psi(E, x)\, dx = -\int_{\mathcal{B}} \frac{1}{2}\kappa E^2\, dx, \tag{19}
$$

where $\kappa_1 = 1$ for $x < 0.5$ and $\kappa_2 = 2$ for $x > 0.5$. Having the decomposition in Equation (6)

$$
E = \overline{E} - \frac{\partial \widetilde{\phi}}{\partial x} \tag{20}
$$

along with the boundary conditions $\widetilde{\phi}(0) = 0$ and $\widetilde{\phi}(1) = 0$, the analytical solution for the electric field reads

$$
E = \begin{cases} \dfrac{2\kappa_2}{\kappa_1 + \kappa_2}\overline{E} & 0 \le x < l/2 \\[2ex] \dfrac{2\kappa_1}{\kappa_1 + \kappa_2}\overline{E} & l/2 < x \le l \end{cases} \tag{21}
$$

and for the electric potential

$$\widetilde{\phi} = \begin{cases} \overline{E}\dfrac{\kappa_1 - \kappa_2}{\kappa_1 + \kappa_2}x & 0 \leq x \leq l/2 \\ \overline{E}\dfrac{\kappa_2 - \kappa_1}{\kappa_1 + \kappa_2}(x - l) & l/2 \leq x \leq l \end{cases} . \tag{22}$$

4.1.1. Global Neural Net Approach on Equidistant Grid

To find the electric scalar potential that optimizes the energy potential in Equation (19), we construct a global trial function according to Equation (10) that automatically fulfills the boundary conditions given above as

$$\widetilde{\phi}_t = A_o + A_1(x)N(x, \boldsymbol{p}) = x(1 - x)N(x, \boldsymbol{p}), \tag{23}$$

where N is a neural network that takes x as an input and has the weights and biases $\boldsymbol{p}$. The derivatives in this case can be computed explicitly as

$$\frac{\partial \widetilde{\phi}_t}{\partial x} = (1 - 2x)N(x, \boldsymbol{p}) + x(1 - x)\frac{\partial N(x, \boldsymbol{p})}{\partial x}. \tag{24}$$

The derivatives then allow us to compute the global potential in terms of the neural network's weights and biases as

$$\Pi(\boldsymbol{p}) = -\int_{\mathcal{B}} \frac{1}{2}\kappa(\overline{E} - \frac{\widetilde{\phi}_t}{\partial x})^2\, dx. \tag{25}$$

Finally, we need a numerical integration scheme to evaluate the integral. In the present work, we use quadrature points in terms of equidistant grid points x_k with distance Δx on the interval of $[0, 1]$ as $\{\Delta x/2, 3\Delta x/2, \ldots, 1 - \Delta x/2\}$, yielding the discrete objective

$$\boxed{\overline{\Pi} = \sup_{\boldsymbol{p}} \frac{1}{l} \sum_{0 < x_k < 1.0} -\frac{1}{2}\kappa(\overline{E} - \frac{\widetilde{\phi}_t}{\partial x})^2 \Delta x.} \tag{26}$$

The maximum of this objective function can be found by means of the gradient descent method. The gradients of the objective with respect to the weights $\boldsymbol{p}$ that are needed for such an iterative solver can be computed using Equation (15). We then have everything at hand to carry out the first numerical example. As for the ANN architecture, we use an SLP with $H = 10$ neurons in the hidden layer. As for the activation function, the logistic sigmoid function $\sigma(z) = 1/(1 + e^{-z})$ is chosen, and the weights and biases $\boldsymbol{p}$ are randomly initalized with a uniform distribution between 0 and 1. Figure 3 shows the result of the numerical experiment (green) compared to the analytical results (black). One can see that the numerical scalar electric potential $\widetilde{\phi}_t$ is close to the analytical solution. However, having a look at the gradients in the form of E reveals the occurrence of oscillations around the discontinuity. The MATLAB [34] code that generates these results can be found in Appendix A. Note that the tolerances for the step size and the optimality as well as the maximum number of function evaluations and iterations is set different from the MATLAB [34] default values to obtain reasonable results. One might decrease the oscillations by having even higher iteration numbers. However, the jump in the solution at the vicinity of the material jump cannot be captured by the global smooth neural network function. The more neurons we use, the better accuracy we obtain, which leads to a trade-off between accuracy and speed.

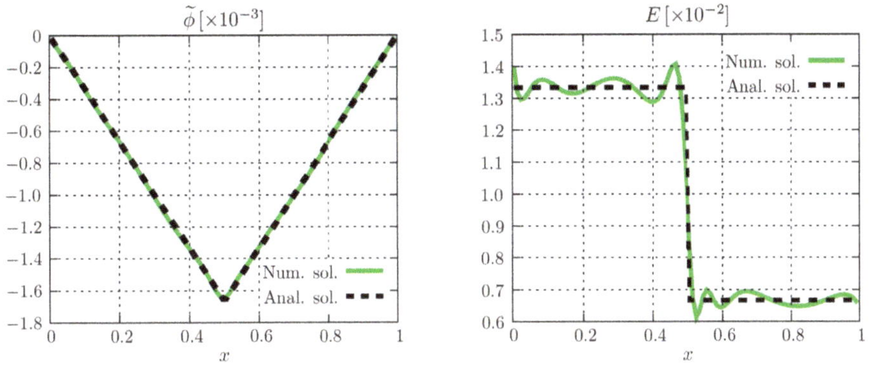

Figure 3. Numerical vs. analytical solution for a global trial function approach using an SLPs with 10 neurons and a random initialization of $p^{(0)} \sim \mathcal{U}(0,1)$, where $\mathcal{U}(0,1)$ denotes a vector of numbers generated from a uniform distribution between 0 and 1. These parameters are statistically independent.

4.1.2. Piecewise-defined Neural Net Approach on Equidistant Grid

To overcome the oscillations observed in the global approach, we next construct a trial function that is piecewise defined for the RVE's different material regions

$$\tilde{\phi}_t = \begin{cases} xN_1(x, \boldsymbol{p}_1) & 0 < x \leq 0.5 \\ (0.5 - x)(1 - x)N_2(x, \boldsymbol{p}_2) + 2(1 - x)\tilde{\phi}_t(0.5) & 0.5 < x \leq 1.0 \end{cases} . \tag{27}$$

Its derivatives can be computed according to

$$\frac{\partial \tilde{\phi}_t}{\partial x} = \begin{cases} N_1(x, \boldsymbol{p}_1) + x\dfrac{\partial N_1(x, \boldsymbol{p}_1)}{\partial x} & 0 < x < 0.5 \\ (2x - 1.5)N_2(x, \boldsymbol{p}_2) + (0.5 - x)(1 - x)\dfrac{\partial N_2(x, \boldsymbol{p}_2)}{\partial x} - 2\tilde{\phi}_t(0.5) & 0.5 < x < 1.0 \end{cases} . \tag{28}$$

The global potential in terms of the neural network's weights and biases then appears as

$$\Pi(\boldsymbol{p}) = \int_B \frac{1}{2}\kappa(\overline{E} - \frac{\tilde{\phi}_t}{\partial x})^2\, dx. \tag{29}$$

In analogy to the global approach, numerical integration is performed using equidistant grid points x_k with distance Δx as quadrature points in the interval $[0, 1]$ to obtain the discrete objective

$$\boxed{\overline{\Pi} = \sup_{\boldsymbol{p}} \frac{1}{l}\left(\sum_{0<x_k<0.5} -\frac{1}{2}\kappa_1(\overline{E} - \frac{\tilde{\phi}_t}{\partial x})^2 - \sum_{0.5<x_k<1.0} \frac{1}{2}\kappa_2(\overline{E} - \frac{\tilde{\phi}_t}{\partial x})^2 \right) \Delta x.} \tag{30}$$

The gradient of the objective function with respect to the parameters $\boldsymbol{p}$ can again be obtained through the derivatives of Equation (15). However, the output and training behavior of artificial neural networks is dependent on the initialization of the weights and biases $\boldsymbol{p}$. In a first run, we set the number of neurons in the two SLPs' hidden layers to 10 and initialize the weights randomly between 0 and 1. As for the activation function, we choose the logistic sigmoid function $\sigma(z) = 1/(1 + e^{-z})$.

Figure 4 shows the numerical results in green compared to the analytical results in black. There is a distinct deviation between them. Having a look at the output layer of the SLP in Equation (13) and its derivative in Equation (14), one can see that the initial output for 10 neurons for an initialization of all weights between 0 and 1 is far from the exact solution. This difference becomes even larger for higher number of neurons. In contrast to the global approach, we use the default values of the unconstrained MATLAB [34] minimizer and, apparently, there are too few iterations.

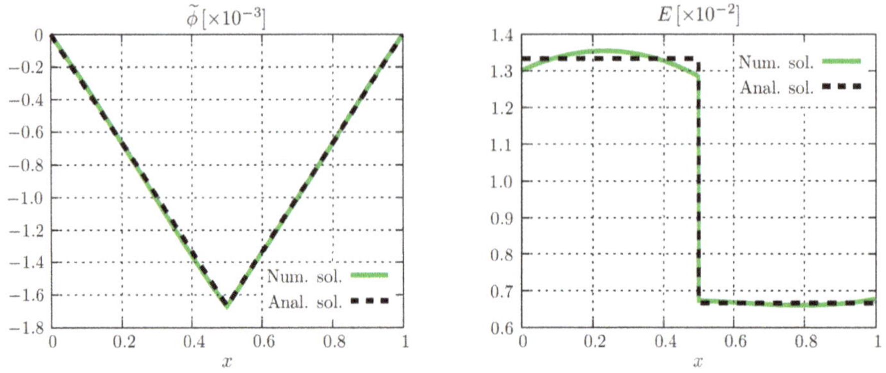

Figure 4. Numerical vs. analytical solution for 2 SLPs with 10 neurons each and a random initialization of $p^{(0)} \sim \boldsymbol{\mathcal{U}}(0,1)$, where $\boldsymbol{\mathcal{U}}(0,1)$ denotes a vector of numbers generated from a uniform distribution between 0 and 1. These parameters are statistically independent.

Next, we want to have fewer iterations of the solver within the default values by using an adaptive way of initializing the weights. The key idea is to initialize the weights in a way that the net and its derivative output values are roughly in the range of the values we would expect with respect to the given macroscopic load. We therefore use a simple modification of the weight initialization given as

$$p^{(0),*} \sim \frac{\overline{E}}{H} * \boldsymbol{\mathcal{U}}(0,1), \tag{31}$$

where $\boldsymbol{\mathcal{U}}(0,1)$ is a vector of random numbers uniformly distributed along 0 and 1. Please note that we use boldface calligraphic $\boldsymbol{\mathcal{U}}$ to indicate the vector notation instead of univariate distribution. The results of the computation using the adaptive initialization method can be seen in Figure 5. The results are closer to the analytical solution and are independent of the number of hidden neurons used, making the overall method much more reliable. The MATLAB [34] code used for generating Figure 5 can be found in Appendix B.

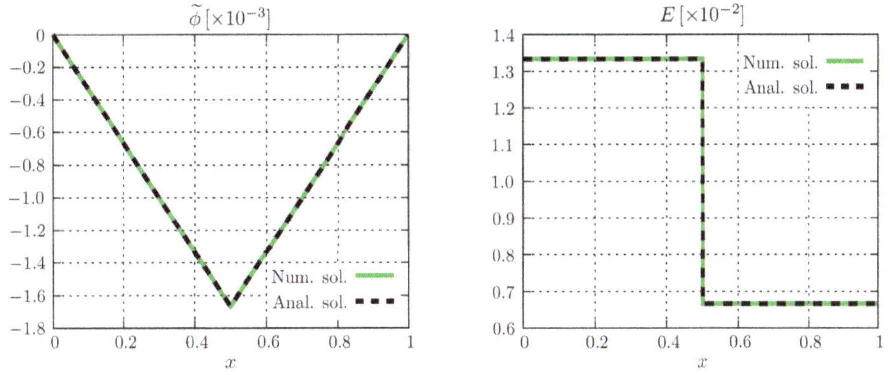

Figure 5. Numerical vs. analytical solution for 2 SLPs with 10 neurons each and a random initialization of $p^{(0),*} \sim \frac{\overline{E}}{H} * \mathcal{U}(0,1)$, where $\mathcal{U}(0,1)$ denotes a vector of numbers generated from a uniform distribution between 0 and 1. These parameters are statistically independent.

4.2. Two-Dimensional Example

Next, we consider a two-dimensional microstructure with a circular inclusion, as shown in Figure 6. The radius of the inclusion is $r_0 = 0.178l$, corresponding to 10% volume fraction. The global energy potential per unit out-of-plane thickness is given as the integral of the internal energy density over the RVE's domain

$$\Pi = \int_{\mathcal{B}} \Psi(\boldsymbol{E})\, dA = -\int_{\mathcal{B}} \frac{1}{2} \kappa \boldsymbol{E} \cdot \boldsymbol{E}\, dA. \tag{32}$$

In this example, the material parameter is set to $\kappa = 1$ in the matrix and $\kappa = 10$ in the inclusion. The square RVE of unit length $l = 1$ is loaded with the constant macroscopic field $\overline{E}_1 = 1$ and $\overline{E}_2 = 0$ under the periodic Dirichlet boundary conditions in Equation $(8)_1$. For reference, a simulation using the finite element method is performed for this optimization problem. The mesh is discretized by linear quadrilateral elements with four Gauss points. The optimized global potential is calculated to $|\Pi|^{\text{FEM}} = 0.588652$. Figure 6 shows the contour plot of the microscopic field E_1. The finite element results serve as a reference for the neural network based approaches in the following examples.

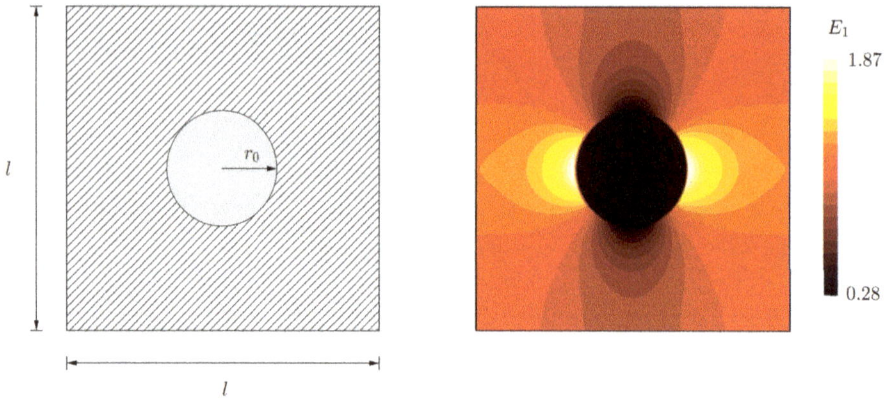

Figure 6. Microstructure of length l with a circular inclusion having the radius $r_0 = 0.178l$. On the right, a finite element solution for a phase contrast of 10 and a loading of $\overline{E}_1 = 1$ and $\overline{E}_2 = 0$ is displayed.

4.2.1. Global Neural Net Approach on Equidistant Grid

First, we want to build a global trial function, which is covering the whole RVE. As we want to implement periodic boundary conditions, a set of three neural networks is used to construct the trial function ϕ_t: $N_{x_1}(x_1)$ acting in the x_1-direction, $N_{x_2}(x_2)$ acting in the x_2-direction and $N_1(\boldsymbol{x})$ acting in both directions. The trial function then takes the form

$$\begin{aligned} \widetilde{\phi}_t(\boldsymbol{x},\boldsymbol{p}) &= A_1(x_1,x_2)N_1(\boldsymbol{x},\boldsymbol{p}_1) + A_2(x_1)N_{x_1}(x_1,\boldsymbol{p}_{x_1}) + A_3(x_2)N_{x_2}(x_2,\boldsymbol{p}_{x_2}) \\ &= x_1(1-x_1)x_2(1-x_2)N_1(\boldsymbol{x},\boldsymbol{p}_1) \\ &\quad + x_1(1-x_1)N_{x_1}(x_1,\boldsymbol{p}_{x_1}) + x_2(1-x_2)N_{x_2}(x_2,\boldsymbol{p}_{x_2}). \end{aligned} \tag{33}$$

Figure 7 shows a visualization of the functions ensuring the boundary conditions. Here, we use SLPs for the boundary networks and a two-layer MLP for the two-dimensional neural network. As for the activation function for the neurons, the hyperbolic tangent $\sigma(z) = \tanh(z)$ is chosen. The negative gradient of the trial function can then be computed to obtain the trial field

$$\boldsymbol{E}_t = \overline{\boldsymbol{E}} - \nabla\widetilde{\phi}_t, \tag{34}$$

where we drop dependencies on $\boldsymbol{x}$ and $\boldsymbol{p}$ in our notation. According to Equation (32), we arrive at the global potential as an objective

$$\overline{\Pi} = \sup_{\boldsymbol{p}} \frac{1}{|\mathcal{B}|}\Pi(\boldsymbol{p}) = \sup_{\boldsymbol{p}} \frac{1}{|\mathcal{B}|}\int_{\mathcal{B}} -\frac{1}{2}\kappa(\boldsymbol{x})\boldsymbol{E}_t\cdot\boldsymbol{E}_t\,dA. \tag{35}$$

The spatial gradient of the trial function $\nabla\widetilde{\phi}_t$ as well as gradients of the global potential $\partial\Pi(\boldsymbol{p})/\partial\boldsymbol{p}$ used in optimization algorithms can be obtained analytically through the derivatives given in Sections 3.2 and 3.3. We use equidistant grid points for establishing a regular mesh of elements and employ nine Gauss points per element as the quadrature points for numerical integration. The objective is optimized without any constraints on the parameters $\boldsymbol{p}$, as we satisfy the boundary conditions a priori by the construction of $\widetilde{\phi}_t$.

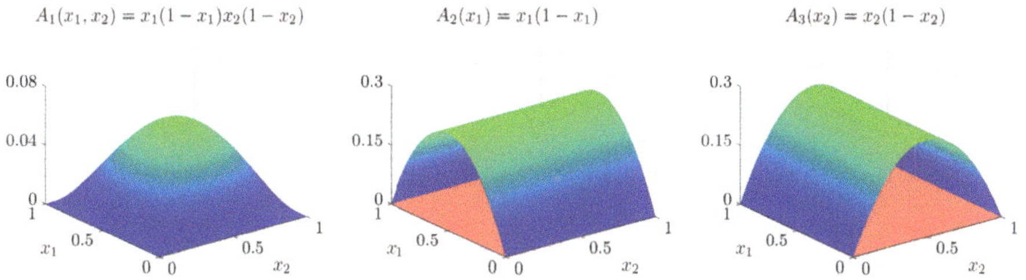

Figure 7. Visualization of the functions A_1, A_2 and A_3 ensuring periodicity of the two-dimensional trial function $\widetilde{\phi}_t$ in a square RVE of unit length $l = 1$. One can see that A_1 covers the volume, A_2 satisfies the periodicity in x_1-direction and A_3 satisfies the periodicity in x_2-direction.

The simulation is carried out for a different number of neurons and integration points as follows: (a) 51 × 51 elements, 15 neurons in each of the two hidden layers of the MLP and 5 neurons each in the layer of the boundary SLPs; (b) 51 × 51 elements, 10 neurons in each of the two hidden layers of the MLP and 5 neurons each in the layer of the boundary SLPs; and (c) 101 × 101 elements, 15 neurons in each of the two hidden layers of the MLP and 5 neurons each in the layer of the boundary SLPs. Additionally, all three set ups are run with a uniform initialization of the weights through

$$\boldsymbol{p}^{(0)} \sim \boldsymbol{\mathcal{U}}(-1,1). \tag{36}$$

Figure 8 shows the contour plot of E_{t1} of the neural network after 20,000 iterations. One can see that the neural network in Case (a) localizes in an unphysical state. This could be a problem related to overfitting. Case (b), with the same amount of integration points but lower neuron count, shows a qualitatively better result. The overall optimization seems to become more reliable as the integration is performed more accurately, as seen in Case (c). Quantitatively, the global potentials are: (a) $|\overline{\Pi}|(\boldsymbol{p}) = 0.213522$; (b) $|\overline{\Pi}|(\boldsymbol{p}) = 0.590266$; and (c) $|\overline{\Pi}|(\boldsymbol{p}) = 0.589887$ compared to the FEM potential of $|\overline{\Pi}|^{\text{FEM}}(\boldsymbol{E}) = 0.588652$, taken from the simulation that creates Figure 6. In two dimensions and with complicated geometries at hand, it can be difficult to estimate the magnitude of the solution fields appearing a priori. A rather simple approach for initializing the weights as described in the one-dimensional case did not seem to significantly improve the method. Here, we test a second approach of initializing the weights according to a normal distribution

$$\boldsymbol{p}^{(0),*} \sim \boldsymbol{\mathcal{N}}(\mu, \sigma^2), \tag{37}$$

where we set the mean $\mu = 0$ and the variance $\sigma^2 = 1$. Figure 8 shows the contour plot of E_{t1} of the neural network after 20,000 iterations for the normal distribution. Case (a) now shows a more physical state. The energies are calculated as: (a) $|\overline{\Pi}|(\boldsymbol{p}) = 0.590054$; (b) $|\overline{\Pi}|(\boldsymbol{p}) = 0.592981$; and (c) $|\overline{\Pi}|(\boldsymbol{p}) = 0.590544$. At this point, there is surely some potential left for improving weight initialization. This is subject to other fields of machine learning as well, especially in the context of training speed and vanishing gradient problems [38].

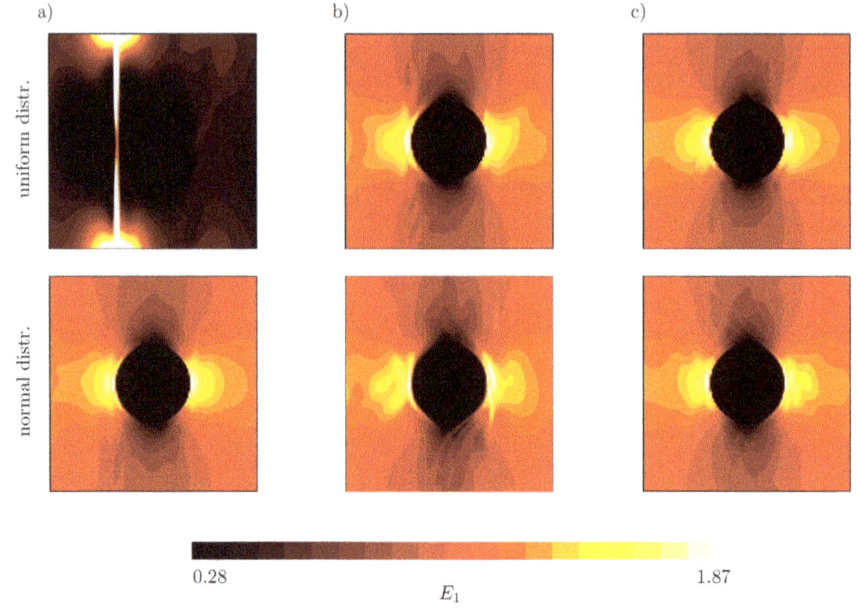

Figure 8. Contour plot of E_{t1} for a set of parameters: (**a**) 51 × 51 elements, 15 neurons per layer in N_1 and 5 neurons each for N_{x_1} and N_{x_2}; (**b**) 51 × 51 elements, 10 neurons per layer in N_1 and 5 neurons each for N_{x_1} and N_{x_2}; and (**c**) 101 × 101 elements, 15 neurons per layer in N_1 and 5 neurons each for N_{x_1} and N_{x_2}.

4.2.2. Piecewise-Defined Neural Net Approach

To further improve the training and approximation properties of the trial functions, we next construct it in a way that it captures the discontinuity of material properties in the microstructure a priori. This is possible due to the simple topology of the microstructure at hand. However, for more complicated microstructures, defining explicit expressions for the trial function might be difficult. We now need a set of four ANNs for the construction of ϕ_t: $N_{x_1}(x_1)$ acting in the x_1-direction, $N_{x_2}(x_2)$ acting in the x_2-direction, $N_1(\boldsymbol{x})$ acting in the matrix and $N_2(\boldsymbol{x})$ acting in the inclusion. The global trial function is defined piecewise in two sub-domains as follows

$$\tilde{\phi}_t = \begin{cases} x_1(1-x_1)x_2(1-x_2)N_1(\boldsymbol{x},\boldsymbol{p}_1) + x_1(1-x_1)N_{x_1}(x_1,\boldsymbol{p}_{x_1}) & r \geq r_0 \\ +x_2(1-x_2)N_{x_2}(x_2,\boldsymbol{p}_{x_2}) & \\ \tilde{\phi}_t(r_0,\varphi) + (r_0-r)N_2(\boldsymbol{r},\boldsymbol{p}_1) & r < r_0 \end{cases}. \tag{38}$$

One can see that the above equations automatically fulfill periodicity at the boundaries as well as the transition condition at the phase interface of the circular inclusion. Here, we implement the trial function in Cartesian coordinates for the matrix and in polar coordinates for the inclusion. Thus, the neural network of the inclusion takes the coordinate vector $\boldsymbol{r} = (r,\varphi)$ in terms of the radius r and the angle φ as its input. Having the origin of our coordinate system in the bottom left corner, the transformation used here takes the form

$$x_1 = 0.5 + r\cos\varphi \quad \text{and} \quad x_2 = 0.5 + r\sin\varphi. \tag{39}$$

With the coordinate transformation at hand, one can calculate the gradient in the matrix and inclusion as

$$\boldsymbol{E}_t = -\nabla(-\overline{\boldsymbol{E}} \cdot \boldsymbol{x} + \widetilde{\phi}_t), \tag{40}$$

where $\overline{\boldsymbol{E}}$ is the applied macroscopic electric field. For a more detailed derivation, see Appendix C. Finally, the potential to optimize in this example takes the form

$$\overline{\Pi} = \sup_{\boldsymbol{p}} \frac{1}{|\mathcal{B}|}\Pi(\boldsymbol{p}) = \sup_{\boldsymbol{p}} \frac{1}{|\mathcal{B}|}\left(-\int_{\mathcal{B}^{\text{matr}}} \frac{1}{2}\kappa_1 \boldsymbol{E}_t \cdot \boldsymbol{E}_t \, dA - \int_{\mathcal{B}^{\text{incl}}} \frac{1}{2}\kappa_2 \boldsymbol{E}_t \cdot \boldsymbol{E}_t \, dA\right). \tag{41}$$

As a numerical integration scheme, we use a simple one-point quadrature rule, where we have an equidistant grid of 3240 integration points in the matrix and 3600 integration points in the inclusion. The MLP $N_1(\boldsymbol{x}, \boldsymbol{p}_1)$ has two layers with five neurons in each, the SLP $N_2(\boldsymbol{x}, \boldsymbol{p}_2)$ has one layer with four neurons and the boundary SLPs $N_{x_1}(x_1, \boldsymbol{p}_{x_1})$ and $N_{x_2}(x_2, \boldsymbol{p}_{x_2})$ have seven neurons in the hidden layer. This sums up to a total of 116 degrees of freedom $\boldsymbol{p}$. In the present example, we use the hyperbolic tangent $\sigma(z) = \tanh(z)$ as the activation function of the neurons. The initial weights are randomly initialized with uniform distribution in the range of -1 to 1. As in the previous examples, the applied macroscopic load in Equation (40) is $\overline{E}_1 = 1.0$ and $\overline{E}_2 = 0.0$.

Figure 9 shows the result for E_{t1} after 10,000 iterations for the local construction of the trial function. Compared with the previous simulations using only one global trial function, the fields have fewer oscillations. Additionally, the maximum energy after 10,000 iterations is $|\overline{\Pi}|(\boldsymbol{p}) = 0.588414$, being lower than in the previous simulations and lower than the maximum energy computed with finite elements. Interestingly, in the first iterations, the learning rates are higher for the global schemes, whether the weights are initialized according to a normal distribution or a uniform distribution (see Figure 9).

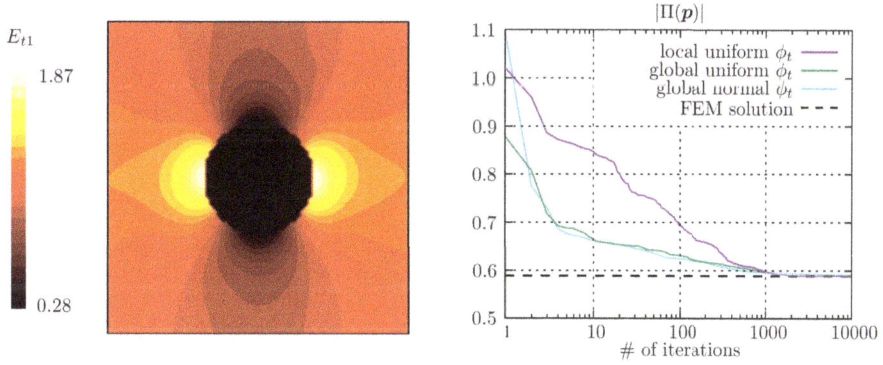

Figure 9. **(left)** Contour plot of E_{t1} for the piecewise-defined trial function in Equation (38) after 10,000 iterations in the optimization procedure; and **(right)** optimization of $\Pi(\boldsymbol{p})$ vs. iteration count for the piecewise-defined trial function with uniformly distributed weight initialization and the global trial function (33) with uniformly and normally distributed weight initialization.

4.3. Three-Dimensional Example

In the present example, we apply the method to a three-dimensional cubic RVE of unit length $l = 1$ with a spherical inclusion of radius $r_0 = 0.178l$. We first construct a global trial function, for which we need a set of seven neural networks: $N_{x_1}(x_1)$ acting in the x_1-direction, $N_{x_2}(x_2)$ acting in the x_2-direction, $N_{x_3}(x_3)$ acting in the x_3-direction, $N_{x_{12}}(x_1, x_2)$ acting in the x_1x_2-plane, $N_{x_{13}}(x_1, x_3)$

acting in the x_1x_3-plane, $N_{x_{23}}(x_2, x_3)$ acting in the x_1x_2-plane and $N_1(\boldsymbol{x})$ acting in the RVE's volume (see Figure 10). The trial function for the matrix then appears as

$$\begin{aligned}\widetilde{\phi}_t(\boldsymbol{x},\boldsymbol{p}) = \quad & A_1\, N_1(\boldsymbol{x},\boldsymbol{p}_1) + A_2\, N_{x_{12}}(x_1, x_2, \boldsymbol{p}_{x_{12}}) + A_3\, N_{x_{13}}(x_1, x_3, \boldsymbol{p}_{x_{13}}) \\ & + A_4\, N_{x_{23}}(x_2, x_3, \boldsymbol{p}_{x_{23}}) + A_5\, N_{x_1}(x_1, \boldsymbol{p}_{x_1}) \\ & + A_6\, N_{x_2}(x_2, \boldsymbol{p}_{x_2}) + A_7\, N_{x_3}(x_3, \boldsymbol{p}_{x_3}),\end{aligned} \tag{42}$$

where the functions A_i take the form

$$\begin{aligned} A_1 &= x_1(1-x_1)x_2(1-x_2)x_3(1-x_3), & A_5 &= x_1(1-x_1), \\ A_2 &= x_1(1-x_1)x_2(1-x_2), & A_6 &= x_2(1-x_2), \\ A_3 &= x_1(1-x_1)x_3(1-x_3), & A_7 &= x_3(1-x_3), \\ A_4 &= x_2(1-x_2)x_3(1-x_3). \end{aligned} \tag{43}$$

By construction, the trial function fulfills the periodic boundary conditions. The negative gradient of the trial function can then be computed analytically according to Equation (40) (see also Appendix D for a more detailed derivation). With the gradient at hand, we are able optimize the global potential in Equation (32) with respect to the weights of the ANNs.

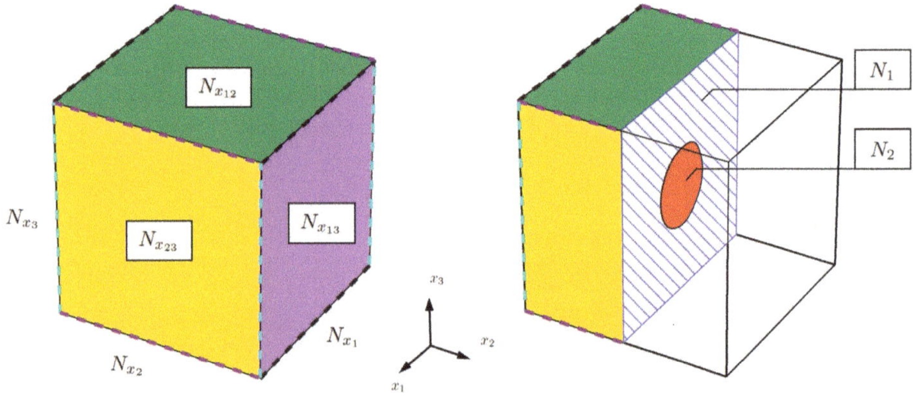

Figure 10. A representation of the artificial neural networks used in the construction of the trial function ϕ_t. There are separate ANNs for the edges, surface boundaries and volumes. In the piecewise-defined approach, there is one additional ANN acting in the inclusion.

As seen in the previous sections, the use of one global trial function might lead to oscillations in the solution field. In line with Section 4.2.2, we additionally construct a piecewise-defined trial function for comparison

$$\widetilde{\phi}_t^{\,*}(\boldsymbol{x},\boldsymbol{p}) = \begin{cases} A_1\, N_1(\boldsymbol{x},\boldsymbol{p}_1) + A_2\, N_{x_{12}}(x_1, x_2, \boldsymbol{p}_{x_{12}}) + A_3\, N_{x_{13}}(x_1, x_3, \boldsymbol{p}_{x_{13}}) & \\ +A_4\, N_{x_{23}}(x_2, x_3, \boldsymbol{p}_{x_{23}}) + A_5\, N_{x_1}(x_1, \boldsymbol{p}_{x_1}) + A_6\, N_{x_2}(x_2, \boldsymbol{p}_{x_2}) & r \geq r_0 \\ +A_7\, N_{x_3}(x_2, \boldsymbol{p}_{x_3}) & \\ \widetilde{\phi}_t(r_0, \varphi, \theta) + A_8\, N_2(\boldsymbol{r}, \boldsymbol{p}_2) & r < r_0. \end{cases} \tag{44}$$

where we add an eighth neural network for the inclusion and the function

$$A_8 = r_0 - r. \tag{45}$$

The transformation between the spherical coordinates $\boldsymbol{r} = (r, \varphi, \theta)$ and the Cartesian coordinates $\boldsymbol{x} = (x_1, x_2, x_3)$ is given as

$$x_1 = 0.5 + r \sin\varphi \cos\theta, \quad x_2 = 0.5 + r \sin\varphi \sin\theta, \quad x_3 = r \cos\varphi. \tag{46}$$

The gradient in Equation (40) can then be computed accordingly in Cartesian coordinates for the matrix and in spherical coordinates for the inclusion, as derived in Appendix D, allowing us to optimize the global potential

$$\overline{\Pi}^* = \sup_{\boldsymbol{p}} \frac{1}{|\mathcal{B}|} \Pi^*(\boldsymbol{p}) = \sup_{\boldsymbol{p}} \frac{1}{|\mathcal{B}|} \left(- \int_{\mathcal{B}^{\mathrm{matr}}} \frac{1}{2} \kappa_1 \boldsymbol{E}_t^* \cdot \boldsymbol{E}_t^* \, dV - \int_{\mathcal{B}^{\mathrm{incl}}} \frac{1}{2} \kappa_2 \boldsymbol{E}_t^* \cdot \boldsymbol{E}_t^* \, dV \right). \tag{47}$$

For our numerical experiment, the RVE depicted in Figure 10 is loaded with a macroscopic field of $\overline{E}_1 = 1.0$, $\overline{E}_2 = 0.0$ and $\overline{E}_3 = 0.0$. The material parameters are chosen as $\kappa_1 = 1$ and $\kappa_2 = 10$. For the global approach, the mesh consists of $43 \times 43 \times 43$ equidistant integration points. The ANNs acting on the edges and surface boundaries are SLPs, having four neurons along each edge ANN and five neurons along each surface ANN. The ANN acting in the volume is a two-layer perceptron with eight neurons in each layer. This trial function totals 256 parameters to optimize in the ANNs. As for the activation function, we choose the softplus function, as it provides the best results in our examples. As for the piecewise-defined trial function in Equation (44), the ANNs are constructed in the same way. The additional ANN acting in the inclusion is an SLP with eight neurons and also has the softplus activation function. The mesh used in this case consists of 32,768 integration points in the matrix and 16,384 integration points in the inclusion. Additionally, an FFT-based simulation [23] using a conjugate gradient solver [24,39–42] is carried out for comparison, where the microstructure is discretized on a $43 \times 43 \times 43$ grid.

Figure 11 shows the contour plot of E_{t1} for the approach using the global and the piecewise-defined trial function after 20,000 iterations. In comparison to the FFT-based solution, both simulations produce qualitatively good results. The global approximation scheme captures the expected jump of E_{t1} across the phase interface a little less distinctly compared with the piecewise-defined one. However, the piecewise-defined approximation is more difficult to construct and will be quite challenging to implement in the case of complicated microstructure geometries. Quantitatively, the optimized global potentials after 20,000 iterations are quite close to each other, with $|\overline{\Pi}|(\boldsymbol{p}) = 0.529926$ for the global approach and $|\overline{\Pi}|^*(\boldsymbol{p}) = 0.529692$ for the piecewise-defined approach, compared to a optimum energy $|\overline{\Pi}|^{\mathrm{FFT}} = 0.527590$ obtained by the FFT-based simulation.

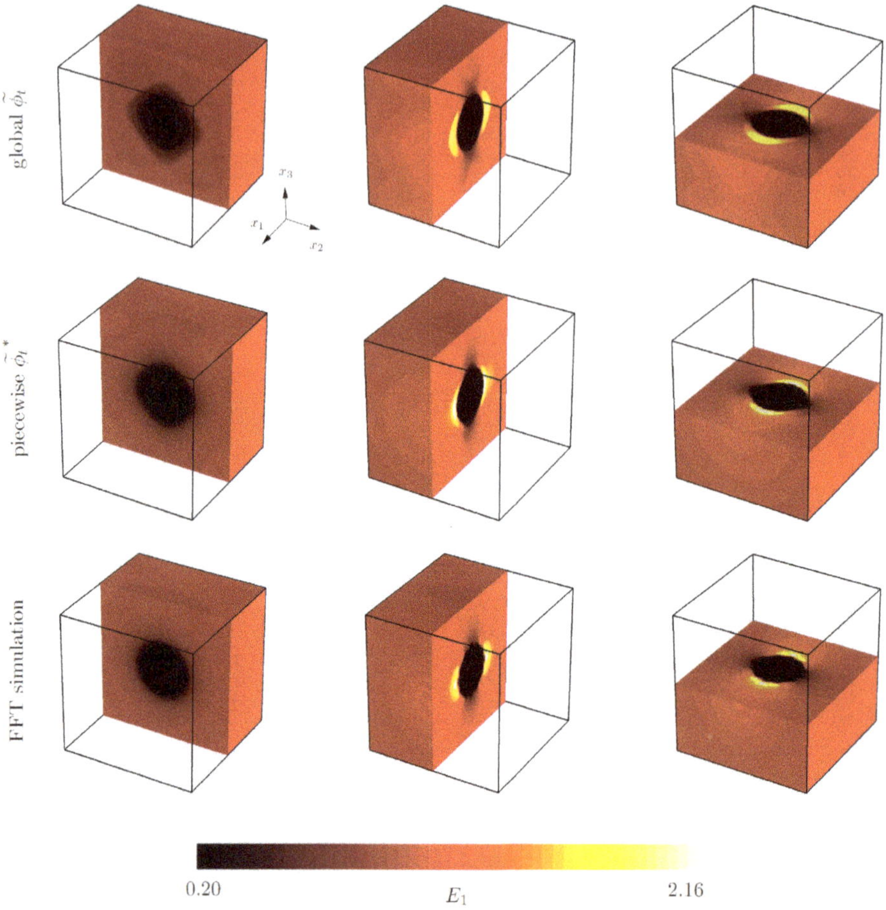

Figure 11. Contour plot of E_1 for a global and a piecewise-defined construction of the trial function as well as an FFT-based simulation for comparison. The results of the ANN simulations are close to the FFT-based simulation. The jump discontinuity is more distinct in the piecewise-defined approach compared with the global approach.

5. Conclusions

We presented a solution scheme to periodic boundary value problems in homogenization based on the training of artificial neural networks. They were employed in tandem with the multiplicative factors in a way that the resulted trial function fulfilled the boundary conditions a priori and thus we arrived at the unconstrained optimization problem. The numerical examples showed that physically reasonable results can be obtained with rather low amounts of neurons, which allows for low memory demand. A construction of trial functions by defining them piecewise in separate sub-domains led to lower oscillations and a generally more stable training behavior compared with a global approach but was geometrically more challenging to construct. The scheme carried over to three dimensions quite well. We assume this to stem from the ratio of neurons compared with the number of integration points: In the considered example of a cube-shaped matrix with spherical inclusion, the solution could be approximated using a quite low neuron count while the number of integration points grew a lot. For future progress, the training speed of the neural network needs to be improved. More sophisticated

ANN structures such as deep nets or recurrent nets might further improve the approximation and training behavior of the method, while methods such as dropout or regularization might assist to avoid problems of overfitting.

Author Contributions: Conceptualization, F.S.G., L.T.K.N. and M.-A.K.; Funding acquisition, M.-A.K.; Investigation, F.S.G.; Methodology, F.S.G., L.T.K.N. and M.-A.K.; Software, F.S.G. and L.T.K.N.; Validation, F.S.G., L.T.K.N. and M.-A.K.; Visualization, F.S.G.; Writing—original draft, F.S.G.; Writing—review & editing, L.T.K.N. and M.-A.K.

Acknowledgments: Funded by Deutsche Forschungsgemeinschaft (DFG, German Research Foundation) under Germany's Excellence Strategy—EXC 2075—390740016.

Conflicts of Interest: The authors declare no conflict of interest.

Appendix A. Matlab Script for Global One-Dimensional Example

```
% Evaluation points in area
dx = 0.01;
x = dx/2:dx:1-dx/2;

% Applied macroscopic electric field
E0 = 0.01;

% Electric permittivity in laminate
KK = zeros(size(x));
KK(x > 1/2) = 2.0;
KK(x < 1/2) = 1.0;

% Number of hidden neurons
nn = 10;

% Initialize weights and biases
wb0 = rand(3*nn+1,1);

% Anonymous function handle
f = @(wb)neuralapprox(wb,KK,E0,x,dx,nn);

% Call unconstrained minimizer
opts = optimoptions(@fminunc,'Algorithm','quasi-newton',...
     'StepTolerance',1e-12,'OptimalityTolerance',1e-12,...
     'MaxFunctionEvaluations',10000,...
     'SpecifyObjectiveGradient',true,'CheckGradients',true,...
     'FiniteDifferenceType','central','MaxIterations',10000);
[wb,f] = fminunc(f,wb0,opts)

function [f, g] = neuralapprox(wb,KK,E0,x,dx,nn)

% restore weights and biases for ANN
u = wb(1:nn);
w = wb(nn+1:2*nn);
b = wb(2*nn+1);
v = wb(2*nn+2:3*nn+1);

% Neural network response for feedforward net (FFN) (Eq.(13))
```

```
z = w*x + u;
sigmoid = (1 + exp(-z)).^(-1);
ANN = sum(v.*sigmoid) + b;

% Spatial derivative response for feedforward net (Eq.(14))
d_sigmoid = sigmoid.*(1 - sigmoid);
d_ANN = sum(v.*w.*d_sigmoid);

% Derivative of 1st FFN and dFFN/dx w.r.t. its weights (Eq.(15))
d2_sigmoid = sigmoid.*(1 - sigmoid).^2 - sigmoid.^2.*(1 - sigmoid);

xmatr = x.*ones(size(z));
ANN_du = v.*d_sigmoid;
ANN_dw = v.*d_sigmoid.*xmatr;
ANN_db = ones(size(x));
ANN_dv = sigmoid;

d_ANN_du = v.*w.*d2_sigmoid;
d_ANN_dw = v.*d_sigmoid + v.*w.*d2_sigmoid.*xmatr;
d_ANN_db = zeros(size(x));
d_ANN_dv = w.*d_sigmoid;

% Derivatives of the trial function phi = x*(1-x)*N (Eq.(24)&(15))
d_phi_t    = (1-2*x).*ANN + x.*(1-x).*d_ANN;
d_phi_t_du   = (1-2*xmatr).*ANN_du + xmatr.*(1-xmatr).*d_ANN_du;
d_phi_t_dw   = (1-2*xmatr).*ANN_dw + xmatr.*(1-xmatr).*d_ANN_dw;
d_phi_t_db   = (1-2*x).*ANN_db + x.*(1-x).*d_ANN_db;
d_phi_t_dv   = (1-2*xmatr).*ANN_dv + xmatr.*(1-xmatr).*d_ANN_dv;

% Cost function and its derivatives w.r.t. the weights (Eq.(26))
JJ = 0.5*KK.*(E0 - d_phi_t).^2;

JJ_du = -KK.*(E0 - d_phi_t).*d_phi_t_du;
JJ_dw = -KK.*(E0 - d_phi_t).*d_phi_t_dw;
JJ_db = -KK.*(E0 - d_phi_t).*d_phi_t_db;
JJ_dv = -KK.*(E0 - d_phi_t).*d_phi_t_dv;

g1 = [JJ_du;JJ_dw;JJ_db;JJ_dv];

f = dx*sum(JJ);
g = dx*sum(g1, 2);

end
```

Appendix B. Matlab Script for Piecewise-Defined One-Dimensional Example

```
% Evaluation points in area
dx = 0.01;
x = dx/2:dx:1-dx/2;

% Applied macroscopic load
E0 = 0.01;

% Electric permittivity in laminate
KK = zeros(size(x));
KK(x > 1/2) = 2.0;
KK(x < 1/2) = 1.0;

% Number of hidden neurons
nn = 10;

% Initialize weights and biases
wb0 = rand(6*nn+2,1)*(E0/nn);

% Anonymous function handle
f = @(wb)neuralapprox(wb,KK,E0,x,dx,nn);

% Call unconstrained minimizer
opts = optimoptions(@fminunc,'Algorithm','quasi-newton',...
    'SpecifyObjectiveGradient',true,'CheckGradients',true,...
    'FiniteDifferenceType','central','MaxIterations',5000);
[wb,f] = fminunc(f,wb0,opts)

function [f, g] = neuralapprox(wb,KK,E0,xx,dx,nn)
%%% Cost function for interval [0,0.5]

% Restore weights and biases for first ANN
u = wb(1:nn);
w = wb(nn+1:2*nn);
b = wb(2*nn+1);
v = wb(2*nn+2:3*nn+1);

% Quadrature points for integration
x = xx(1:floor(size(xx,2)/2));
K = KK(1:floor(size(KK,2)/2));

% Neural network response for 1st feedforward net (FFN) (Eq.(13))
z = w*x + u;
sigmoid = (1 + exp(-z)).^(-1);
ANN = sum(v.*sigmoid) + b;

% Spatial derivative response for 1st feedforward net (Eq.(14))
d_sigmoid = sigmoid.*(1 - sigmoid);
d_ANN = sum(v.*w.*d_sigmoid);
```

```

% Derivative of 1st FFN and dFFN/dx w.r.t. its weights (Eq.(15))
d2_sigmoid = sigmoid.*(1 - sigmoid).^2 - sigmoid.^2.*(1 - sigmoid);

xmatr = x.*ones(size(z));

ANN_du = v.*d_sigmoid;
ANN_dw = v.*d_sigmoid.*xmatr;
ANN_db = ones(size(x));
ANN_dv = sigmoid;

d_ANN_du = v.*w.*d2_sigmoid;
d_ANN_dw = v.*d_sigmoid + v.*w.*d2_sigmoid.*xmatr;
d_ANN_db = zeros(size(x));
d_ANN_dv = w.*d_sigmoid;

% FFN and derivatives evaluated at the discontinuity (Eq.(28)&(15))
z_disc = w*0.5 + u;

sigmoid_disc = (1 + exp(-z_disc)).^(-1);
d_sigmoid_disc = sigmoid_disc.*(1 - sigmoid_disc);

ANN_disc = sum(v.*sigmoid_disc) + b;
ANN_disc_du = v.*d_sigmoid_disc;
ANN_disc_dw = v.*d_sigmoid_disc*0.5;
ANN_disc_db = 1.0;
ANN_disc_dv = sigmoid_disc;

% Derivatives of the trial function phi = x*N_1 (Eq.(28)&(15))
phi_t_disc      = 0.5.*ANN_disc;
d_phi_t_disc_du = ANN_disc_du;
d_phi_t_disc_dw = ANN_disc_dw;
d_phi_t_disc_db = ANN_disc_db;
d_phi_t_disc_dv = ANN_disc_dv;

d_phi_t  = ANN + x.*d_ANN;

d_phi_t_du  = ANN_du + xmatr.*d_ANN_du;
d_phi_t_dw  = ANN_dw + xmatr.*d_ANN_dw;
d_phi_t_db  = ANN_db + x.*d_ANN_db;
d_phi_t_dv  = ANN_dv + xmatr.*d_ANN_dv;

% Cost function and its derivatives w.r.t. the weights (Eq.(30))
JJ = 0.5*K.*(E0 - d_phi_t).^2;

JJ_du = -K.*(E0 - d_phi_t).*d_phi_t_du;
JJ_dw = -K.*(E0 - d_phi_t).*d_phi_t_dw;
JJ_db = -K.*(E0 - d_phi_t).*d_phi_t_db;
JJ_dv = -K.*(E0 - d_phi_t).*d_phi_t_dv;

```

```
g1 = [JJ_du;JJ_dw;JJ_db;JJ_dv];

f = dx*sum(JJ);
g = dx*sum(g1, 2);

%%% Cost function for interval [0.5,1.0]

% Restore weights and biases for second ANN
u = wb(3*nn+2:4*nn+1);
w = wb(4*nn+2:5*nn+1);
b = wb(5*nn+2);
v = wb(5*nn+3:6*nn+2);

% Quadrature points for integration
x = xx(floor(size(xx,2)/2)+1:end);
K = KK(floor(size(KK,2)/2)+1:end);

% Neural network response for 2nd feedforward net (FFN) (Eq.(13))
z = w*x + u;
sigmoid = (1 + exp(-z)).^(-1);
ANN = sum(v.*sigmoid) + b;

% Spatial derivative response for 2nd feedforward net (Eq.(14))
d_sigmoid = sigmoid.*(1 - sigmoid);
d_ANN = sum(v.*w.*d_sigmoid);

% Derivative of end FFN and dFFN/dx w.r.t. its weights (Eq.(28)&(15))
d2_sigmoid = sigmoid.*(1 - sigmoid).^2 - sigmoid.^2.*(1 - sigmoid);

xmatr = x.*ones(size(z));

ANN_du = v.*d_sigmoid;
ANN_dw = v.*d_sigmoid.*xmatr;
ANN_db = ones(size(x));
ANN_dv = sigmoid;

d_ANN_du = v.*w.*d2_sigmoid;
d_ANN_dw = v.*d_sigmoid + v.*w.*d2_sigmoid.*xmatr;
d_ANN_db = zeros(size(x));
d_ANN_dv = w.*d_sigmoid;

% Derivatives of the trial function
% phi = (0.5-x)*(1-x)*N_2 + 2*(1-x)*phi(0.5) (Eq.(28)&(15))
d_phi_t  = (2*x-1.5).*ANN + (0.5-x).*(1-x).*d_ANN - 2*phi_t_disc;

d_phi_t_du  = (2*xmatr-1.5).*ANN_du + (0.5-xmatr).*(1-xmatr).*d_ANN_du;
d_phi_t_dw  = (2*xmatr-1.5).*ANN_dw + (0.5-xmatr).*(1-xmatr).*d_ANN_dw;
d_phi_t_db  = (2*x-1.5).*ANN_db     + (0.5-x).*(1-x).*d_ANN_db;
d_phi_t_dv  = (2*xmatr-1.5).*ANN_dv + (0.5-xmatr).*(1-xmatr).*d_ANN_dv;
```

```
% Cost function and its derivatives w.r.t. the weights (Eq.(30))
JJ = 0.5*K.*(E0 - d_phi_t).^2;

JJ_du = -K.*(E0 - d_phi_t).*d_phi_t_du;
JJ_dw = -K.*(E0 - d_phi_t).*d_phi_t_dw;
JJ_db = -K.*(E0 - d_phi_t).*d_phi_t_db;
JJ_dv = -K.*(E0 - d_phi_t).*d_phi_t_dv;

JJ_disc_du = -K.*(E0 - d_phi_t).*d_phi_t_disc_du;
JJ_disc_dw = -K.*(E0 - d_phi_t).*d_phi_t_disc_dw;
JJ_disc_db = -K.*(E0 - d_phi_t).*d_phi_t_disc_db;
JJ_disc_dv = -K.*(E0 - d_phi_t).*d_phi_t_disc_dv;

g0 = [JJ_disc_du;JJ_disc_dw;JJ_disc_db;JJ_disc_dv];
g1 = [JJ_du;JJ_dw;JJ_db;JJ_dv];

f = f + dx*sum(JJ);
g = g - dx*sum(g0, 2);
g = [g;dx*sum(g1, 2)];
end
```

Appendix C. Two-Dimensional Trial Function and Derivatives

Recalling Section 4.2, we have the trial function in the matrix and the inclusion defined as

$$\widetilde{\phi}_t = \begin{cases} x_1(1-x_1)x_2(1-x_2)N_1(\boldsymbol{x},\boldsymbol{p}_1) + x_1(1-x_1)N_{x_1}(x_1,\boldsymbol{p}_{x_1}) \\ +x_2(1-x_2)N_{x_2}(x_2,\boldsymbol{p}_{x_2}) & r \geq r_0 \\ \widetilde{\phi}_t(r_0,\varphi) + (r_0 - r)N_2(\boldsymbol{r},\boldsymbol{p}_1) & r < r_0 \end{cases}, \tag{A1}$$

where the coordinate transformation between Cartesian and polar coordinates is performed as

$$x_1 = 0.5 + r\cos\varphi \quad \text{and} \quad x_2 = 0.5 + r\sin\varphi. \tag{A2}$$

The negative gradient of the trial function in the matrix ($r \geq r_0$) in Cartesian coordinates then appears as

$$\begin{aligned} \widetilde{\boldsymbol{E}}_t = -\nabla\widetilde{\phi}_t = &-\begin{bmatrix} (1-2x_1)x_2(1-x_2)N_1 + x_1(1-x_1)x_2(1-x_2)\frac{\partial N_1}{\partial x_1} \\ (1-2x_2)x_1(1-x_1)N_1 + x_1(1-x_1)x_2(1-x_2)\frac{\partial N_1}{\partial x_2} \end{bmatrix} \\ &-\begin{bmatrix} (1-2x_1)N_{x_1} + x_1(1-x_1)\frac{\partial N_{x_1}}{\partial x_1} \\ (1-2x_2)N_{x_2} + x_2(1-x_2)\frac{\partial N_{x_2}}{\partial x_2} \end{bmatrix}. \end{aligned} \tag{A3}$$

The gradient of the trial function in the inclusion ($r < r_0$) in polar coordinates can be computed through

$$\widetilde{\boldsymbol{E}}_t = -\nabla_r\widetilde{\phi}_t = -\boldsymbol{J}_0^T \cdot \begin{bmatrix} \frac{\partial\widetilde{\phi}_t}{\partial x_1} \\ \frac{\partial\widetilde{\phi}_t}{\partial x_2} \end{bmatrix}_{r_0,\varphi} - \begin{bmatrix} -N_2 + (r_0 - r)\frac{\partial N_2}{\partial r} \\ \frac{r_0 - r}{r}\frac{\partial N_2}{\partial\varphi} \end{bmatrix}, \tag{A4}$$

where the derivatives with respect to x_1 and x_2 are evaluated at $x_1(r_0, \varphi)$ and $x_2(r_0, \varphi)$. The Jacobian for a radius of r_0 takes the form

$$J_0 = \begin{bmatrix} 0 & -\frac{r_0}{r}\sin\varphi \\ 0 & \frac{r_0}{r}\cos\varphi \end{bmatrix}. \tag{A5}$$

The integration of the global potential in Equation (41) is then piecewise carried out

$$\Pi(\boldsymbol{p}) = \int_{\mathcal{B}^{\text{matr}}} \frac{1}{2}\kappa_1 \boldsymbol{E}_t \cdot \boldsymbol{E}_t \, dx\, dy + \int_{\mathcal{B}^{\text{incl}}} \frac{1}{2}\kappa_2 \boldsymbol{E}_t \cdot \boldsymbol{E}_t \, r\, dr\, d\varphi, \tag{A6}$$

where the integration for the matrix is carried out in Cartesian coordinates and the integration for the inclusion is carried out in polar coordinates.

Appendix D. Three-Dimensional Trial Function and Derivatives

Recalling Section 4.3, the piecewise-defined trial function is constructed as

$$\widetilde{\phi}_t^{\,*} = \begin{cases} A_1\, N_1(\boldsymbol{x}, \boldsymbol{p}_1) + A_2\, N_{x_{12}}(x_1, x_2, \boldsymbol{p}_{x_{12}}) + A_3\, N_{x_{13}}(x_1, x_3, \boldsymbol{p}_{x_{13}}) & \\ +A_4\, N_{x_{23}}(x_2, x_3, \boldsymbol{p}_{x_{23}}) + A_5\, N_{x_1}(x_1, \boldsymbol{p}_{x_1}) + A_6\, N_{x_2}(x_2, \boldsymbol{p}_{x_2}) & r \geq r_0 \\ +A_7\, N_{x_3}(x_3, \boldsymbol{p}_{x_3}) & \\ \widetilde{\phi}_t(r_0, \varphi, \theta) + A_8\, N_2(\boldsymbol{r}, \boldsymbol{p}_2) & r < r_0 \end{cases}, \tag{A7}$$

where the factors

$$\begin{aligned} A_1 &= x_1(1-x_1)x_2(1-x_2)x_3(1-x_3), & A_5 &= x_1(1-x_1), \\ A_2 &= x_1(1-x_1)x_2(1-x_2), & A_6 &= x_2(1-x_2), \\ A_3 &= x_1(1-x_1)x_3(1-x_3), & A_7 &= x_3(1-x_3), \\ A_4 &= x_2(1-x_2)x_3(1-x_3). & A_8 &= r - r_0 \end{aligned} \tag{A8}$$

ensure the satisfaction of the boundary conditions. As the matrix material ($r \geq r_0$) is described in Cartesian coordinates, the gradient can be straightforward computed as

$$\begin{aligned} \widetilde{\boldsymbol{E}}_t^* = -\nabla\widetilde{\phi}_t = &- \begin{bmatrix} \frac{\partial A_1}{\partial x_1} N_1 + A_1 \frac{\partial N_1}{\partial x_1} \\ \frac{\partial A_1}{\partial x_2} N_1 + A_1 \frac{\partial N_1}{\partial x_2} \\ \frac{\partial A_1}{\partial x_3} N_1 + A_1 \frac{\partial N_1}{\partial x_3} \end{bmatrix} - \begin{bmatrix} \frac{\partial A_2}{\partial x_1} N_{x_{12}} + A_2 \frac{\partial N_{x_{12}}}{\partial x_1} \\ \frac{\partial A_2}{\partial x_2} N_{x_{12}} + A_2 \frac{\partial N_{x_{12}}}{\partial x_2} \\ 0 \end{bmatrix} - \begin{bmatrix} \frac{\partial A_3}{\partial x_1} N_{x_{13}} + A_3 \frac{\partial N_{x_{13}}}{\partial x_1} \\ 0 \\ \frac{\partial A_3}{\partial x_3} N_{x_{13}} + A_3 \frac{\partial N_{x_{13}}}{\partial x_3} \end{bmatrix} \\ &- \begin{bmatrix} 0 \\ \frac{\partial A_4}{\partial x_2} N_{x_{23}} + A_4 \frac{\partial N_{x_{23}}}{\partial x_2} \\ \frac{\partial A_4}{\partial x_3} N_{x_{23}} + A_4 \frac{\partial N_{x_{23}}}{\partial x_3} \end{bmatrix} - \begin{bmatrix} \frac{\partial A_5}{\partial x_1} N_{x_1} + A_5 \frac{\partial N_{x_1}}{\partial x_1} \\ \frac{\partial A_6}{\partial x_2} N_{x_2} + A_6 \frac{\partial N_{x_2}}{\partial x_2} \\ \frac{\partial A_7}{\partial x_3} N_{x_3} + A_7 \frac{\partial N_{x_3}}{\partial x_3} \end{bmatrix}, \end{aligned} \tag{A9}$$

where the derivatives of the factors A_i take the form

$$\begin{aligned}
\frac{\partial A_1}{\partial x_1} &= (1-2x_1)x_2(1-x_2)x_3(1-x_3), & \frac{\partial A_1}{\partial x_2} &= (1-2x_2)x_1(1-x_1)x_3(1-x_3),\\
\frac{\partial A_1}{\partial x_3} &= (1-2x_3)x_1(1-x_1)x_2(1-x_2), & \frac{\partial A_2}{\partial x_1} &= (1-2x_1)x_2(1-x_2),\\
\frac{\partial A_2}{\partial x_2} &= (1-2x_2)x_1(1-x_1), & \frac{\partial A_3}{\partial x_1} &= (1-2x_1)x_3(1-x_3),\\
\frac{\partial A_3}{\partial x_3} &= (1-2x_3)x_1(1-x_1), & \frac{\partial A_4}{\partial x_2} &= (1-2x_2)x_3(1-x_3),\\
\frac{\partial A_4}{\partial x_3} &= (1-2x_3)x_2(1-x_2), & \frac{\partial A_5}{\partial x_1} &= (1-2x_1),\\
\frac{\partial A_6}{\partial x_2} &= (1-2x_2), & \frac{\partial A_7}{\partial x_3} &= (1-2x_3).
\end{aligned} \tag{A10}$$

As for the inclusion, the transformation between the spherical coordinates $\boldsymbol{r} = (r, \varphi, \theta)$ and the Cartesian coordinates $\boldsymbol{x} = (x_1, x_2, x_3)$ is given as

$$x_1 = 0.5 + r\sin\varphi\cos\theta, \quad x_2 = 0.5 + r\sin\varphi\sin\theta \quad \text{and} \quad x_3 = r\cos\varphi. \tag{A11}$$

One could now either directly substitute the latter transformation into the trial function in Equation (A7) and take the derivatives with respect to $\boldsymbol{r}$. Alternatively, the Jacobian for the matrix can be computed as

$$J_0 = \begin{bmatrix} 0 & \frac{r_0}{r}\cos\varphi\cos\theta & -\frac{r_0}{r}\sin\theta \\ 0 & \frac{r_0}{r}\cos\varphi\sin\theta & \frac{r_0}{r}\cos\theta \\ 0 & -\frac{r_0}{r}\sin\varphi & 0 \end{bmatrix}, \tag{A12}$$

which allows for the computation of the gradient of ϕ_t^* in spherical coordinates through

$$\widetilde{E}_t = -\nabla_r \widetilde{\phi}_t = -J_0^T \cdot \begin{bmatrix} \frac{\partial \widetilde{\phi}_t}{\partial x_1} \\ \frac{\partial \widetilde{\phi}_t}{\partial x_2} \\ \frac{\partial \widetilde{\phi}_t}{\partial x_3} \end{bmatrix}_{r_0,\varphi,\theta} - \begin{bmatrix} \frac{\partial A_8}{\partial r} N_2 + A_8 \frac{\partial N_2}{\partial r} \\ \frac{A_8}{r}\frac{\partial N_2}{\partial \varphi} \\ \frac{A_8}{r\sin\varphi}\frac{\partial N_2}{\partial \theta} \end{bmatrix}. \tag{A13}$$

Here, the partial derivative is simply $\partial A_8/\partial r = -1$. The integration of the global potential in Equation (41) is then piecewise carried out

$$\Pi(\boldsymbol{p}) = \int_{\mathcal{B}^{\text{matr}}} \frac{1}{2}\kappa_1 \boldsymbol{E}_t \cdot \boldsymbol{E}_t \, dx_1\, dx_2\, dx_3 + \int_{\mathcal{B}^{\text{incl}}} \frac{1}{2}\kappa_2 \boldsymbol{E}_t \cdot \boldsymbol{E}_t \, r^2 \sin\varphi \, dr\, d\varphi\, d\theta, \tag{A14}$$

where one has to appropriately add the constant macroscopic loads to the fluctuations in Equations (A9) and (A13) in Cartesian and polar coordinates.

References

1. Hebb, D.O. *The Organization of Behavior*; John Wiley & Sons: New York, NY, USA, 1949.
2. McCulloch, W.S.; Pitts, W. A logical calculus of the ideas immanent in nervous activity. *Bull. Math. Biophys.* **1943**, *5*, 115–133. [CrossRef]
3. Krizhevsky, A.; Sutskever, I.; Hinton, G.E. Imagenet classification with deep convolutional neural networks. In *Advances in Neural Information Processing Systems 25*; Pereira, F., Burges, C.J.C., Bottou, L., Weinberger, K.Q., Eds.; Curran Associates, Inc.: New York, NY, USA, 2012; pp. 1097–1105.

4. Srivastava, N.; Hinton, G.; Krizhevsky, A.; Sutskever, I.; Salakhutdinov, R. Dropout: A simple way to prevent neural networks from overfitting. *J. Mach. Learn. Res.* **2014**, *15*, 1929–1958.
5. Denning, D.E. An intrusion-detection model. *IEEE Trans. Softw. Eng.* **1987**, *2*, 222–232. [CrossRef]
6. Steinwart, I.; Christmann, A. *Support Vector Machines*; Springer Science & Business Media: Cham, Switzerland, 2008.
7. Raissi, M.; Yazdani, A.; Karniadakis, G.E. Hidden fluid mechanics: A Navier–Stokes informed deep learning framework for assimilating flow visualization data. Available online: http://arxiv.org/abs/1808.04327 (accessed on 12 April 2019).
8. Ghaboussi, J.; Garrett, J.H.; Wu, X. Knowledge-based modeling of material behavior with neural networks. *J. Eng. Mech.* **1991**, *117*, 132–153. [CrossRef]
9. Huber, N.; Tsakmakis, C. A neural network tool for identifying the material parameters of a finite deformation viscoplasticity model with static recovery. *Comput. Methods Appl. Mech. Eng.* **2001**, *191*, 353–384. [CrossRef]
10. Le, B.; Yvonnet, J.; He, Q.-C. Computational homogenization of nonlinear elastic materials using neural networks. *Int. J. Numer. Methods Eng.* **2015**, *104*, 1061–1084. [CrossRef]
11. Bessa, M.; Bostanabad, R.; Liu, Z.; Hu, A.; Apley, D.W.; Brinson, C.; Chen, W.; Liu, W. A framework for data-driven analysis of materials under uncertainty: Countering the curse of dimensionality. *Comput. Methods Appl. Mech. Eng.* **2017**, *320*, 633–667. [CrossRef]
12. Liu, Z.; Bessa, M.; Liu, W.K. Self-consistent clustering analysis: An efficient multi-scale scheme for inelastic heterogeneous materials. *Comput. Methods Appl. Mech. Eng.* **2016**, *306*, 319–341. [CrossRef]
13. Voigt, W. *Theoretische Studien über die Elastizitastsverhältnisse der Cristalle*; Königliche Gesellschaft der Wissenschaften zu Göttingen: Göttingen, Germany, 1887.
14. Reuss, A. Berechnung der fließgrenze von mischkristallen auf grund der plastizitätsbedingung für einkristalle. *Zeitschrift für Angewandte Mathematik und Mechanik* **1929**, *9*, 49–58. [CrossRef]
15. Hill, R. The elastic behaviour of a crystalline aggregate. *Proc. Phys. Soc.* **1952**, *65*, 349. [CrossRef]
16. Hashin, Z.; Shtrikman, S. A variational approach to the theory of the elastic behaviour of multiphase materials. *J. Mech. Phys. Solids* **1963**, *11*, 127–140. [CrossRef]
17. Willis, J. Bounds and self-consistent estimates for the overall properties of anisotropic composites. *J. Mech. Phys. Solids* **1977**, *25*, 185–202. [CrossRef]
18. Budiansky, B. On the elastic moduli of some heterogeneous materials. *J. Mech. Phys. Solids* **1965**, *13*, 223–227. [CrossRef]
19. Hill, R. A self-consistent mechanics of composite materials. *J. Mech. Phys. Solids* **1965**, *13*, 213–222. [CrossRef]
20. Mori, T.; Tanaka, K. Average stress in matrix and average elastic energy of materials with misfitting inclusions. *Acta Mech.* **1973**, *21*, 571–574. [CrossRef]
21. Miehe, C.; Schröder, J.; Schotte, J. Computational homogenization analysis in finite plasticity. Simulation of texture development in polycrystalline materials. *Comput. Methods Appl. Mech. Eng.* **1999**, *171*, 387–418. [CrossRef]
22. Schröder, J. A numerical two-scale homogenization scheme: The FE^2–method. In *Plasticity and Beyond*; Schröder, J., Hackl, K., Eds.; Springer: Cham, Switzerland, 2014; pp. 1–64.
23. Moulinec, H.; Suquet, P. A fast numerical method for computing the linear and nonlinear mechanical properties of composites. *Académie des Sciences* **1994**, *2*, 1417–1423.
24. Zeman, J.; Vondřejc, J.; Novak, J.; Marek, I. Accelerating a fft-based solver for numerical homogenization of periodic media by conjugate gradients. *J. Comput. Phys.* **2010**, *229*, 8065–8071. [CrossRef]
25. Fritzen, F.; Leuschner, M. Reduced basis hybrid computational homogenization based on a mixed incremental formulation. *Comput. Methods Appl. Mech. Eng.* **2013**, *260*, 143–154. [CrossRef]
26. Ryckelynck, D. A priori hyperreduction method: An adaptive approach. *J. Comput. Phys.* **2005**, *202*, 346–366. [CrossRef]
27. Lagaris, I.E.; Likas, A.; Fotiadis, D.I. Artificial neural networks for solving ordinary and partial differential equations. *IEEE Trans. Neural Netw.* **1998**, *9*, 987–1000. [CrossRef] [PubMed]
28. Hill, R. Elastic properties of reinforced solids: Some theoretical principles. *J. Mech. Phys. Solids* **1963**, *11*, 357–372. [CrossRef]
29. Jackson, J.D. *Classical Electrodynamics*; Wiley: Hoboken, NJ, USA, 1999.
30. Bensoussan, A.; Lions, J.-L.; Papanicolaou, G. *Asymptotic Analysis for Periodic Structures*; American Mathematical Society: Providence, RI, USA, 2011; Volume 374.

31. Nemat-Nasser, S.; Lori, M.; Datta, S. Micromechanics: Overall properties of heterogeneous materials. *J. Appl. Mech.* **1996**, *63*, 561. [CrossRef]
32. Terada, K.; Hori, M.; Kyoya, T.; Kikuchi, N. Simulation of the multi-scale convergence in computational homogenization approaches. *Int. J. Solids Struct.* **2000**, *37*, 2285–2311. [CrossRef]
33. Goodfellow, I.; Bengio, Y.; Courville, A. *Deep Learning*; MIT Press: Cambridge, MA, USA, 2016.
34. MATLAB R2017b (Version 9.3.0). The MathWorks Inc.: Natick, MA, USA, 2017. Available online: https://www.mathworks.com/products/matlab.html (accessed on 12 April 2019).
35. Rumelhart, D.E.; McClelland, J.L. *Parallel Distributed Processing: Explorations in the Microstructure of Cognition, Volume 1: Foundations*; MIT Press: Cambridge, MA, USA, 1986.
36. Byrd, R.H.; Lu, P.; Nocedal, J.; Zhu, C. A limited memory algorithm for bound constrained optimization. *SIAM J. Sci. Comput.* **1995**, *16*, 1190–1208. [CrossRef]
37. Zhu, C.; Byrd, R.H.; Lu, P.; Nocedal, J. Algorithm 778: L-bfgs-b: Fortran subroutines for large-scale bound-constrained optimization. *ACM Trans. Math. Softw.* **1997**, *23*, 550–560. [CrossRef]
38. Glorot, X.; Bengio, Y. Understanding the difficulty of training deep feedforward neural networks. In Proceedings of the Thirteenth International Conference on Artificial Intelligence and Statistics, Sardinia, Italy, 13–15 May 2010; pp. 249–256.
39. Gelebart, L.; Mondon-Cancel, R. Non-linear extension of fft-based methods accelerated by conjugate gradients to evaluate the mechanical behavior of composite materials. *Comput. Mat. Sci.* **2013**, *77*, 430–439. [CrossRef]
40. Göküzüm, F.S.; Keip, M.-A. An algorithmically consistent macroscopic tangent operator for FFT-based computational homogenization. *Int. J. Numer. Methods Eng.* **2018**, *113*, 581–600. [CrossRef]
41. Göküzüm, F.S.; Nguyen, L.T.K.; Keip, M.A. A multiscale fe-fft framework for electro-active materials at finite strains. *Comput. Mech.* **2019**, 1–22. [CrossRef]
42. Kabel, M.; Böhlke, T.; Schneider, M. Efficient fixed point and newton-krylov solvers for fft-based homogenization of elasticity at large deformations. *Comput. Mech.* **2014**, *54*, 1497–1514. [CrossRef]

Mathematical and Computational Applications

Article

Data-Driven Microstructure Property Relations

Julian Lißner and Felix Fritzen *

Efficient Methods for Mechanical Analysis, Institute of Applied Mechanics (CE), University of Stuttgart, 70569 Stuttgart, Germany; lissner@mechbau.uni-stuttgart.de

* Correspondence: fritzen@mechbau.uni-stuttgart.de; Tel.: +49-711-685 66283

Received: 26 March 2019; Accepted: 28 May 2019; Published: 31 May 2019

Abstract: An image based prediction of the effective heat conductivity for highly heterogeneous microstructured materials is presented. The synthetic materials under consideration show different inclusion morphology, orientation, volume fraction and topology. The prediction of the effective property is made exclusively based on image data with the main emphasis being put on the 2-point spatial correlation function. This task is implemented using both unsupervised and supervised machine learning methods. First, a snapshot proper orthogonal decomposition (POD) is used to analyze big sets of random microstructures and, thereafter, to compress significant characteristics of the microstructure into a low-dimensional feature vector. In order to manage the related amount of data and computations, three different incremental snapshot POD methods are proposed. In the second step, the obtained feature vector is used to predict the effective material property by using feed forward neural networks. Numerical examples regarding the incremental basis identification and the prediction accuracy of the approach are presented. A Python code illustrating the application of the surrogate is freely available.

Keywords: microstructure property linkage; unsupervised machine learning; supervised machine learning; neural network; snapshot proper orthogonal decomposition

MSC: 74-04, 74A40, 74E30, 74Q05, 74S30

1. Introduction

In material analysis and design of heterogeneous materials, multiscale modeling can be used for the discovery of microstructured materials with tuned properties for engineering applications. Thereby, it contributes to the improvement of the technical capabilities, reduces the amount of resources invested into the construction and enhances the reliability of the description of the material behavior. However, the discovery of materials with the desired material property, which is characterized by the microstructure of the solid, constitutes a highly challenging inverse problem.

The basis for all multiscale models and simulations is information on the microstructure and on the microscale material behavior. If at hand, physical experiments can be replaced by—often costly—computations in order to determine the material properties by virtual testing [1–3]. Separation of structural and microstructural length scales can often be assumed. This enables the use of the representative volume element (RVE) [4] equipped with the preferable periodic fluctuation boundary conditions [5]. The RVE characterizes the highly heterogeneous material using a single frame (or image) and the (analytical or numerical) computation can be conducted on this frame.

The concurrent simulation of the underlying microstructure (e.g., through nested FE simulations, cf., e.g., [6,7], or considering microstructure behavior in the constitutive laws, e.g., [8]) and of the problem on the structural scale is computationally intractable. In view of the correlation between computational complexity and energy consumption, nested FE simulations should be limited in application for ecological reasons, too. Therefore, efficient methods giving reliable prediction of the material property are an active field of research: POD-driven reduced order models with hyper-reduction (e.g., [9,10]), with multiple reduced bases spanning also internal variable [11,12] and for finite strains [13,14] are a selection of recent examples. We refer also to general review articles on the topic such as [15,16].

Supposing that two similar images representing microstructured materials are considered, it is natural to expect similar effective properties in many physically relevant problems such as elasticity, thermal and electric conduction to mention only two applications. The main task, thus, persists in finding low-dimensional parameterizations of the images that capture the relevant information, use these parameterizations to compress the image information and build a surrogate model operating only on the reduced representation. A black-box approach, exploiting precomputed data for the construction of the surrogate to link features to characteristics and using established machine learning methods, is the topic of this paper.

As the no free lunch theorem [17] states, an algorithm can not be arbitrarily fast and arbitrarily accurate at the same time. Hence, there has to be a compromise either in accuracy, computational speed or in versatility. At the cost of generality, i.e., by focusing on subclasses of microstructures, fast and accurate models can be deployed while still allowing for considerable variations of the microstructures. This does not mean that these subclasses must be overly confined: For instance, inclusion volume fractions ranging from 20 up to 80% are considered in this work. Using a limited number of computations performed on relevant microstructure images, machine learned methods can be trained for the subclass under consideration. The sampling of the data, the feature extraction and the training of the machine learning (ML) algorithm constitutes the offline phase in which the surrogate model is built. Typically, the evaluation of the surrogate can be realized almost in real-time (at least this is the aspired and ambitious objective), thereby enabling previously infeasible applications in microstructure tayloring, interactive user interfaces and computations on mobile devices.

To have a reliable prediction for a broader range of considered microstructures, the material knowledge system (MKS) framework [18] is currently actively researched. Many branches thereof exist, all trying to attain low-dimensional microstructure descriptors from the truncation of selected n-point correlation functions. For instance, a principal component analysis (PCA) of the 2-point correlation functions is performed, using the principal scores in a polynomial regression model, in order to predict material properties. The MKS is actively researched for different material structures [19–21]. For instance, [19,20] successfully predict the elastic strain and yield stress for the underlying microstructure using the MKS approach, however they confine their focus on either the topological features of the microstructure or a confined range of allowed volume fractions (0–20%), often held constant in individual studies.

A different approach for target driven microstructure tayloring deploys reconstruction techniques [22,23] to generate similar microstructures which fulfill certain criteria. In order to explicitly find the optimal microstructure geometry, sensitivities of descriptors, as, e.g., the number of inclusions, with respect to material property are obtained with machine learning [24,25]. With the sensitivities at hand, target driven construction enables the generation of optimal microstructure topology for the desired material property, even when considering a broad design space [26].

The goal of the present study is to make accurate image based predictions for RVEs spanning large subclasses of all possible microstructured materials: Substantial variations of the volume fraction, the morphology and of the topology are considered.

Similarly to key ideas of the MKS approach, a reduced basis is deployed to reduce the dimensionality of the microstructural features contained in the n-point correlation functions. With the sheer amount of samples required, conventional methods fail to capture the key features of all considered microstructures. Therefore, we propose three novel incremental reduced basis updates to make the computation possible. Combining these techniques with the use of synthetic microstructure data, the costly training of the reduced basis and of the artificial neural network (e.g., [27]) become feasible, thereby allowing the creation of a surrogate model for the image-property linkage. The surrogate accepts binarized image representations of bi-phasic materials as inputs. The outputs constitute the effective heat conductivity tensor of the considered material.

In Section 2 the microstructure classification and the three different incremental snapshot POD procedures used during feature extraction are presented (unsupervised learning). In Section 3 the use of feedforward artificial neural networks for the processing of the extracted features is discussed. Numerical examples are presented in Section 4 including different inclusion morphologies and an investigation of the relaxation of the microstructure subclass confinement, of the procedure by using mixed data sets, is made. A Python code illustrating the application of the surrogate is freely available via Github.

2. Materials and Methods

2.1. Microstructure Classification

The microstructure is defined by the representative volume element (RVE) [4], which is one periodic frame (or image) characterizing the heterogeneous material under consideration, see Figure 1 for examples of the microstructure and its 2-point spatial correlation function (see below for its definition). Due to their favorable properties regarding the needed size of the RVE, periodic fluctuation boundary conditions, e.g., [5], are used for the computations during the offline phase.

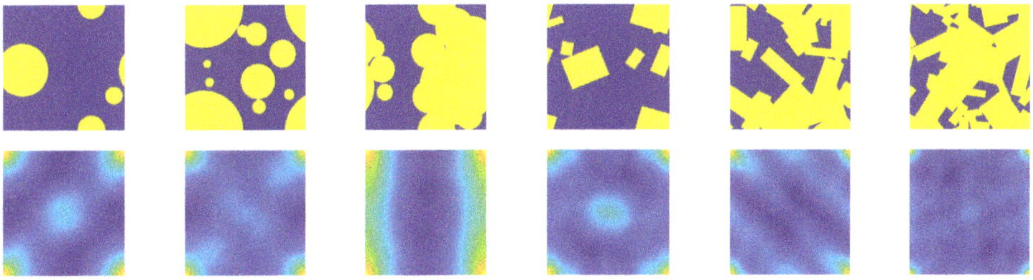

Figure 1. Depicting some exemplary microstructures with their respective 2-point spatial correlation functions $c_2(\boldsymbol{r}; b, b)$ below.

The n-point spatial correlation functions represent a widely used mathematical framework for microstructural characterization [28,29]. Roughly described, the n-point correlation is obtained by placing a polyline consisting of $(n-1)$ nodes defined relative to the first point by vectors $\boldsymbol{r}_1, \boldsymbol{r}_2, \ldots$.

By placing the first point uniformly randomly into the microstructure and computing the mean probability of finding a prescribed sequence of material phases at the nodes of the polyline (including the initial point) denotes the n-point correlation $c_n(\boldsymbol{r}_1, \boldsymbol{r}_2, \ldots, \boldsymbol{r}_{n-1}; m_1, m_2, \ldots, m_n)$, where m_k is the material label expected to be found at the kth node.

For example, the 1-point spatial correlation function, i.e., the probability of finding phase m ($m \in \{a, b, \ldots\}$), yields the phase volume fraction f_m of phase m. In the present study bi-phasic materials are considered. Here $m = a$ corresponds to the matrix material (drawn blue in Figure 1) and $m = b$ to the inclusion phase (drawn yellow in Figure 1). The trivial relation

$$f_a = 1 - f_b \tag{1}$$

holds. The 2-point spatial correlation function (2PCF) $c_2(\boldsymbol{r}; a, b)$ places the vector $\boldsymbol{r}$ in each pixel/voxel x of the RVE and states the probability of starting in the matrix phase a and ending in the inclusion phase b. Mathematically we have

$$c_2(\boldsymbol{r}; a, b) = \left\langle \chi^{(a)}(\boldsymbol{x}) \, \chi^{(b)}(\boldsymbol{x} + \boldsymbol{r}) \right\rangle_x \tag{2}$$

with $\chi^{(m)}$ being the indicator function of phase m, $\boldsymbol{r}$ the point offset and $\langle \bullet \rangle_x$ denoting the averaging operator over the RVE. The 2PCF is efficiently computed in Fourier space by making use of the algorithmically sleek fast Fourier transform (FFT) [30,31]

$$c_2(\boldsymbol{r};a,b) = \mathscr{F}^{-1}\left(\overline{\mathscr{F}(\chi^{(a)})} \odot \mathscr{F}(\chi^{(b)})\right), \tag{3}$$

where $\mathscr{F}$ and $\mathscr{F}^{-1}$ denote the forward and backward FFT, $\overline{\bullet}$ is the complex conjugate and $\odot$ denotes the point-wise multiplication, respectively. For bi-phasic materials the three different two-point functions $c_2(\bullet;a,b)$, $c_2(\bullet;a,a)$, $c_2(\bullet;b,b)$ are related via

$$c_2(\boldsymbol{r};a,a) = f_a - c_2(\boldsymbol{r};a,b), \qquad c_2(\boldsymbol{r};b,b) = f_b - c_2(\boldsymbol{r};a,b). \tag{4}$$

In view of computational efficiency, this redundancy can be exploited. Some key characteristics of the non-negative 2PCF are

$$c_2(\boldsymbol{0};a,a) = f_a = \max_{\boldsymbol{r}\in\Omega} c_2(\boldsymbol{r};a,a), \tag{5}$$

$$c_2(\boldsymbol{0};b,b) = f_b = \max_{\boldsymbol{r}\in\Omega} c_2(\boldsymbol{r};b,b), \tag{6}$$

$$c_2(\boldsymbol{0};a,b) = 0, \tag{7}$$

$$c_2(\boldsymbol{r};a,b) = c_2(\boldsymbol{r};b,a) = c_2(-\boldsymbol{r};a,b), \tag{8}$$

$$\langle c_2(\boldsymbol{x};m,m)\rangle_x = f_m^2 \qquad (m = a,b). \tag{9}$$

In addition to that, a key property of the 2PCF is its invariance with respect to translations of the periodic microstructure. This property is of essential importance when it comes to the comparison of several images under consideration, i.e., during the evaluation of similarities within images.

Examples of $c_2(\boldsymbol{r};b,b)$ (referred to also as auto-correlation of the inclusion phase) are depicted by the lower set of images in Figure 1. By the metric of vision, the following characteristics can be observed:

- The maximum of $c_2(\boldsymbol{r};b,b)$ occurs at the corners of the domain (corresponding to $\boldsymbol{r} = \boldsymbol{0}$);
- Preferred directions of the inclusion placement and/or orientation correspond to laminate-like images (best seen in the third microstructure from the left);
- The domain around $\boldsymbol{r} = \boldsymbol{0}$ partially reflects the average inclusion shape;
- Some similarities are found, particularly with respect to shape of the 2PCF at the corners and in the center.

These observations hint at the existence of a low-dimensional parameterization of relevant microstructural features. In the following this property is exploited by using a snapshot proper orthogonal decomposition (snapshot POD) in order to capture reoccurent patterns of the 2PCF. By working on the two-point function the afore-mentioned elimination of possible translations of the images is an important feature.

The influence of higher order spatial correlation functions has been investigated in the literature, e.g., [28,32]. These considerations often yield minor gains relative to the additional computations and the increased dimensionality (for instance, the 3PCF takes to vectors $\boldsymbol{r}_1, \boldsymbol{r}_2 \in \Omega$ as inputs. Hence, the full 3PCF is basically inaccessible in practice but only after major truncation). While it has been demonstrated that the two point function does not suffice to uniquely describe the microstructure in periodic domains [33], there is evidence that the level of microstructural ambiguity for identical 2PCF can be considered low. Therefore, only the n-point correlation functions up to second order are accounted for in the present study.

2.2. Unsupervised Learning via Snapshot Proper Orthogonal Decomposition

The snapshot POD [34] can be used to construct a reduced basis (RB) [35–37] that provides an optimal subspace for approximating a given snapshot matrix $\underline{\underline{S}} \in \mathbb{R}^{n \times n_s}$. The matrix $\underline{\underline{S}}$ consists of n_s individual snapshots $\underline{s}_i \in \mathbb{R}^n$ with the size n being the dimension of the discrete representation of the unreduced field information. In the case of the 2PCF n denotes the total number of pixels within the RVE, i.e., the discrete two-dimensional 2PCF (represented as image data) is recast into vector format for further processing ($\underline{c}_2^0(m,m) \in \mathbb{R}^n$). In the present study, the constructed RB is used for information compression, i.e., for the extraction of relevant microstructural features from the image data. The reduced basis $\underline{\underline{B}} \in \mathbb{R}^{n \times N}$ retains the N most salient features of the data contained in $\underline{\underline{S}}$ in a few eigenmodes represented by the orthonormal columns of $\underline{\underline{B}}$.

The actual snapshot data stored in $\underline{\underline{S}}$ is constructed from the discrete 2-point function data $\underline{s}_i^0 \in \mathbb{R}^n$ via scaling and shifting according to

$$\underline{s}_i = \frac{1}{f_b}\left(\underline{s}_i^0 - f_b^2\,\underline{1}\right), \tag{10}$$

where $\underline{1} \in \mathbb{R}^n$ is a vector containing ones at all entries. This shift ensures a peak value of 1 in the corner and the mean of 0 for every snapshot.

The reduced basis is computed under the premise to minimize the overall relative projection error

$$\mathcal{P}_\delta = \frac{||\underline{\underline{S}} - \underline{\underline{B}}\,\underline{\underline{B}}^\mathsf{T}\,\underline{\underline{S}}||_F}{||\underline{\underline{S}}||_F} \tag{11}$$

with respect to the Frobenius norm $||\bullet||_F$. The RB can be constructed with multiple methods, e.g., with the snapshot correlation matrix $\underline{\underline{C}}_S$ and its eigenvalue decomposition, which is given by

$$\underline{\underline{C}}_S = \underline{\underline{S}}^\mathsf{T}\,\underline{\underline{S}} = \underline{\underline{V}}\,\underline{\underline{\Theta}}\,\underline{\underline{V}}^\mathsf{T}. \tag{12}$$

The following properties of the sorted eigenvalue decomposition hold

$$\underline{\underline{V}}^\mathsf{T}\,\underline{\underline{V}} = \underline{\underline{I}} \qquad \mathbb{R}^{n_s \times n_s}, \qquad \Theta_{ij} = \theta_i \delta_{ij}, \qquad \theta_1 \geq \theta_2 \geq \ldots \geq \theta_{n_s} \geq 0, \tag{13}$$

and δ_{ij} denotes the Kronecker delta. The dimension of the reduced basis is determined by the POD threshold, i.e., the truncation criterion is given by

$$\delta_N = \sqrt{\frac{\sum_{j=N+1}^{n_s}\theta_j}{\sum_{i=1}^{n_s}\theta_i}} = \sqrt{\frac{\sum_{j=N+1}^{n_s}\theta_j}{||\underline{\underline{S}}||_F^2}} = \sqrt{||\underline{\underline{S}}||_F^2 - \sum_{j=1}^{N}\theta_j} \overset{!}{\leq} \varepsilon, \tag{14}$$

where $\varepsilon > 0$ is a given tolerance denoting the admissible approximation error. Then, the reduced basis is computed via

$$\underline{\underline{B}} = \underline{\underline{S}}\,\underline{\underline{\widetilde{V}}}\,\underline{\underline{\widetilde{\Theta}}}^{-\frac{1}{2}} \tag{15}$$

after truncation of the eigenvalue and eigenvector matrices to reduced dimension N represented by $\underline{\underline{\widetilde{\Theta}}} \in \mathbb{R}^{N \times N}$ and $\underline{\underline{\widetilde{V}}} \in \mathbb{R}^{n \times N}$, respectively. The sorting of the eigenvalues with their corresponding eigenvectors leads to the property that the least recurrent information given in $\underline{\underline{S}}$ is omitted. Hence, the first eigenmode in $\underline{\underline{B}}$ has the most dominant pattern, the second eigenmode the second most, etc. The properties of the reduced basis computed with the snapshot correlation matrix remain the same as for the singular value decomposition (SVD) introduced below.

The SVD [38] of the snapshot matrix is given by

$$\underline{\underline{S}} = \underline{\underline{U}}\,\underline{\underline{\Sigma}}\,\underline{\underline{W}}^\mathsf{T} \tag{16}$$

with the following properties (asserting $n_s \geq n$)

$$\underline{\underline{U}} \in \mathbb{R}^{n \times n_s} : \underline{\underline{U}}^\mathsf{T} \underline{\underline{U}} = \underline{\underline{I}}, \qquad \underline{\underline{W}} \in \mathbb{R}^{n_s \times n_s} : \underline{\underline{W}}^\mathsf{T} \underline{\underline{W}} = \underline{\underline{I}}, \qquad \underline{\underline{\Sigma}} \in \mathbb{R}^{n_s \times n_s} : \underline{\underline{\Sigma}} = \mathrm{diag}(\sigma_i) \tag{17}$$

and the sorted non-negative singular values σ_i such that $\sigma_1 \geq \sigma_2 \geq \cdots \geq \sigma_{n_s} \geq 0$. The criterion for determining the reduced dimension N matching Equation (14) takes the form

$$\delta_N = \sqrt{\frac{\sum_{j=N+1}^{n_s} \sigma_j^2}{\|\underline{\underline{S}}\|_F^2}} = \sqrt{\frac{\sum_{j=N+1}^{n_s} \sigma_j^2}{\sum_{i=1}^{n_s} \sigma_i^2}} = \sqrt{\|\underline{\underline{S}}\|_F^2 - \sum_{j=1}^{N} \sigma_j^2} \overset{!}{\leq} \varepsilon. \tag{18}$$

Then the reduced basis is given by truncation of the columns of $\underline{\underline{U}}$ yielding $\underline{\underline{\widetilde{U}}} \in \mathbb{R}^{n \times N}$

$$\underline{\underline{B}} = \underline{\underline{\widetilde{U}}}. \tag{19}$$

More specifically, the left subspace associated with the leading singular values represents the RB. Both introduced methods yield the exact same result for the same snapshot matrix $\underline{\underline{S}}$.

2.3. Incremental Generation of the Reduced Basis $\underline{\underline{B}}$

The RB is deployed in order to compress the information contained in n_s snapshots into an N-dimensional set of eigenmodes stored in the columns of $\underline{\underline{B}} \in \mathbb{R}^{n \times N}$, where $N \ll n_s$ is asserted. Since the RB is computed with the snapshot matrix alone, the information contained in $\underline{\underline{S}}$ needs to contain data representing the relevant microstructure range, i.e., covering the parameter range used in the generation of the synthetic materials, in order for $\underline{\underline{B}}$ to be representative for the problem under consideration.

In the case of bi-phasic microstructural images containing n pixels, a ludicrous amount of 2^n states could theoretically be considered when allowing for fully arbitrary microstructures. When limiting attention to certain microstructure classes, then less information is needed. Still, thousands of snapshots are usually required, at least. In the following, attention is limited to synthetic materials generated using random sequential adsorption of morphological prototypes with variable size, orientation, aspect ratio, overlap and randomized phase volume fraction. Due to the high variability of such microstructures (see, e.g., Figure 1), a large number of snapshots exceeding available memory would be needed, i.e., a monolithic snapshot matrix $\underline{\underline{S}}$ is not at hand in practice. While attention is limited to two-dimensional model problems in this study, the problem aggravates considerably for three-dimensional images which imply technical challenges of various sort (storage, processing time, data management, etc.).

In order to be able to generate a rich RB accounting for largely varying microstructural classes, the incremental basis generation represents a core concept within the present work. It enables the RB generation based on a sequence of input snapshots but without the need to store previously considered data except for the current RB. Three different methods are proposed, two of which rely on approximations of the snapshot correlation matrix $\underline{\underline{C}}_S$, and one of which relies on the SVD of an approximate snapshot matrix. The general incremental scheme depicted in Figure 2 remains the same for all the procedures, i.e., the only difference is found during the step labeled 'adjust'.

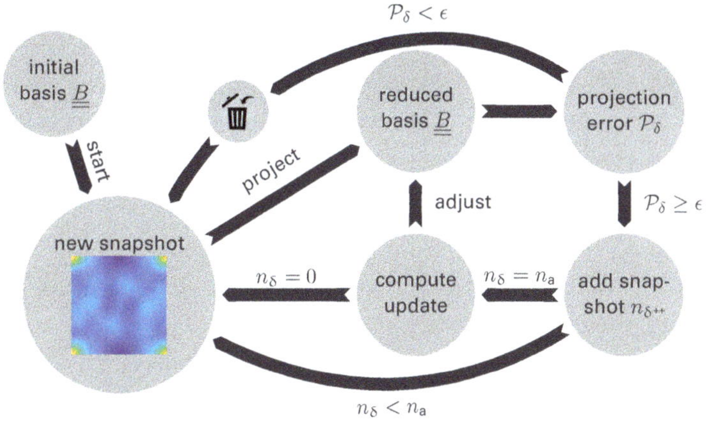

Figure 2. Graphical overview of the incremental update of the reduced basis.

The algorithm is initialized by a small sized set of initial snapshots of the shifted and scaled 2-point correlation function (cf. Equation (10) in Section 2.2). Further, the algorithmic variables $n_\delta = 0$ and $\underline{\underline{\Delta S}} = \emptyset$ are set. The initial RB is computed classically using either the correlation matrix or the SVD (see previous section for details). After computation of the RB, the snapshots are stored neither in memory nor on a hard drive. The algorithm then takes input snapshots in the order of appearance, i.e., the data gets abandoned. For each newly generated snapshot $\underline{s}_i$ the relative projection error with respect to the current RB is computed

$$\mathcal{P}_\delta = \frac{||\underline{s}_i - \underline{\underline{B}}\,\underline{\underline{B}}^\mathsf{T}\,\underline{s}_i||_F}{||\underline{s}_i||_F}. \tag{20}$$

If $\mathcal{P}_\delta$ is greater than the tolerance $\varepsilon > 0$ the snapshot is considered as inappropriately represented by the existing RB. Consequently, $\underline{s}_i$ is appended to a buffer $\underline{\underline{\Delta S}}$ containing candidates for the next basis enrichment and the counter n_δ is incremented. Once the buffer contains a critical number of n_a elements the actual enrichment is triggered and the buffer is emptied thereafter. Thereby the computational overhead is reduced. The three different update procedures are described later on in detail. The procedure is continued until $n_\mathrm{c} > 0$ consecutive snapshots were found to be approximated up to the relative tolerance ε. Then the basis is considered as converged for the microstructure class under consideration.

In the following three methods for the update procedure are described. Formally, the update of an existing basis $\underline{\underline{B}}$ with a block of snapshots contained in the buffer $\underline{\underline{\Delta S}}$ is sought-after. The new basis is required to remain orthonormal.

2.3.1. Method A: Append Eigenmodes to $\underline{\underline{B}}$

A trivial enrichment strategy is given in terms of appending new modes to the existing basis while preserving orthonormality of the basis. Therefore, the projection of $\underline{\underline{\Delta S}}$ onto the existing RB is subtracted in a first step

$$\underline{\underline{\Delta \widehat{S}}} = \underline{\underline{\Delta S}} - \underline{\underline{B}}\,\underline{\underline{B}}^\mathsf{T}\underline{\underline{\Delta S}}. \tag{21}$$

It is readily seen that $\underline{\underline{\Delta \widehat{S}}}$ is orthogonal to $\underline{\underline{B}}$. Then the correlation matrix of the additional data and its eigen-decomposition are computed according to

$$\underline{\underline{\Delta C}} = \underline{\underline{\Delta \widehat{S}}}^\mathsf{T}\underline{\underline{\Delta \widehat{S}}} = \underline{\underline{V}}\,\underline{\underline{\Theta}}\,\underline{\underline{V}}^\mathsf{T}. \tag{22}$$

Eventually, the enrichment is given through the truncated matrices $\underline{\underline{\widetilde{V}}}$ and $\underline{\underline{\widetilde{\Theta}}}$

$$\underline{\underline{\Delta B}} = \underline{\underline{\Delta \widehat{S}}}\,\underline{\underline{\widetilde{V}}}\,\underline{\underline{\widetilde{\Theta}}}^{-\frac{1}{2}}. \tag{23}$$

The new basis is then obtained by appending the newly computed modes $\underline{\underline{\Delta B}}$

$$\underline{\underline{B}} \leftarrow \begin{bmatrix} \underline{\underline{B}} & \underline{\underline{\Delta B}} \end{bmatrix}. \tag{24}$$

Method A simply adds modes generated from the projection residual $\underline{\underline{\Delta \widehat{S}}}$ in a decoupled way, i.e., the existing basis is not modified. In order to compute the basis update, only the existing RB $\underline{\underline{B}}$ and the temporarily stored snapshots $\underline{\underline{\Delta S}}$ are required.

Remarks on Method A

A.1 The truncation parameter δ_N must be chosen carefully such that

$$\frac{\|\underline{\underline{\Delta \widehat{S}}} - \underline{\underline{\Delta B}}\,\underline{\underline{\Delta B}}^{\mathsf{T}}\underline{\underline{\Delta \widehat{S}}}\|_F}{\|\underline{\underline{\Delta S}}\|_F} \leq \delta_N. \tag{25}$$

In particular, the normalization with respect to the original data prior to projection onto the existing RB must be taken.

A.2 By appending orthonormal modes to the existing basis it is a priori guaranteed that the accuracy of previously considered snapshots cannot worsen, i.e., an upper bound for the relative projection error of all snapshots considered until termination of the algorithm is given by the truncation parameter δ_N and n_{a}:

$$\max \frac{|\underline{s}_i - \underline{\underline{B}}\,\underline{\underline{B}}^{\mathsf{T}}\underline{s}_i|}{|\underline{s}_i|} \leq \sqrt{n_{\mathrm{a}}}\,\delta_N. \tag{26}$$

This estimate is, however, overly pessimistic and it must be noted that the enrichment will guarantee a drop in the residual for all snapshots contained in $\underline{\underline{\Delta S}}$

2.3.2. Method B: Approximate Reconstruction of the Snapshot Correlation Matrix

This update scheme is based on an approximation of the new correlation matrix

$$\underline{\underline{C}} = \begin{bmatrix} \underline{\underline{S}}^{\mathsf{T}}\underline{\underline{S}} & \underline{\underline{S}}^{\mathsf{T}}\underline{\underline{\Delta S}} \\ \underline{\underline{\Delta S}}^{\mathsf{T}}\underline{\underline{S}} & \underline{\underline{\Delta S}}^{\mathsf{T}}\underline{\underline{\Delta S}} \end{bmatrix} = \begin{bmatrix} \underline{\underline{C}}_0 & \underline{\underline{S}}^{\mathsf{T}}\underline{\underline{\Delta S}} \\ \underline{\underline{\Delta S}}^{\mathsf{T}}\underline{\underline{S}} & \underline{\underline{\Delta S}}^{\mathsf{T}}\underline{\underline{\Delta S}} \end{bmatrix}. \tag{27}$$

Here $\underline{\underline{S}}$ denotes all snapshots considered in the RB so far and $\underline{\underline{\Delta S}}$ contains the candidate snapshots. However, the previously used snapshots formally written as $\underline{\underline{S}}$ are no longer available since they can not be stored due to storage limitations. Using the previously computed matrices $\underline{\underline{B}}, \underline{\underline{\widetilde{V}}}, \underline{\underline{\widetilde{\Theta}}}$ the following approximations are available

$$\underline{\underline{S}}^{\mathsf{T}}\,\underline{\underline{S}} = \underline{\underline{C}}_0 \approx \underline{\underline{\widetilde{C}}}_0 = \underline{\underline{\widetilde{V}}}\,\underline{\underline{\widetilde{\Theta}}}\,\underline{\underline{\widetilde{V}}}^{\mathsf{T}}, \qquad \underline{\underline{B}} = \underline{\underline{S}}\,\underline{\underline{\widetilde{V}}}\,\underline{\underline{\widetilde{\Theta}}}^{-\frac{1}{2}}, \qquad \underline{\underline{S}} \approx \underline{\underline{B}}\,\underline{\underline{B}}^{\mathsf{T}}\,\underline{\underline{S}}, \tag{28}$$

where the accuracy of the approximation is governed by the truncation threshold δ_N. Using these approximations and using intrinsic properties of the spectral decomposition, the snapshot matrix $\underline{\underline{S}}$ up to the last basis adjustment is approximated by

$$\underline{\underline{S}} \approx \underline{\underline{B}}\,\underline{\underline{\widetilde{\Theta}}}^{\frac{1}{2}}\,\underline{\underline{\widetilde{V}}}^{\mathsf{T}}. \tag{29}$$

Note that $\underline{\underline{B}} \in \mathbb{R}^{n\times N}$ is stored anyway, $\underline{\underline{\widetilde{\Theta}}} \in \mathbb{R}^{N\times N}$ is diagonal and $\underline{\underline{\widetilde{V}}} \in \mathbb{R}^{n_S\times N}$ is of manageable size (here $n_S \ll n$ is the number of snapshots with $\mathcal{P}_\delta \geq \epsilon$ considered in the basis generation up to now). The snapshot correlation matrix $\underline{\underline{C}}$ that considers the additional snapshots can be approximated as

$$\underline{\underline{C}} \approx \begin{bmatrix} \underline{\underline{\widetilde{C}}}_0 & \underline{\underline{\widetilde{V}}}\,\underline{\underline{\widetilde{\Theta}}}^{\frac{1}{2}}\,\underline{\underline{B}}^\mathsf{T}\,\underline{\underline{\Delta S}} \\ \underline{\underline{\Delta S}}^\mathsf{T}\,\underline{\underline{B}}\,\underline{\underline{\widetilde{\Theta}}}^{\frac{1}{2}}\,\underline{\underline{\widetilde{V}}}^\mathsf{T} & \underline{\underline{\Delta S}}^\mathsf{T}\,\underline{\underline{\Delta S}} \end{bmatrix} = \underbrace{\begin{bmatrix} \underline{\underline{\widetilde{V}}} & \underline{\underline{0}} \\ \underline{\underline{0}} & \underline{\underline{I}} \end{bmatrix}}_{\underline{\underline{V}}_*} \underbrace{\begin{bmatrix} \underline{\underline{\widetilde{\Theta}}} & \underline{\underline{\widetilde{\Theta}}}^{\frac{1}{2}}\,\underline{\underline{B}}^\mathsf{T}\,\underline{\underline{\Delta S}} \\ \text{sym.} & \underline{\underline{\Delta S}}^\mathsf{T}\,\underline{\underline{\Delta S}} \end{bmatrix}}_{\underline{\underline{C}}_1} \underbrace{\begin{bmatrix} \underline{\underline{\widetilde{V}}}^\mathsf{T} & \underline{\underline{0}} \\ \underline{\underline{0}} & \underline{\underline{I}} \end{bmatrix}}_{\underline{\underline{V}}_*^\mathsf{T}} . \tag{30}$$

In order to compute the updated basis, the inexpensive eigenvalue decomposition of $\underline{\underline{C}}_1 \in \mathbb{R}^{(N+n_a)\times(N+n_a)}$ is computed

$$\underline{\underline{C}}_1 = \underline{\underline{V}}_1\,\underline{\underline{\Theta}}_1\,\underline{\underline{V}}_1^\mathsf{T} . \tag{31}$$

Analogously to the previous RB computation in Equation (15), the adjusted and enriched basis is computed by

$$\underline{\underline{B}} = \begin{bmatrix} \underline{\underline{S}} & \underline{\underline{\Delta S}} \end{bmatrix} \underline{\underline{\widetilde{V}}}\,\underline{\underline{\widetilde{\Theta}}}^{-\frac{1}{2}} \approx \begin{bmatrix} \underline{\underline{B}}\,\underline{\underline{\widetilde{\Theta}}}^{\frac{1}{2}}\,\underline{\underline{\widetilde{V}}}^\mathsf{T} & \underline{\underline{\Delta S}} \end{bmatrix} \underbrace{\underline{\underline{V}}_*\,\underline{\underline{\widetilde{V}}}_1}_{\underline{\underline{\widetilde{W}}} \in \mathbb{R}^{(n_S+n_a)\times N}} \underline{\underline{\widetilde{\Theta}}}_1^{-\frac{1}{2}} . \tag{32}$$

To update the RB the truncated eigenvector matrix ($\underline{\underline{B}}, \underline{\underline{\widetilde{V}}} \leftarrow \underline{\underline{\widetilde{W}}} \in \mathbb{R}^{(n_S+n_a)\times N}$) need to be stored as well as the diagonal eigenvalue matrix $\underline{\underline{\widetilde{\Theta}}}$.

Remarks on Method B

B.1 The existing RB is not preserved but it is updated using the newly available information. Thereby, the accuracy of the RB for the approximation of the previous snapshots is not guaranteed a priori. However, numerical experiments have shown no increase in the approximation errors of previously well-approximated snapshots.

B.2 In contrast to Method A the dimension of the RB can remain constant, i.e., a mere adjustment of existing modes is possible. The average number of added modes per enrichment is well below that of Method A.

B.3 The additional storage requirements are tolerable and the additional computations are of low algorithmic complexity. In particular, the correlation matrix $\underline{\underline{C}}_1$ consists of a diagonal block complemented by a dense rectangular block, rendering the eigenvalue decomposition more affordable.

2.3.3. Method C: Incremental SVD

Method C is closely related to Method B. However, instead of building on the use of the correlation matrix, it relies on the use of an updated SVD, i.e., an approximate truncated SVD is sought after

$$\text{trunc svd}\left(\begin{bmatrix} \underline{\underline{S}} & \underline{\underline{\Delta S}} \end{bmatrix} \right) \approx \underline{\underline{B}}\,\underline{\underline{\Sigma}}\,\underline{\underline{W}}^\mathsf{T} . \tag{33}$$

Since the original snapshot matrix $\underline{\underline{S}}$ can not be stored, only an approximation of the actual truncated SVD in (33) can be computed. Methods to compute an incremental SVD were, e.g., introduced in [39,40], with the latter referring to Brand's incremental algorithm [41] which is used in the present study with minor modifications. With the previously computed basis $\underline{\underline{B}}$ at hand, the approximation of $\underline{\underline{S}}$ is known

$$\underline{\underline{S}} \approx \underline{\underline{B}}\,\underline{\underline{\Sigma}}\,\underline{\underline{W}}^\mathsf{T} . \tag{34}$$

First, the projection residual $\underline{\underline{\Delta\widehat{S}}}$ of the enrichment snapshots $\underline{\underline{\Delta S}}$ and its SVD

$$\underline{\underline{\Delta\widehat{S}}} = \underline{\underline{\Delta S}} - \underline{\underline{B}}\,\underline{\underline{B}}^\mathsf{T}\,\underline{\underline{\Delta S}} \quad = \underline{\underline{U}}_S\,\underline{\underline{\Sigma}}_S\,\underline{\underline{W}}_S^\mathsf{T} , \tag{35}$$

are computed. By using the truncated SVD to approximate the previous snapshots cf. Equation (34) and accounting for the newly added snapshots via Equation (35), the new snapshot matrix including the candidate snapshots can be approximated by

$$[\underline{\underline{S}} \quad \Delta\underline{\underline{S}}] \approx [\underline{\underline{B}}\,\underline{\underline{\Sigma}}\,\underline{\underline{W}}^{\mathsf{T}} \quad \Delta\underline{\underline{S}}] = [\underline{\underline{B}} \quad \underline{\underline{U}}_S] \underbrace{\begin{bmatrix} \underline{\underline{\Sigma}} & \underline{\underline{B}}^{\mathsf{T}} \Delta\underline{\underline{S}} \\ \underline{\underline{0}} & \underline{\underline{\Sigma}}_S \underline{\underline{W}}_S^{\mathsf{T}} \end{bmatrix}}_{\underline{\underline{\Gamma}}} \begin{bmatrix} \underline{\underline{W}} & \underline{\underline{0}} \\ \underline{\underline{0}} & \underline{\underline{I}} \end{bmatrix}^{\mathsf{T}}. \tag{36}$$

The matrix $\underline{\underline{\Gamma}}$ consists of a $N \times N$ diagonal block and a rectangular matrix of size $(N + n_{\mathrm{a}}) \times n_{\mathrm{a}}$. Due to this sparsity pattern, the SVD $\underline{\underline{\Gamma}} = \underline{\underline{U}}_\Gamma \underline{\underline{\Sigma}}_\Gamma \underline{\underline{W}}_\Gamma^{\mathsf{T}} \in \mathbb{R}^{(N+n_{\mathrm{a}})\times(N+n_{\mathrm{a}})}$ is inexpensive to compute. It allows to rewrite Equation (36) as

$$[\underline{\underline{S}} \quad \Delta\underline{\underline{S}}] \approx \Big(\underbrace{[\underline{\underline{B}} \quad \underline{\underline{U}}_S]\,\underline{\underline{U}}_\Gamma}_{\underline{\underline{U}}_*} \Big) \underbrace{\underline{\underline{\Sigma}}_\Gamma}_{\underline{\underline{\Sigma}}_*} \Big(\underbrace{\begin{bmatrix} \underline{\underline{W}} & \underline{\underline{0}} \\ \underline{\underline{0}} & \underline{\underline{I}} \end{bmatrix} \underline{\underline{W}}_\Gamma}_{\underline{\underline{W}}_*} \Big)^{\mathsf{T}}. \tag{37}$$

It is easily shown that the matrices $\underline{\underline{U}}_*$ and $\underline{\underline{W}}_*$ are column-orthogonal and that $\underline{\underline{\Sigma}}_*$ is diagonal and non-negative. Therefore, the three matrices constitute an approximate SVD of the enlarged snapshot matrix at low computational expense. This implies the following updates after the enrichment step

$$\underline{\underline{B}} \leftarrow \begin{bmatrix} \underline{\underline{B}} & \underline{\underline{U}}_S \end{bmatrix} \underline{\underline{U}}_\Gamma \qquad \underline{\underline{\Sigma}} \leftarrow \tilde{\underline{\underline{\Sigma}}}_\Gamma \qquad \underline{\underline{W}} \leftarrow \begin{bmatrix} \underline{\underline{W}} & \underline{\underline{0}} \\ \underline{\underline{0}} & \underline{\underline{I}} \end{bmatrix} \underline{\underline{W}}_\Gamma \tag{38}$$

after truncation of $\underline{\underline{B}}$, where the truncation criteria needs to ensure that $\underline{\underline{B}}$ does not decrease in size. To compute the enrichment of the RB, $\underline{\underline{B}} \in \mathbb{R}^{n\times N}$ and the sparse singular values $\underline{\underline{\Sigma}} \in \mathbb{R}^{N\times N}$ after truncation need to be stored.

Remarks on Method C

C.1 As highlighted for Method B (see remark **B.1**), the existing RB is not preserved but adjusted by considering the newly added information. A priori guarantees regarding the subset approximation accuracy can not be made, i.e., the approximation error of the previous snapshots $\underline{\underline{S}}$ could theoretically worsen. However, our numerical experiments did not exhibit such behavior at any point.

C.2 In contrast to Method A the dimension of the RB can remain constant, i.e., a mere adjustment of existing modes is possible. The average number of added modes per enrichment is well below that of Method A.

C.3 Each update step in (38) is computed separately and, consequently, storing $\underline{\underline{W}}$ is not required since only the RB $\underline{\underline{B}}$ is of interest.

C.4 The diagonal matrix $\underline{\underline{\Sigma}}$ has low storage requirements corresponding to that of a vector in $\mathbb{R}^N$.

3. Supervised Learning Using Feed Forward Neural Network

During the supervised learning phase, the machine is provided with data sets consisting of inputs and the related outputs: We aim at learning an unknown function relating inputs (here: Image data compressed into a low dimensional feature vector) to outputs (here: Effective thermal conductivity tensors) without or with limited prior knowledge of the structure of this function. Artificial Neural Networks (ANN) are a powerful machine learning tool which has gained wide popularity in the recent years due to the surge in computational power [27,42] and the availability of easy to use software packages (as a frontend in Python: Keras, Pytorch, TensorFlow or as graphical user interfaces Neuraldesigner amongst many others).

The functionality of the ANN is inspired by the (human) brain, propagating a signal (input) through multiple neurons where it is lastly transformed into an action (output). Various types of neural networks have been invented, e.g., feedforward, recurrent or convolutional networks, being applicable to almost any field of interest [43–46].

In the present study a regression model from the input, i.e., the feature vector $\underset{\sim}{\xi}$ which is derived with the converged basis $\underline{\underline{B}}$, to the output, i.e., the effective heat conduction tensor $\underline{\underline{\bar{\kappa}}}$, is deployed with a dense feedforward ANN.

In a dense feedforward ANN (Figure 3) a signal is propagated through the hidden layers where every output of the previous layer $\underline{a}^{l-1}$ affects the activation $\underline{z}^{l}$ of the current layer l $(l = 1, \ldots, L+1)$. The activation of each layer gets wrapped into an activation function f where the output of each neuron in the layers is computed, i.e., $\underline{a}^{l} = f(\underline{z}^{l})$. Note that matrix/vector notation is used, where each entry in the vectors denotes one neuron in the respective layer.

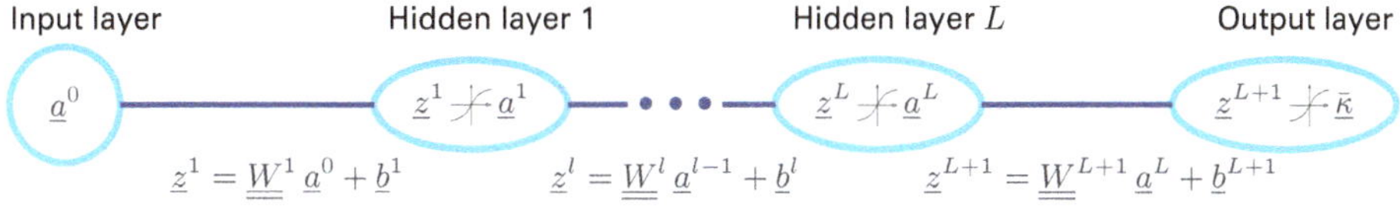

Figure 3. The basic functionality of a dense feedforward neural network is depicted in simplified form.

The basic learning algorithm/optimizer usually employed for a feedforward ANN is the back propagation algorithm [47] and modifications thereof. The learning of the network consists in the numerical identification of the unknown weights $\underline{\underline{W}}^{l}$ and biases $\underline{b}^{l}$ minimizing a given cost function, where a random initialization defines the initial guess for all parameters. The cost function gives an indication of the quality of the ANN prediction. The gradient back propagation computes suitable corrections for the parameters of the ANN by evaluating the gradients of the cost function to the weights.

The learning itself is an iterative procedure in which the training data is cycled multiple times through the ANN (one run called an 'epoch'). In each epoch the internal parameters are updated with the aim of improving the mapping relating input and output data, aiming at reduction of the cost function. The optimization problem itself is (usually) high-dimensional. In most situations it is not well-posed and local minima and maxima can hinder convergence to the global minimum. Therefore, multiple random instantations of the network parameters are usually required to assure that a good set of parameters is found, even if the network layout remains unaltered.

The training requires a substantial input data set as input-output tuples in order to allow for robust and accurate predictions.

It is important to note that the (repeated) training of the ANN usually results in a parameter set that is able to approximate the training data with high accuracy under the given meta-parameters describing the network architecture (number of layers, number of neurons per layer, type of activation function). However, the approximation quality of the ANN may be different for query points not contained in the training set. Thus, it is important to validate the generality of the discovered surrogate for the underlying problem setting. Therefore, an additional validation data set is introduced, where only the evaluation of the cost function is tracked over the epochs. Generally, when overfitting occurs (overfitting relates to the fact that a subset of the data is nicely matched but small variations in the inputs can lead to substantial loss in accuracy, similar to oscillating higher-order polynomial interpolation functions), the errors for the validation set increase whereas the errors of the training set decrease. The training should be halted if such a scenario is detected.

Since the choice of activation function as well as the number of hidden layers and the number of neurons within the individual layers are arbitrary (describing the ANN architecture), these meta-parameters should be tailored specifically for the desired mapping. Finding the best neural network architecture is not straight-forward and usually relies on intuition, experience and a substantial amount of numerical experiments. As mentioned earlier, the identification of a well-suited ANN requires various random realizations (corresponding to different initial biases and weights) for each ANN architecture under consideration. The optimum is then found as the best ANN over all realizations over all tested architectures.

In the present study the ANN training is performed using TensorFlow in Python [48]. TensorFlow is an open source project by the Google team, providing highly efficient algorithms for ANN implementation. The ADAM [49] optimizer, which is a modification of the gradient back propagation, has been deployed for the learning.

4. Results

4.1. Generation of Synthetic Microstructures

All of the used synthetic microstructures have been generated by a random sequential adsortion algorithm with some examples shown in Figure 1. Two morphological prototypes were used: spheres and rectangles. The deployed microstructure generation algorithm ensures a broad variability in the resulting microstructure geometry. Indeed, any bi phasic microstructure image can be considered. The parameters used to instantiate the generation of a new microstructure were modeled as uniformly distributed variables:

M.1 The phase volume fraction f_b of the inclusions (0.2–0.8);
M.2 The size of each inclusion (0.0–1.0);
M.3 For rectangles: The orientation (0–π) and the aspect ratio (1.0–10.0);
M.4 The admissible relative overlap ϱ for each inclusion (0.0–1.0).

For $\varrho = 0$ and the spherical inclusion, a boolean model of hard spheres is obtained. Setting $\varrho = 1$ induces a boolean model without placement restrictions, i.e., new inclusions can be placed independent of the existing ones. The generated microstructures were stored as images with resolution 400×400. After the generation of the RVE, the 2-point spatial correlation function was computed for the RVE. This was then shifted and scaled, see Equation (10) in Section 2.2, and used as a snapshot $\underline{s}_i$ for the identification of the reduced basis.

Additionally, a smaller random set of RVEs used for the supervised learning phase was simulated using the recent Fourier-based solver FANS [3] in order to compute the effective heat conduction tensor $\underline{\underline{\bar{\kappa}}}$. The heat conductivity of the matrix and of the inclusion phase are prescribed as

$$\kappa_a = 1.0\left[\frac{\mathrm{W}}{\mathrm{m}\cdot\mathrm{K}}\right], \qquad \kappa_b = \frac{\kappa_a}{R}\left[\frac{\mathrm{W}}{\mathrm{m}\cdot\mathrm{K}}\right]. \tag{39}$$

Here $R > 0$ denotes the material contrast. In the present study, $R = 5$ was considered, i.e., the matrix of the microstructure has a five times higher conductivity than the inclusions. These values can be seen as typical values for metal ceramic composites (Figure 4).

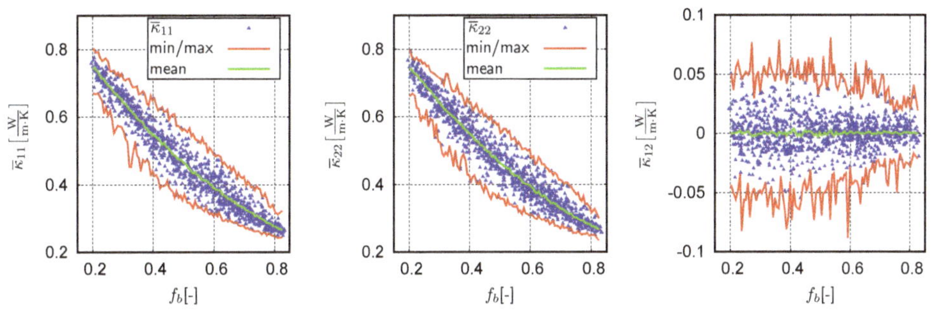

Figure 4. The range of each $\underline{\underline{\bar{\kappa}}}$ entry computed with 15,000 microstructures of the mixed set is shown. Only 1000 discrete values are shown in each plot.

An inverse phase contrast has exemplarily been studied, i.e., inclusions with $\kappa_b = 1\,\mathrm{W\,m^{-1}\,K^{-1}}$ and $\kappa_a = \frac{\kappa_b}{5}$ (corresponding to $R = \frac{1}{5}$) $R = 1/5$) has also been investigated. Qualitatively, the results for the inverse phase contrast did not show any new findings or qualitative differences. Therefore, the following results focus on $R = 5$, corresponding to rather insulating inclusions.

The symmetric tensor $\underline{\underline{\bar{\kappa}}}$ can be represented as a three-dimensional vector $\underline{\bar{\kappa}}$ using the normalized Voigt notation

$$\underline{\underline{\bar{\kappa}}} = \begin{bmatrix} \bar{\kappa}_{11} & \bar{\kappa}_{12} \\ \bar{\kappa}_{21} & \bar{\kappa}_{22} \end{bmatrix} \quad \rightarrow \quad \underline{\bar{\kappa}}_{\mathrm{V}} = \begin{bmatrix} \bar{\kappa}_{11} \\ \bar{\kappa}_{22} \\ \sqrt{2}\,\bar{\kappa}_{12} \end{bmatrix}. \tag{40}$$

For the supervised learning of the ANNs (see Section 3), multiple files each containing 1,500 data sets for different inclusion morphologies were generated (circle only; rectangle only; mixed; see following section). Each data set contains the image of the microstructure, the respective autocorrelation of the inclusion phase $c_2(\bullet; b, b)$ and the effective heat conductivity $\underline{\bar{\kappa}}_{\mathrm{V}}$.

4.2. Unsupervised Learning

First, the reduced basis is identified using the iterative procedure presented in Section 2.3. All three proposed methods were considered and for each of these, three different sets of microstructures were used as inputs: The first set of microstructures consisted of RVEs with only circular inclusions, the second set consisted of RVEs with only rectangular inclusions, and the third set was divided into equal parts, each part consisting of RVEs with either circular or rectangular inclusions (i.e., each structure contained exclusively one of the two morphological prototypes and the same number of realizations for each prototype was enforced), respectively. Each type of microstructure was processed using each of the three incremental RB schemes introduced in Section 2.3. Hence, a total of nine different trainings were conducted, each using different randomly generated snapshots.

For the iterative enrichment process, the initial RB was computed from 200 snapshots $\underline{\underline{S}}_0$. Thereafter, snapshots were randomly generated and processed by the enrichment algorithm sketched in Figure 2. The number of snapshots per enrichment step has been set to $n_{\mathrm{a}} = 75$ and the number of consecutive snapshots with $\mathcal{P}_\delta < \varepsilon$, used to indicate convergence, has been set to $n_{\mathrm{c}} = 100$. The relative projection tolerance $\varepsilon = 0.025$ was chosen. Note that this corresponds to the maximum value of the mean relative $\|\cdot\|_{L^2}$-error that is considered exact for the shifted and scaled snapshots. The actual accuracy in the reproduction of the 2PCF $c_2(\boldsymbol{r}; b, b)$ is significantly lower than this (results are given in Figure 7).

Key attributes for each of the nine trainings are provided in Table 1. There is an obvious discrepancy between Method A and the remaining methods in basically all outputs. While Method A claims the lowest computing times, it yields approximately twice the number of modes. However, the number of snapshots needed is substantially lower which can be relevant if the generation of the synthetic microstructures is computationally involved.

Table 1. Data of the unsupervised learning (incremental reduced basis (RB) identification) for the nine considered scenarios; the parameters $\varepsilon = 0.025$, $n_c = 100$ and $n_a = 75$ were used. Some numbers are rounded for easier readability.

Method	Final Basis Size	Snapshots with $\mathcal{P}_\delta \geq \varepsilon$	Snapshots with $\mathcal{P}_\delta \leq \varepsilon$	Enrichment Steps	Time [s]	Used Microstructures
A	143	150	730	4	20	
B	80	400	2400	7	70	
C	96	800	7700	12	200	
A	596	670	4500	11	150	
B	294	2400	12,700	34	500	
C	312	2600	16,500	37	550	
A	464	560	2900	9	150	 
B	274	2000	16,100	29	500	
C	244	1540	8000	22	280	

Note that methods B and C yield similar results, although for the rectangular and circular training Method C needed significantly more snapshots, Method B needed significantly more snapshots for the mixed training. The outliers between methods B and C in the number of snapshots needed are due to the randomness of the materials and the chosen convergence criterion. The resulting basis size of methods B and C indicate very similar results from these methods. Note that methods B and C yield identical results when operating on an identical sequence of microstructures used as inputs when leaving aside perturbations due to numerical truncation.

In addition, note that the computational effort for the relative projection error $\mathcal{P}_\delta$ grows linearly with the dimension of the RB, i.e., the faster offline time of Method A can quickly be compensated by the costly online procedure induced by the high dimension of the RB in comparison to the competing techniques.

To compare the accuracy of the resulting basis as well as during the training, the relative projection error $\mathcal{P}_\delta$ of the snapshots used for the original basis construction $\underline{\underline{S}}_0$ are plotted in Figure 5.

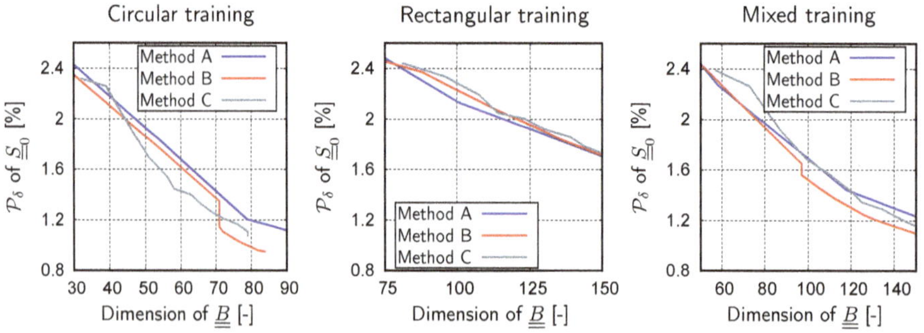

Figure 5. Development of the relative projection error $\mathcal{P}_\delta$ of the snapshots $\underline{\underline{S}}_0$ with respect to the current basis size N over the enrichment.

While methods B and C do not, unlike Method A, a priori guarantee an improvement of the relative projection error of $\underline{\underline{S}}_0$ over the enrichment, a strict downward trend is observed. The adjustment of already existing eigenmodes in methods B and C allow for an improvement of the relative projection error of $\underline{\underline{S}}_0$ for a constant basis size.

Method B and C seem to outperform Method A in most cases; however, the basis of Method A achieves a lower projection error on convergence (not shown in the plot), but at the expense of a considerably larger dimension of the RB.

Since there seems to be an obvious correlation between resulting accuracy and the final basis size for the initial snapshots $\underline{\underline{S}}_0$ (see Figure 5, Table 1), the general quality for arbitrary stochastic inputs must be investigated. In order to quantify the quality of the RB, the accuracy can be expressed in terms of the relative projection error of approximating additional, newly generated snapshot data $\underline{\underline{S}}$ as a function of the Method (A, B, C) and the number of modes $N \geq 1$ via

$$\mathcal{P}_\delta(N) = \sqrt{\frac{||\underline{\underline{S}} - \underline{\underline{B}}(:,1:N)\,\underline{\underline{B}}^\mathsf{T}(:,1:N)\,\underline{\underline{S}}||_F^2}{||\underline{\underline{S}}||_F^2}} \tag{41}$$

in Matlab notation.

This measure captures to what extend the first N basis functions represent the 2PCF of the underlying microstructure class. In the current work sets of 1500 newly generated snapshots assure an unbiased validation, i.e., the data was used in neither of the three training procedures. The results are stated in Figure 6. Again, Method B and C yield similar results, achieving lower projection errors with fewer eigenmodes compared to Method A, i.e., the basis produced by Method A cannot catch up with its two competitors. On a side note, the rectangular inclusions apparently lead to significantly richer microstructure information which can be seen by direct comparison of the left to the middle plot in Figure 6. For methods B and C and for circular inclusions the relative error of 5% is reached for approximately 15 modes while rectangular inclusions require more than 60 modes to attain a similar accuracy. This is supported also by the rightmost plot determined from a sort of blend of the two microstructural types.

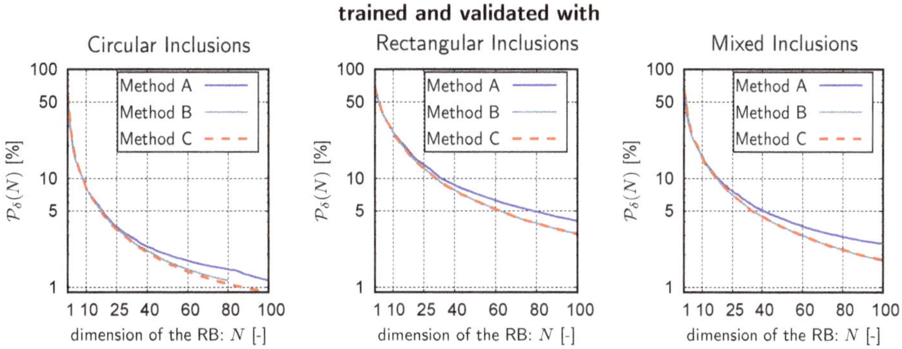

Figure 6. Relative projection error for three different microstructure classes as a function of the number of eigenmodes. The relative projection error is determined for a validation set of 1,500 newly generated microstructures for each class.

Since all of the previous error measures are given on the shifted snapshot according to Equation (10), the true relative projection error on the unshifted snapshot is also investigated as a function of the basis size. It describes the actual relative accuracy of the approximation of the 2PCF $c_2(\boldsymbol{r}; b, b)$ as a function of the basis size. The errors in the shifted data (Figure 7, left) and the corresponding reconstructed 2PCF (Figure 7, right) for five randomly selected snapshots show that the actual relative error in the 2PCF reconstruction is below 5% for 10 reduced coefficients even for the challenging rectangular inclusion morphology, while the error in the shifted and scaled snapshots is on the order of 50%. This highlights the statement made earlier regarding the choice of ε which is not directly the accepted mean error in the 2PCF, but only after application of the shift. The high discrepancy in the two relative projection errors is due to the fact that the shifted snapshots fluctuate closely around 0, i.e., the homogeneous part of the 2PCF is obviously of high relevance.

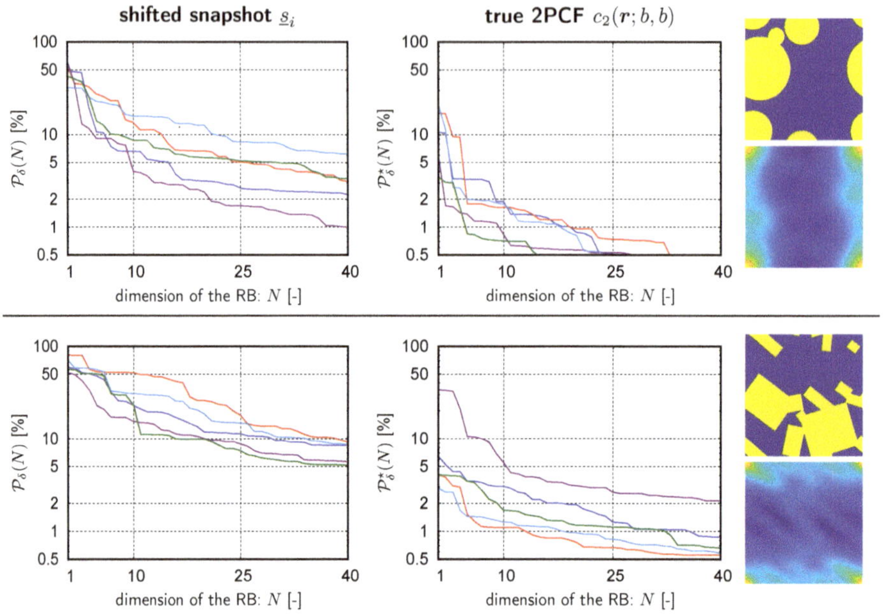

Figure 7. Using the RB of Method C, the relative projection error on the shifted snapshot $\mathcal{P}_\delta$ is given on the left for five random samples. For comparison the relative projection error of the reconstruction of the actual 2-point correlation function $\mathcal{P}^*_\delta$ is given on the right for the same five samples.

The development, i.e., the stabilization of the mode shapes over the enrichment steps, of a few selected eigenmodes is shown in Figure 8 using RVEs with circular inclusions for training of Method C. Similar results are expected for Method B, whereas for Method A the eigenmodes would remain unconditionally unchanged over the enrichment steps, i.e., a pure enlargement of the basis takes place. The faster stabilization of the leading eigenmodes indicates a quick stabilization of the lower order statistics of the microstructure ensemble, while the tracking of higher order fluctuations is more involved.

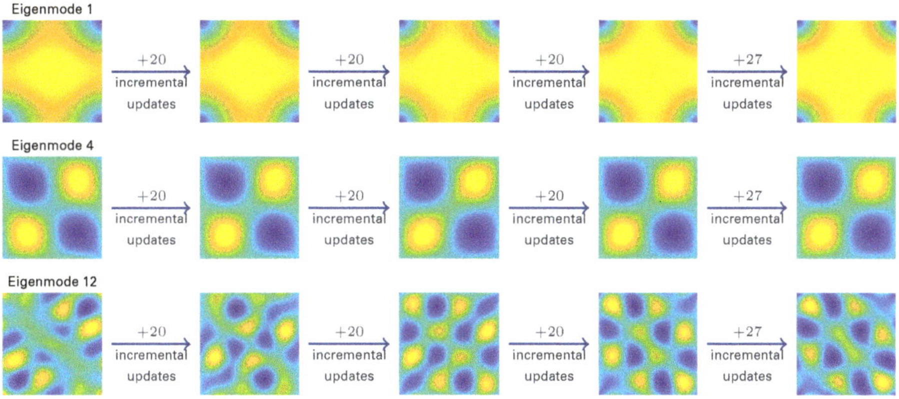

Figure 8. The development of a few selected eigenmodes over the enrichment are shown for the circular inclusion morphology. Note that these results are generated with $n_a = 15$ and $\varepsilon = 0.01$ using Method C. The procedure comprised a total of 87 basis enrichments/adjustment.

4.3. Supervised Learning

After the training of the RB, the input for the neural network, the feature vector $\underline{\xi}$ was derived using the 1- and 2-point spatial correlation functions of the ith RVE as

$$\underline{\xi}_i = \begin{bmatrix} f_{b,i} \\ \underline{\underline{B}}^\mathsf{T} \underline{s}_i \end{bmatrix} \qquad \in \mathbb{R}^{(h+1)} . \tag{42}$$

The size of the feature vector is determined by the amount of reduced coefficients $1 \le h \le N$, i.e., the snapshot is projected onto the leading h eigenmodes of $\underline{\underline{B}}$.

Since the inputs and outputs have a highly varying magnitude, they need to be shifted such that they are equally representative. Therefore, each entry of the feature vector is separately shifted and scaled such that its distribution of all samples has zero mean and a standard deviation of one. The output is shifted combinedly such that the mean of $\underline{\bar{\kappa}}_V$ is $\underline{0}$. The transformed inputs and outputs are then given to the ANN for the training phase. Thus, the outputs of the ANN need to undergo an inverse scaling in order to yield the sought-after vector representation of the heat conduction tensor. These shifts and scalings need to be extracted from the available training data. Hence, every data set used for training purposes has its own parameters.

The training for the neural network has been conducted for all of the three microstructure classes, i.e., using only RVEs with circular inclusions, only RVEs with rectangular inclusions and lastly using RVEs with either circular or rectangular inclusions with equal number of realizations of each shape within the mixed set. In order to derive the feature vector, the converged basis of Method C has been used. Note that depending on the training set, either circular or rectangular or both inclusion shapes (for the mixed set) contributed to the RB.

In order to find a good overall ANN, the network architecture has been intensely studied: The accuracy of the prediction after the training has been evaluated with various sizes of the feature vector, different network layouts and for different activation functions (Figure 9).

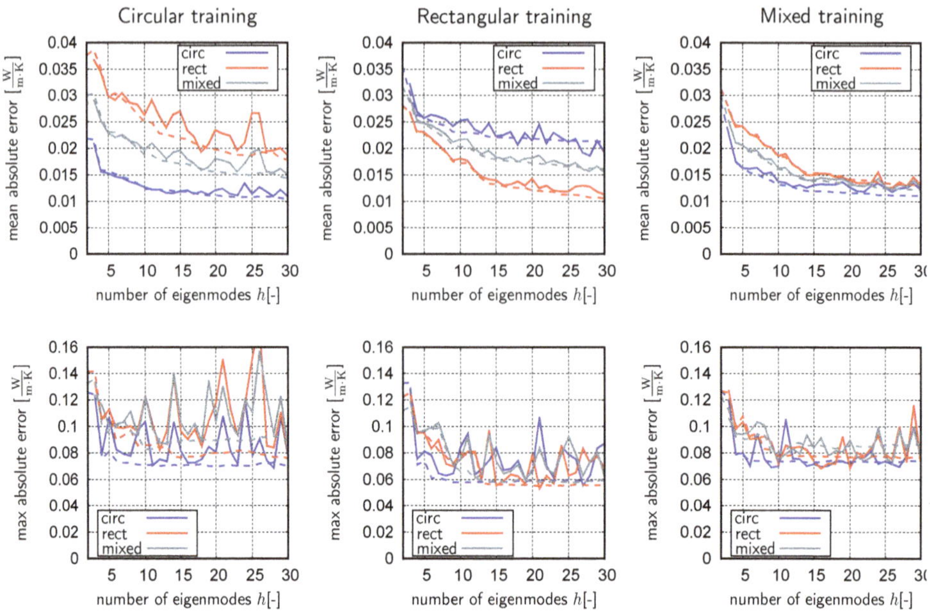

Figure 9. The given error measures over the test sets are shown for the Gaussian Process Model (GPM) (dashed lines) and the Artificial Neural Networks (ANN) (full lines) which achieved the lowest MSE (cost) on the validation set for each number of reduced coefficients and training type.

The training of the ANN was conducted with an early stop algorithm, stopping the training after 500 consecutive epochs of no improvement of the cost function with respect to the validation set. The learning rate of the ANN has been held constant during the training, being randomly initialized between 0.01 and 0.05. A network depth of up to 6 hidden layers and a network width of up to 100 hidden neurons have been considered and the number of neurons was chosen on a per layer basis. Recall that a vanilla dense feedforward ANN has been deployed. In order to find the best ANN architecture, 35 randomly initialized ANN trainings have been considered for each size of the feature vector. A total amount of 1500 samples have been considered for each ANN training. These were shuffled randomly and split into the training set ($n_t = 1000$) and the validation set ($n_v = 500$).

In the following, the error measurements used and the term of unbiased testing refers to the prediction of 7500 unseen data points for each of the three microstructure classes named 'test sets'.

The prediction error is given by the 2-norm, i.e.,

$$e^{\mathrm{P}} = ||\underline{\overline{\kappa}}_{\mathrm{V}} - \underline{\overline{\kappa}}_{\mathrm{V}}^{\mathrm{P}}||_2, \tag{43}$$

with $\underline{\overline{\kappa}}_{\mathrm{V}}^{\mathrm{P}}$ denoting the prediction of the regression model. The mean and maximum errors of the prediction error for all test sets are shown in Figure 9. For comparison of the regression model, we have deployed a Gaussian Process Model (GPM) [50], which reliably finds the global minimum of the optimization for the kernel regression. The ANN is given with full lines and the GPM model is given with dashed lines in Figure 9. Note that each ANN realization refers to a randomly initialized ANN architecture.

The GPM model seems to achieve slightly lower errors than the ANN, however, in the interest of computational speed the ANN regressor is preferred. Not only is the training significantly faster, the prediction times for the GPM highly depend on the size of the input vector, whereas the prediction times of the ANN mostly depend on the ANN architecture. More details are given in Section 5.

The spikes in Figure 9 regarding the maximum error, are explained by each depicted ANN having the lowest overall MSE of the validation set, which did not consider the maximum errors directly. Though, only a few outliers yielded a high prediction error, as can be seen in Figures 10 and 11.

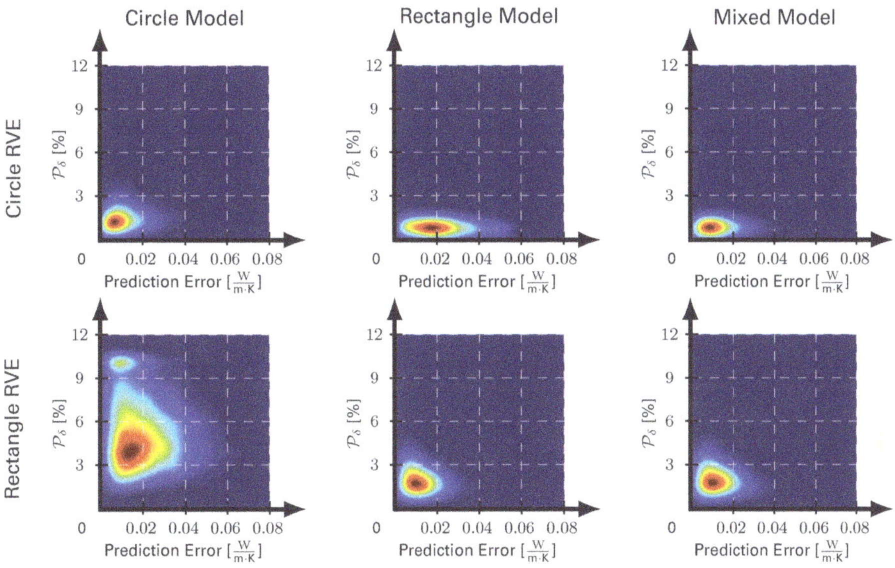

Figure 10. A density map of the projection error of the reduced basis compared to the prediction error of the ANN is given for all of the three training variants, for the prediction of the circle and rectangle test set, respectively. Each with one exemplary ANN (basis dimensions are 23, 25 and 25, respectively, from left to right).

Note that the ANN trained with rectangular RVE achieved lower maximum errors, whereas the circular RVE training achieved lower mean errors (Figure 9). A possible explanation is that rectangular inclusions allow for more complex geometries in the microstructure than perfectly, yet overlapping spherical inclusions. This possibly allows the RB as well as the ANN to better learn about microstructure geometries which, usually, lead to a high prediction error. The ANN trained with both microstructure classes manages to nicely capture both training advantages of the RVE classes and achieves a good mean accuracy as well as low maximum errors across the board.

The conductivity $\overline{\kappa}_{12}$ fluctuates mildly around zero for all inputs. In order to accurately capture this fluctuation, only the specific training and RB dimensions of four or higher ($h \geq 4$) are required cf. Figure 12. Albeit the values can be considered small in comparison to the $\overline{\kappa}_{11}$ and $\overline{\kappa}_{22}$ errors.

The overall downward trend of the prediction errors validate our approach, implying that a higher amount of reduced coefficients leads to more detailed information about the microstructure geometry, allowing for a better prediction of the regression model. However, the prediction errors do not seem to completely vanish, therefore the 2PCF alone does not suffice to perfectly describe the microstructure geometry.

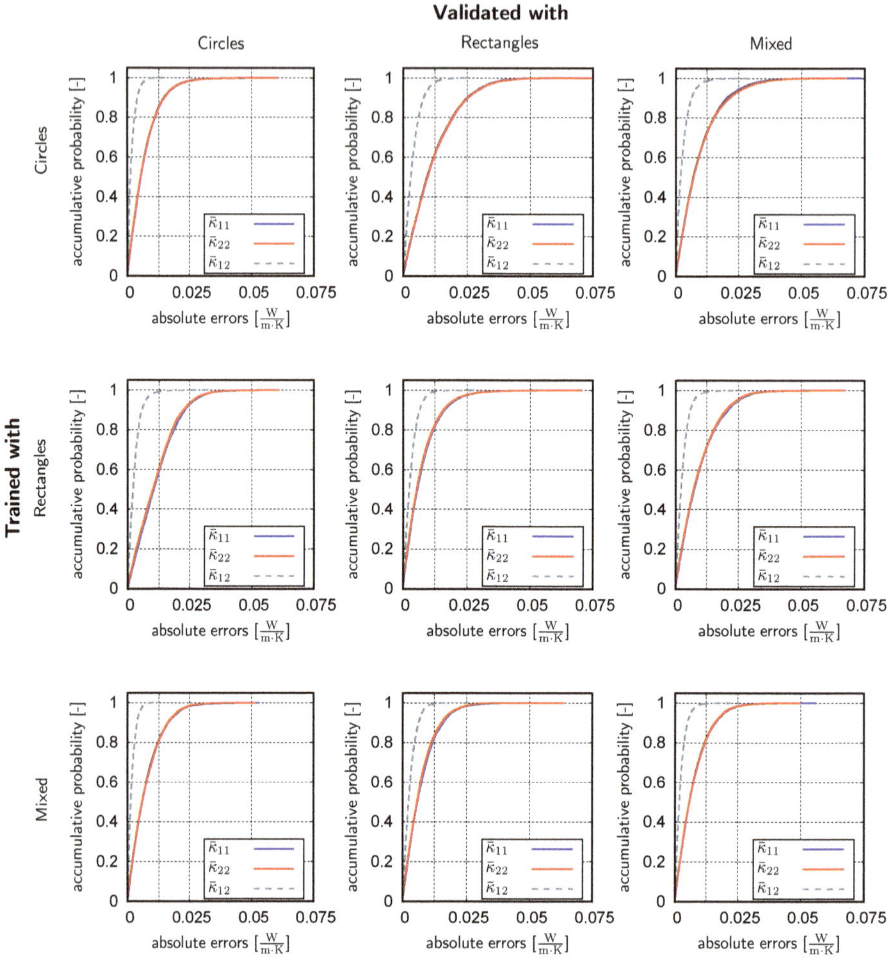

Figure 11. Results for the best of all tested ANN for the test sets. The graphs represent a probability distribution of the absolute error in each component of $\bar{\kappa}$.

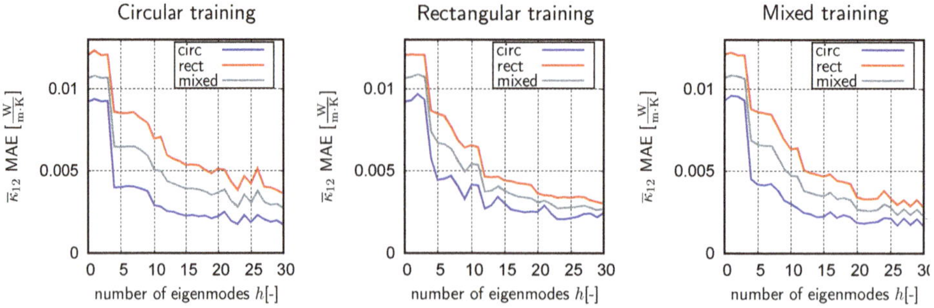

Figure 12. The mean absolute error (MAE) of $\bar{\kappa}_{12}$ is given for each of the training types and test sets.

To further study the accuracy of our surrogate model, which is divided into two processes namely the feature extraction with the RB and thereafter the prediction of the ANN, the error committed in each step is examined in Figure 10. Intuitively, a high projection error of the reduced basis is expected to yield poor knowledge of the microstructure geometry and, as a consequence, lead to a high prediction error of the ANN. On the contrary, microstructures with the highest projection errors still allowed for accurate ANN predictions and the highest ANN prediction errors occurred for relatively small projection errors. The comparison of the RB relative projection error plotted against the GPM prediction error yielded very similar results. Note that the relatively high projection errors on the circle-trained RB are due to the fact that the basis is significantly smaller, leading to an overall higher projection error (Table 1). The relative projection errors have been measured on the shifted and scaled 2PCF which is a more pessimistic prediction than the actual 2PCF cf. the results shown in Figure 7.

An observation of the worst predictions for each ANN (Figure 13) shows, that the inclusions of each RVE either just barely do not overlap, leaving a small gap for the matrix phase, or the inclusions just barely perculate. This phenomena has a pronounced impact on the resulting effective heat conductivity. Hence, a miniscule change in the image data can result in notable variation of the conductivity tensor, which can lead to high prediction errors of the surrogate.

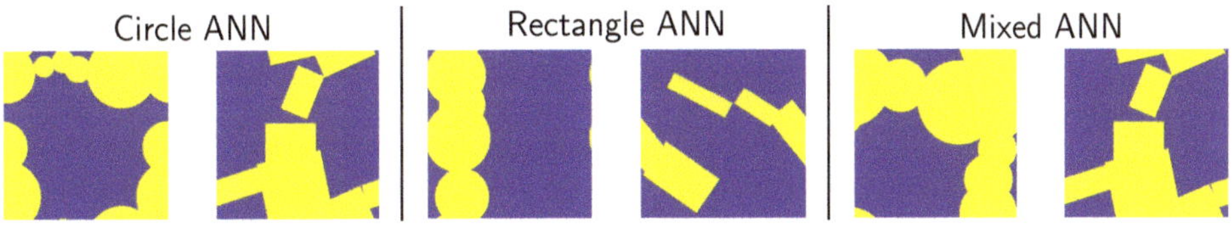

Figure 13. Representative volume element (RVE) with the highest prediction error for each of the ANN models given in Figure 10.

A detailed study of various ANN architectures revealed, that almost every architecture was suitable for the regression problem, e.g., an ANN with 2 hidden layers and a total of 13 hidden neurons had almost identical prediction errors as an ANN with 5 hidden layers and roughly 230 hidden neurons. The used activation functions were the sigmoid, relu, tanh and softplus, where only some combinations delivered poor results. Not a clear trend of ANN architecture and quality of prediction could be seen and, consequently, the best ANN were randomly found based on the lowest error on the test set.

The prediction accuracies for each test set of three differently trained ANNs, which have been deemed the *best*, is given in Figure 11. The training and architecture of the best ANNs in Figure 11 had the following properties:

- Circular training: $h = 23$; $11,206$ epochs; 5 hidden layers
  $\{5, 40, 77, 75, 74\}$ hidden neurons
  {sigm, softplus, sigm, softplus, softplus} activation functions
- Rectangular training: $h = 29$; $1,054$ epochs; 6 hidden layers
  $\{10, 42, 56, 18, 63, 59\}$ hidden neurons
  {relu, sigm, relu, softplus, tanh, tanh} activation functions
- Mixed training: $h = 26$; $6,177$ epochs; 2 hidden layer
  $\{6, 7\}$ hidden neurons
  {softplus, softplus} activation function

The shown error measures (Figure 11) are evaluated for each point in the whole test set, yielding a kind of probability distribution for the prediction error. For an easier readability, the percentage mean and max errors for each of the explicitly depicted ANN are given in Table 2. Note that since the values of $\bar{\kappa}_{12}$ vary closely around 0 (Figure 4), relative errors are not sensible for the quantity of interest.

Table 2. Percentage errors for $\bar{\kappa}_{11}$ and $\bar{\kappa}_{22}$ given for each of the best ANNs, evaluated over the complete test set (7500 data samples).

		Validated with					
		Circles		Rectangles		Mixed	
Trained with	Error Measures	κ_{11}	κ_{22}	κ_{11}	κ_{22}	κ_{11}	κ_{22}
Circles	Mean [%]	1.58	1.57	2.60	2.62	2.11	2.14
	Max [%]	12.8	12.5	13.9	13.0	14.7	11.7
Rectangles	Mean [%]	2.68	2.57	1.60	1.58	2.14	2.09
	Max [%]	12.9	11.7	12.5	12.0	13.8	13.0
Mixed	Mean [%]	1.77	1.76	1.65	1.60	1.72	1.71
	Max [%]	11.7	14.1	11.6	10.5	10.4	12.4

PARAGRAPH MOVED (after the table) As a side note, a descriptor based GPM has been trained for RVEs with circular inclusions, using the average minimum distance of inclusions, average inclusion radius, number of inclusions and volume fraction as an input, achieving mean relative errors of around 5% on the circle set.

A GUI code is provided in Github, where the user can choose between the three proposed surrogate model, the input for the prediction is a 400×400 image in matrix format written in a text file or a TIFF image and the output is the prediction for the heat conduction tensor as described above. In order to compile the code, Python3 with TensorFlow is required, additional required modules are pillow, numpy and matplotlib, as well as the default modules os and tkinter. Some exemplary RVE with their respective heat conductivity are uploaded in a subfolder.

5. Computational Effort

For the training and the deployment of the proposed surrogate model, the computational effort can be split into online and offline part. The offline phase describes the building of the surrogate model and is obviously computationally expensive due to the iterative nature of the supervised as well as the unsupervised learning. However, since the cost of the offline phase has no impact on the actual evaluation, i.e., prediction of the surrogate model, its impact is neglectable. All of the following measured times have been documented while computing with only an AMD Ryzen Threadripper 2920X 12-Core Processor, unless stated otherwise. In order to evaluate the surrogate model in the online phase, firstly, the 2PCF of the RVE has to be computed. Therefore a FFT, complex point-wise multiplication and lastly an IFFT is performed, summing up to a computational complexity of $\mathcal{O}(2\,n\,\log n + n)$. Recall that n is the dimension of the unreduced problem, i.e., the total number of voxels in the present study.

To derive the input for the ANN, the complexity for the computation of the reduced coefficients is $\mathcal{O}(n\,h)$ together with a computation of the volume fraction with an additional effort of $\mathcal{O}(n)$. This mounts up to a total computation effort of $\mathcal{O}(n\,(2+h) + 2\,n\log n\,)$ just to derive the input of the regression model. To give sensitivity to the computational effort, the computation of the feature vectors for one test set, i.e., 7500 images, took roughly 95 s.

As has been mentioned earlier, the ANN has been significantly faster than the GPM in the online, as well as offline phase. The training of the regression model for each number of reduced coefficients (i.e., 1–30) took roughly 12 hours for the ANN and about 31 hours for the GPM. Note that GPM has been trained in R with the code provided by [50], whereas the ANN has been implemented in Python with TensorFlow, using a Pali6GB D6 RTX 2060 GamingPro OC graphics card as well. As it is more important, in the online phase the ANN has been significantly faster than the GPM. Each prediction refers to the prediction of the three test sets, i.e., 3×7500 data points with each output being a three-dimensional vector. The prediction times for the GPM highly depends on the size of the input vector and takes from 0.82 s (with one reduced coefficient) up to 4.1 s to predict the test sets for an input dimension of 31. In comparison, the ANN took on average roughly 0.24 s for any dimension of the feature vector.

The computational complexity of the forward propagation in the ANN is governed by the matrix multiplication of a complexity of $\mathcal{O}(n_{\text{neuron}}^2)$ and the element wise evaluation of the activation function for each neuron with the complexity $\mathcal{O}(n_{\text{neuron}})$. For a quick overview, assume that the ANN has the same number of neurons in each layer, the computational complexity amounts to $\mathcal{O}(n_{\text{layer}}\,(n_{\text{neuron}}^2 + n_{\text{neuron}}))$. Therefore, we have an a priori estimate of the prediction time required for the ANN.

To compute the effective heat conductivity for 7500 images using the FANS solver [3], ≈ 4000 s were required. Note that the deployed FFT solver for the heat conductivity is intrinsically fast. The proposed method could be easily expanded to different material properties, yielding an even more significant computational speedup. Since usually $n \gg n_{\text{neuron}}$, the main computational effort lies within the computation of the feature vector, especially when considering the extension to the 3D case.

6. Conclusions

6.1. Summary and Concluding Remarks

The computational homogenization of highly heterogeneous microstructures is a challenging procedure with massive computational requirements. In the present study a method to efficiently and accurately predict the heat conductivity for any RVE with the image and no further information is proposed. Key ideas of the Materials Knowledge System (MKS) [21,32] have been adopted in the sense that a subset of the POD compressed 2-point correlation function is used to identify a low-dimensional microstructure description. In contrast to [32] the 2PCF is not truncated to a small neighborhood, but the full field information is considered. Similar to other works related to the MKS [18], a truncated PCA of the 2-point information is used to extract microstructural key features.

However, the classical truncated PCA used, e.g., in [18] is not applicable to the considered rich class of microstructures due to the high number of needed samples and the related unmanageable computational resources. Therefore, our proposal is founded on a novel incremental procedure for the generation of the RB of the 2PCF. Similar techniques have not been considered in the literature to the best of the authors' knowledge. The shifting and scaling of the images of 2PCF before entering the POD is another feature that can help in reducing the impact of the inclusion volume fraction, i.e., the shifted function has zero mean and a peak value of one. The authors would like to emphasize that such scaling is relevant in the present study where the phase volume fractions varies in a wide range.

Other than in [32] no higher-order statistics are used. This is by purpose as the selection of the relevant entries of the higher order PCF is ambiguous and a challenge in itself. Most notably it is based on a priori selections of the relevant components of the higher spatial correlations which allows for very limited insights to our understanding. Instead, the present study focuses on the variability of the input images in terms phase volume fractions in a broad range (20–80%) alongside topological variations (impenetrable, partial overlap, unrestricted placement) and different morphologies (circles and rectangles). Generally speaking, a much higher microstructural variation is accounted for, than in many previous studies. Therefore, the current study also investigates how the proposed technique and similar MKS related approach can possibly generalize towards truly arbitrary input images (e.g., stemming from 3D micrographs of real materials) and for databases containing millions of snapshots in order to build a powerful tool for material analysis and design.

In order to cope with the variability of the 2PCF, the classical truncated PCA or snapshot POD operating on a monolithic snapshot matrix during the unsupervised learning phase is replaced by novel incremental procedures for the construction of small-sized reduced microstructure parameterization. Three incremental POD methods are proposed and their results are compared regarding the computational effort, the projection accuracy of the snapshots and the quality of the basis in view of capturing random inputs.

The learned reduced bases are used to extract low-dimensional feature vectors. These are used as inputs for fully connected feedforward Artificial Neural Networks. The ANN is used to predict the homogenized heat conductivity of the material defined by the microstructure. The mean relative error of the surrogate is well below 2% for the majority of the considered test data. This is remarkable in view of the phase contrast $R = 5$ and the particle volume fractions ranging from 0.2–0.8, as well as morphological and topological variations. Further, an immense speedup in computing time is achieved by the surrogate over FE or FFT simulations (factors around 40 without tweaking the projection operation).

Importantly, the presented methodology can immediately be adopted to different physical settings such as thermo-elastic properties, fluid permeability, dielectricity constants, etc. The same holds for three-dimensional problems. However, the limited number of samples in 3D could be problematic as more features are likely required to attain a sufficiently accurate RB.

6.2. Discussion and Outlook

A weakness of the current approach remains the computational complexity of the method: Although the feature vector is rather low-dimensional, it requires the evaluation of the 2PCF using the FFT which is of complexity $\mathcal{O}(n\log(n))$ where n is the number of pixels/voxels in the image. In order to extract the reduced coefficient vector from the 2PCF, the latter must be projected onto the RB. This operation scales with $\mathcal{O}(n\,h)$. These two operations are at least linear to the number of pixels or voxels of the image which can be critical, especially in three-dimensional settings. Consequently, the computational effort of the feature vector computation heavily out-weights the computational complexity of the regression model as can readily be seen from the provided timings (95 s vs. 0.08 s for the ANN for 7500 predictions). In the future, optimizations, e.g., in the spirit of reduced cubature rules [51], will be explored to render the overall computation more efficient in view of 3D microstructures at resolutions of 512^3 and beyond.

Another extension of the current scheme could account for variable phase contrast R which was fixed as $R = 5$ in this work. In particular, higher phase contrasts should be explored. Preliminary investigations state that the accuracy of the machine learned surrogate deteriorates considerably for a high phase contrast of $R = 1/100$. The source of error and the possible measures to cope with extreme contrasts ($R \ll 1$ and $R \gg 1$) in the data-driven model should be studied in the future. Thereby, the dimension of the feature vector must increase, even beyond the 2PCF. This could possibly lead to a data scarcity dilemma: The number of input samples for the supervised learning should grow exponentially with the dimension of the feature vector. However, this is not realizable in practice due to limited computational resources. With the goal of predictions for nearly arbitrary 3D microstructures in mind, in the authors' opinion this dependence is the most pronounced short-coming of the method and future studies should focus on limiting the number of required input samples in order to fight the curse of dimensionality as more reduced coefficients require an exponential growth in the available data, making the offline procedure unaffordable, today.

Advantages of the current scheme comprise the independence of the underlying simulation scheme. This does allow for heterogeneous simulation environments, the use of commercial software, multi-fidelity input data and blended sources of information (e.g., in silico data supported by experimental results).

Author Contributions: Conceptualization, J.L. and F.F.; Data curation, J.L.; Formal analysis, F.F.; Funding acquisition, F.F.; Investigation, J.L. and F.F.; Methodology, J.L. and F.F.; Project administration, F.F.; Resources, F.F.; Software, J.L.; Supervision, F.F.; Validation, J.L.; Visualization, J.L. and F.F.; Writing—original draft, J.L. and F. F.; Writing—review & editing, J.L. and F.F.

Funding: This research was funded by Deutsche Forschungsgemeinschaft (DFG) within the Emmy-Noether programm under grant DFG-FR2702/6 (contributions of F.F.).

Acknowledgments: Support from Mauricio Fernández on the implementation and layout for the training of the Artificial Neural Networks using Google's TensorFlow is highly appreciated. Stimulating discussions within the Cluster of Excellence SimTech (DFG EXC2075) on machine learning and reduced basis methods are highly acknowledged. Further, the authors would like to thank the three anonymous reviewers for their remarks (particularly in view of the GPM method) which helped in improving the quality of the manuscript.

Conflicts of Interest: The authors declare no conflict of interest.

References

1. Ghosh, S.; Lee, K.; Moorthy, S. Multiple scale analysis of heterogeneous elastic structures using homogenization theory and Voronoi cell finite element method. *Int. J. Solids Struct.* **1995**, *32*, 27–62. [CrossRef]
2. Dhatt, G.; Lefrançois, E.; Touzot, G. *Finite Element Method*; John Wiley & Sons: Hoboken, NJ, USA, 2012.
3. Leuschner, M.; Fritzen, F. Fourier-Accelerated Nodal Solvers (FANS) for homogenization problems. *Comput. Mech.* **2018**, *62*, 359–392. [CrossRef]
4. Torquato, S. *Random Heterogeneous Materials: Microstructure and Macroscopic Properties*; Springer Science & Business Media: Berlin, Germany, 2013.
5. Jiang, M.; Alzebdeh, K.; Jasiuk, I.; Ostoja-Starzewski, M. Scale and boundary condition effects in elastic properties of random composites. *Acta Mech.* **2001**, *148*, 63–78. [CrossRef]
6. Feyel, F. Multiscale FE^2 elastoviscoplastic analysis of composite structures. *Comput. Mater. Sci.* **1999**, *16*, 344–354. [CrossRef]
7. Miehe, C. Strain-driven homogenization of inelastic microstructures and composites based on an incremental variational formulation. *Int. J. Numer. Methods Eng.* **2002**, *55*, 1285–1322. [CrossRef]
8. Beyerlein, I.; Tomé, C. A dislocation-based constitutive law for pure Zr including temperature effects. *Int. J. Plast.* **2008**, *24*, 867–895. [CrossRef]
9. Ryckelynck, D. Hyper-reduction of mechanical models involving internal variables. *Int. J. Numer. Methods Eng.* **2009**, *77*, 75–89. [CrossRef]
10. Hernández, J.; Oliver, J.; Huespe, A.; Caicedo, M.; Cante, J. High-performance model reduction techniques in computational multiscale homogenization . *Comput. Methods Appl. Mech. Eng.* **2014**, *276*, 149–189. [CrossRef]
11. Fritzen, F.; Hodapp, M. The Finite Element Square Reduced (FE^{2R}) method with GPU acceleration: Towards three-dimensional two-scale simulations. *Int. J. Numer. Methods Eng.* **2016**, *107*, 853–881. [CrossRef]
12. Leuschner, M.; Fritzen, F. Reduced order homogenization for viscoplastic composite materials including dissipative imperfect interfaces. *Mech. Mater.* **2017**, *104*, 121–138. [CrossRef]
13. Yvonnet, J.; He, Q.C. The reduced model multiscale method (R3M) for the non-linear homogenization of hyperelastic media at finite strains. *J. Comput. Phys.* **2007**, *223*, 341–368. [CrossRef]
14. Kunc, O.; Fritzen, F. Finite strain homogenization using a reduced basis and efficient sampling. *Math. Comput. Appl.* **2019**, *24*, 56. [CrossRef]
15. Kanouté, P.; Boso, D.; Chaboche, J.; Schrefler, B. Multiscale Methods For Composites: A Review. *Arch. Comput. Methods Eng.* **2009**, *16*, 31–75. [CrossRef]
16. Matouš, K.; Geers, M.G.; Kouznetsova, V.G.; Gillman, A. A review of predictive nonlinear theories for multiscale modeling of heterogeneous materials. *J. Comput. Phys.* **2017**, *330*, 192–220. [CrossRef]
17. Wolpert, D.H.; Macready, W.G. No free lunch theorems for optimization. *IEEE Trans. Evol. Comput.* **1997**, *1*, 67–82. [CrossRef]
18. Brough, D.B.; Wheeler, D.; Kalidindi, S.R. Materials knowledge systems in python—A data science framework for accelerated development of hierarchical materials. *Integr. Mater. Manuf. Innov.* **2017**, *6*, 36–53. [CrossRef] [PubMed]
19. Paulson, N.H.; Priddy, M.W.; McDowell, D.L.; Kalidindi, S.R. Reduced-order structure-property linkages for polycrystalline microstructures based on 2-point statistics. *Acta Mater.* **2017**, *129*, 428–438. [CrossRef]
20. Gupta, A.; Cecen, A.; Goyal, S.; Singh, A.K.; Kalidindi, S.R. Structure–property linkages using a data science approach: Application to a non-metallic inclusion/steel composite system. *Acta Mater.* **2015**, *91*, 239–254. [CrossRef]
21. Kalidindi, S.R. Computationally efficient, fully coupled multiscale modeling of materials phenomena using calibrated localization linkages. *ISRN Mater. Sci.* **2012**, *2012*, 1–13. [CrossRef]

22. Bostanabad, R.; Bui, A.T.; Xie, W.; Apley, D.W.; Chen, W. Stochastic microstructure characterization and reconstruction via supervised learning. *Acta Mater.* **2016**, *103*, 89–102. [CrossRef]
23. Kumar, A.; Nguyen, L.; DeGraef, M.; Sundararaghavan, V. A Markov random field approach for microstructure synthesis. *Model. Simul. Mater. Sci. Eng.* **2016**, *24*, 035015. [CrossRef]
24. Xu, H.; Liu, R.; Choudhary, A.; Chen, W. A machine learning-based design representation method for designing heterogeneous microstructures. *J. Mech. Des.* **2015**, *137*, 051403. [CrossRef]
25. Xu, H.; Li, Y.; Brinson, C.; Chen, W. A descriptor-based design methodology for developing heterogeneous microstructural materials system. *J. Mech. Des.* **2014**, *136*, 051007. [CrossRef]
26. Bessa, M.; Bostanabad, R.; Liu, Z.; Hu, A.; Apley, D.W.; Brinson, C.; Chen, W.; Liu, W.K. A framework for data-driven analysis of materials under uncertainty: Countering the curse of dimensionality. *Comput. Methods Appl. Mech. Eng.* **2017**, *320*, 633–667. [CrossRef]
27. Basheer, I.A.; Hajmeer, M. Artificial neural networks: Fundamentals, computing, design, and application. *J. Microbiol. Methods* **2000**, *43*, 3–31. [CrossRef]
28. Torquato, S.; Stell, G. Microstructure of two-phase random media. I. The *n*-point probability functions. *J. Chem. Phys.* **1982**, *77*, 2071–2077. [CrossRef]
29. Berryman, J.G. Measurement of spatial correlation functions using image processing techniques. *J. Appl. Phys.* **1985**, *57*, 2374–2384. [CrossRef]
30. Cooley, J.W.; Tukey, J.W. An Algorithm for the Machine Calculation of Complex Fourier Series. *AMS Math. Comput.* **1965**, *19*, 297–301. [CrossRef]
31. Frigo, M.; Johnson, S.G. FFTW: An adaptive software architecture for the FFT. In Proceedings of the 1998 IEEE International Conference on Acoustics, Speech and Signal Processing, Seattle, WA, USA, 15 May 1998; Volume 3, pp. 1381–1384. [CrossRef]
32. Fast, T.; Kalidindi, S.R. Formulation and calibration of higher-order elastic localization relationships using the MKS approach. *Acta Mater.* **2011**, *59*, 4595–4605. [CrossRef]
33. Fullwood, D.T.; Niezgoda, S.R.; Kalidindi, S.R. Microstructure reconstructions from 2-point statistics using phase-recovery algorithms. *Acta Mater.* **2008**, *56*, 942–948. [CrossRef]
34. Sirovich, L. Turbulence and the Dynamics of Coherent Structures. Part 1: Coherent Structures. *Q. Appl. Math.* **1987**, *45*, 561–571. [CrossRef]
35. Liang, Y.; Lee, H.; Lim, S.; Lin, W.; Lee, K.; Wu, C. Proper Orthogonal Decomposition and Its Applications—Part I: Theory. *J. Sound Vib.* **2002**, *252*, 527–544. [CrossRef]
36. Camphouse, R.C.; Myatt, J.; Schmit, R.; Glauser, M.; Ausseur, J.; Andino, M.; Wallace, R. A snapshot decomposition method for reduced order modeling and boundary feedback control. In Proceedings of the 4th Flow Control Conference, Seattle, WA, USA, 23–26 June 2008; p. 4195. [CrossRef]
37. Quarteroni, A.; Manzoni, A.; Negri, F. *Reduced Basis Methods for Partial Differential Equations: An Introduction*; Springer: Berlin, Germany, 2016.
38. Klema, V.; Laub, A. The singular value decomposition: Its computation and some applications. *IEEE Trans. Autom. Control* **1980**, *25*, 164–176. [CrossRef]
39. Gu, M.; Eisenstat, S.C. A Stable and Fast Algorithm for Updating the Singular Value Decomposition. Available online: http://citeseerx.ist.psu.edu/viewdoc/summary?doi=10.1.1.46.9767 (accessed on 30 May 2019).
40. Fareed, H.; Singler, J.; Zhang, Y.; Shen, J. Incremental proper orthogonal decomposition for PDE simulation data. *Comput. Math. Appl.* **2018**, *75*, 1942–1960. [CrossRef]
41. Horn, R.A.; Johnson, C.R. *Matrix Analysis*; Cambridge University Press: Cambridge, UK, 1985.
42. Widrow, B.; Lehr, M.A. 30 years of adaptive neural networks: Perceptron, madaline, and backpropagation. *Proc. IEEE* **1990**, *78*, 1415–1442. [CrossRef]
43. Kimoto, T.; Asakawa, K.; Yoda, M.; Takeoka, M. Stock market prediction system with modular neural networks. In Proceedings of the 1990 IJCNN International Joint Conference on Neural Networks, San Diego, CA, USA, 17–21 June 1990; pp. 1–6. [CrossRef]
44. Sundermeyer, M.; Schlüter, R.; Ney, H. LSTM neural networks for language modeling. In Proceedings of the Thirteenth Annual Conference of the International Speech Communication Association, Portland, OR, USA, 9–13 September 2012; pp. 194–197.
45. Simonyan, K.; Zisserman, A. Very deep convolutional networks for large-scale image recognition. *arXiv* **2014**, arXiv:1409.1556.

46. Angermueller, C.; Pärnamaa, T.; Parts, L.; Stegle, O. Deep learning for computational biology. *Mol. Syst. Biol.* **2016**, *12*, 878. [CrossRef]
47. Hecht-Nielsen, R. Theory of the backpropagation neural network. In *Neural Networks for Perception*; Elsevier: Cambridge, MA, USA, 1992; pp. 65–93.
48. Abadi, M.; Barham, P.; Chen, J.; Chen, Z.; Davis, A.; Dean, J.; Devin, M.; Ghemawat, S.; Irving, G.; Isard, M.; et al. Tensorflow: A system for large-scale machine learning. *OSDI* **2016**, *16*, 265–283.
49. Kingma, D.P.; Ba, J. Adam: A method for stochastic optimization. *arXiv* **2014**, arXiv:1412.6980.
50. Bostanabad, R.; Kearney, T.; Tao, S.; Apley, D.W.; Chen, W. Leveraging the nugget parameter for efficient Gaussian process modeling. *Int. J. Numer. Methods Eng.* **2018**, *114*, 501–516. [CrossRef]
51. An, S.; Kim, T.; James, D.L. Optimizing cubature for efficient integration of subspace deformations. *ACM Trans. Graph.* **2009**, *27*, 165. [CrossRef] [PubMed]

Mathematical and Computational Applications

MDPI

Article

Multiple Tensor Train Approximation of Parametric Constitutive Equations in Elasto-Viscoplasticity

Clément Olivier [1,2], David Ryckelynck [2,*] and Julien Cortial [1]

1 Safran Tech, Modelling & Simulation, Rue des Jeunes Bois, Châteaufort, 78114 Magny-Les-Hameaux, France; clement.olivier@safrangroup.com (C.O.); julien.cortial@safrangroup.com (J.C.)

2 MAT-Centre des Matériaux, MINES ParisTech, PSL Research University, CNRS UMR 7633, 10 rue Desbruères, 91003 Evry, France

* Correspondence: david.ryckelynck@mines-paristech.fr

Received: 9 November 2018; Accepted: 23 January 2019; Published: 28 January 2019

Abstract: This work presents a novel approach to construct surrogate models of parametric differential algebraic equations based on a tensor representation of the solutions. The procedure consists of building simultaneously an approximation given in tensor-train format, for every output of the reference model. A parsimonious exploration of the parameter space coupled with a compact data representation allows alleviating the curse of dimensionality. The approach is thus appropriate when many parameters with large domains of variation are involved. The numerical results obtained for a nonlinear elasto-viscoplastic constitutive law show that the constructed surrogate model is sufficiently accurate to enable parametric studies such as the calibration of material coefficients.

Keywords: parameter-dependent model; surrogate modeling; tensor-train decomposition; gappy POD; heterogeneous data; elasto-viscoplasticity

1. Introduction

Predictive numerical simulations in solid mechanics require material laws that involve systems of highly nonlinear Differential Algebraic Equations (DAEs). These models are essential in challenging industrial applications, for instance to study the effects of the extreme thermo-mechanical loadings that turbine blades may sustain in helicopter engines ([1,2]), as well as in biomechanical analyses [3,4].

These DAE systems are referred to as constitutive laws in the material science community. They express, for a specific material, the relationship between the mechanical quantities such as the strain, the stress, and miscellaneous internal variables and stand as the closure relations of the physical equations of mechanics. When the model aims to reproduce physically-complex behaviors, constitutive equations are often tuned through numerous parameters called material coefficients.

An appropriate calibration of these coefficients is necessary to ensure that the numerical model mimics the actual physical behavior [5]. Numerical parametric studies, consisting of analyzing the influence of the parameter values on the solutions, are typically used to perform the identification. However, when the number of parameters increases and unless the computational effort required for a single numerical simulation is negligible, the exploration of the parameter domain turns into a tedious task, and exhaustive analyses become unfeasible. Moreover, defining an unambiguous criterion measuring the fidelity of the model to experimental data is a challenge for models with complex behaviors.

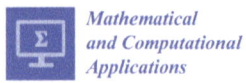

A common technique to mitigate the aforementioned challenges is to build surrogate models (or metamodels) mapping points of a given parameter space (considered as the inputs of the model) to the outputs of interest of the model. The real-time prediction of DAE solutions for arbitrary parameter values, enabled by the surrogate model, helps the comprehension of constitutive laws and facilitates the conducting of parametric studies. In particular, the robustness of the calibration process can be dramatically improved using surrogate model approaches.

The idea of representing the set of all possible parameter-dependent solutions of ODEs and PDEs as a multiway tensor was pioneered with the introduction of the Proper Generalized Decomposition (PGD) [6–8]. In this representation, each dimension corresponds to a spatial/temporal coordinate or a parameter coefficient. The resulting tensor is never assembled explicitly, but instead remains an abstract object for which a low-rank approximation based on a canonical polyadic decomposition [9] is computed. The PGD method further alleviates the curse of dimensionality by introducing a multidimensional weak formulation over the entire parameter space, and the solutions are sought in a particular form where all variables are separated. When differential operators admit a tensor decomposition, the PGD method is very efficient because the multiple integrals involved in the multidimensional weak form of the equations can be rewritten as a sum of products of simple integrals.

Unfortunately, realistic constitutive equations or even less sophisticated elasto-viscoplastic models admit no tensor decomposition with respect to the material coefficients and the time variables. An extension of the PGD to highly nonlinear laws is therefore non-trivial. However, many other tensor decomposition approaches have been successfully proposed to approximate functions or solutions of differential equations defined over high-dimensional spaces. We refer the reader to [10–12] for detailed reviews on tensor decomposition techniques and their applications.

Among the existing formats—CP decomposition [9,13,14], Tucker decomposition [11,15], hierarchical Tucker decomposition [11,16]—this work investigates the Tensor-Train (TT) decomposition [17,18]. The TT-cross algorithm, introduced in [17] and further developed in [19,20], is a sampling procedure to build an approximation of a given tensor under the tensor-train format. Sampling procedures in parameter space have proven their ability to reduce nonlinear and non-separable DAEs by using the Proper Orthogonal Decomposition (POD) [21], the gappy POD [22], or the Empirical Interpolation Method (EIM) [23,24]. These last methods are very convenient when the solutions have only two variables; hence, they are considered as second-order tensors.

This paper aims to extend the sampling procedure of the TT-cross method to DAEs having heterogeneous and time-dependent outputs. A common sampling of the parameter space is proposed, though several TT-cross approximations are computed to cope with heterogeneous outputs. These outputs can be scalars, vectors, or tensors, with various physical units. In the proposed algorithm, sampling points are not specific to any output, although parameters do not affect equally each DAE output. The proposed method is named multiple TT-cross approximation. Similarly to the construction of a reduced integration domain for the hyper-reduction of partial differential equations [25] or for the GNATmethod [26], the set of sampling points is the union of contributions from the various outputs of the DEA. In this paper, the multiple TT-cross incorporates the gappy POD method, and the developments are focused on the numerical outputs obtained through a numerical integration scheme applied to the DAE.

2. Materials and Methods

The parametrized material model generates several time-dependent Quantities of Interest (QoI). These quantities can be scalar-, vector-, or even tensor-valued (e.g., stress) and are generally of distinct natures, namely expressed with different physical units and/or have different magnitudes. For instance, in the physical model described in Appendix A, the outputs of the model are $\underset{\sim}{\varepsilon}(t)$, $\underset{\sim}{\varepsilon}_{vp}(t)$, $\underset{\sim}{\sigma}(t)$, and $p(t)$, where t is the time variable, $\underset{\sim}{\varepsilon}$, $\underset{\sim}{\varepsilon}_{vp}$, $\underset{\sim}{\sigma}$ have six components each, and p is a scalar. Therefore, the generated data will be segregated according to the QoI to which they relate. This will also be structured in a tensor-like fashion to make it amenable to the numerical methods presented in this paper. We restrict our attention to discrete values of $d-1$ parameters and discrete values of time instants, related to indices $i_1 \dots i_d$, $i_k \in \{1, \dots n_k\}$ for $k = 1 \dots d$. For instance, all the computable scalar outputs p will be considered as a tensor $\boldsymbol{\mathcal{A}}^1 \in \mathbb{R}^{n_1 \times \cdots \times n_{d-1} \times n_d^1}$.

For a given $\chi = 1,\ldots,N$ denoting an arbitrary QoI, the tensor of order d, $\boldsymbol{\mathcal{A}}^\chi \in \mathbb{R}^{n_1\times\cdots\times n_d^\chi}$ (denoted with bold calligraphic letters) refers to a multidimensional array (also called a multiway array). Each element of $\boldsymbol{\mathcal{A}}^\chi$ identified by the indices $(i_1,\ldots i_d) \in D_1 \times \cdots \times D_{d-1} \times D_d^\chi$ is denoted by:

$$\boldsymbol{\mathcal{A}}^\chi(i_1,\ldots,i_d) \in \mathbb{R}$$

where $D_k = [1 : n_k]$ for $k < d$ is the set of natural numbers from one to n_k (inclusive) and $D_d^\chi = [1 : n_d^\chi]$. The last index accounts for the number of components in each QoI. Therefore, the last index is specific to each $\boldsymbol{\mathcal{A}}^\chi$, while the others are common to all tensors for $\chi = 1,\ldots,N$. Hence, a common sampling of the parameter space $D_1 \times \ldots \times D_{d-1}$ can be achieved. The vector $\boldsymbol{\mathcal{A}}^\chi(i_1,\ldots,i_{d-1},:) \in \mathbb{R}^{n_d^\chi}$ contains all the components of output χ at all time instants used for the numerical solution of the DAE and for a given point in the parameter space.

Matricization designates a special case of tensor reshaping that allows representing an arbitrary tensor as a matrix. The qth matricization of $\boldsymbol{\mathcal{A}}^\chi$ denoted by $\langle\boldsymbol{\mathcal{A}}^\chi\rangle_q$ consists of dividing the dimensions of $\boldsymbol{\mathcal{A}}^\chi$ into two groups, the q leading dimensions and the $(d-q)$ trailing dimensions, such that the newly-defined multi-indices enumerate respectively the rows and columns of the matrix $\langle\boldsymbol{\mathcal{A}}^\chi\rangle_q$. For instance, $\langle\boldsymbol{\mathcal{A}}^\chi\rangle_1$ and $\langle\boldsymbol{\mathcal{A}}^\chi\rangle_2$ are matrices of respective sizes n_1-by-$n_2\ldots n_{d-1}\, n_d^\chi$ and $n_1 n_2$-by-$n_3\ldots n_{d-1}\, n_d^\chi$. Their elements are given by:

$$\langle\boldsymbol{\mathcal{A}}^\chi\rangle_1(i_1, j^\star) = \boldsymbol{\mathcal{A}}^\chi(i_1,\ldots,i_d)$$
$$\langle\boldsymbol{\mathcal{A}}^\chi\rangle_2(i_1 + (i_2-1)n_1, j^{\star\star}) = \boldsymbol{\mathcal{A}}^\chi(i_1,\ldots,i_d)$$

where $j^\star = 1 + \sum_{k=2}^{d}[(i_k-1)\prod_{l=2}^{k} n_l]$ enumerates the multi-index $(i_2,\ldots,i_d)$ and $j^{\star\star} = 1 + \sum_{k=3}^{d}[(i_k - 1)\prod_{l=3}^{k} n_l]$ enumerates the multi-index $(i_3,\ldots,i_d)$. Here again, these matricizations are purely formal because of the curse of dimensionality.

The Frobenius norm is denoted by $\|.\|$ without the usual subscript $\mathcal{F}$. For $\boldsymbol{\mathcal{A}}^\chi \in \mathbb{R}^{n_1\times\ldots n_d^\chi}$, it reads:

$$\|\boldsymbol{\mathcal{A}}^\chi\| = \sqrt{\sum_{i_1,\ldots,i_d \in D_1\times\cdots\times D_{d-1}\times D_d^\chi} \boldsymbol{\mathcal{A}}^\chi(i_1,\ldots,i_d)^2}$$

The Frobenius norm of a tensor is invariant under all matricizations of a given tensor.

In [17], Singular-Value Decomposition (SVD) is considered in the algorithm called TT-SVD. Because of the curse of dimensionality, the TT-SVD has no practical use, even if tensors have a low rank. More workable approaches aim to sample the entries of tensors.

For instance, in the snapshot Proper Orthogonal Decomposition (POD) [21], the sampling procedure aims to estimate the rank and an orthogonal reduced basis for the approximation of a matrix A. The method consists of applying the truncated SVD on the submatrix $\tilde{A} = A(:, \mathcal{J}_{pod})$ constituted by a selection of columns $\mathcal{J}_{pod}$ of A. Hence, the accuracy of the resulting POD reduced basis relies on the quality of the sampling procedure that generally introduces a sampling error. This sampling procedure seems to be convenient when considering the first matricizations $\langle\boldsymbol{\mathcal{A}}^\chi\rangle_q$ if the product $n_1\, n_2\ldots n_q$ and $\text{Card}(\mathcal{J}_{pod})$ are reasonably small regarding the available computing resources. However,, for large values of q, the curse of dimensionality makes the snapshot POD, alone, intractable.

A more practical approach to construct an approximate TT decomposition effectively, called the TT-cross method, is proposed in [17]. The TT-cross consists of dropping the concept of a POD basis and using the Pseudo-Skeleton Decomposition (PSD) introduced in [27] as the low-rank approximation.

Unlike the TT-SVD, the TT-cross enables building an approximation based on a sparse exploration of a reference tensor. The PSD can be used to approximate any matrix $A \in \mathbb{R}^{n\times m}$ and is written as:

$$A = \underbrace{A(:, \mathcal{J}_{psd}) \left[A(\mathcal{I}_{psd}, \mathcal{J}_{psd})\right]^{-1} A(\mathcal{I}_{psd}, :)}_{= T_{psd}} + E_{psd} \tag{1}$$

where the sets $\mathcal{I}_{psd}$ and $\mathcal{J}_{psd}$ are respectively a selection of row and column indices. The definition is valid only when the matrix $A(\mathcal{I}_{psd}, \mathcal{J}_{psd})$ is non-singular. In particular, the number s of rows and columns has to be identical.

This approximation (1) features an interpolation property at the selected rows and columns:

$$T_{psd}(\mathcal{I}_{psd}, :) = A(\mathcal{I}_{psd}, :) \quad \text{and} \quad T_{psd}(:, \mathcal{J}_{psd}) = A(:, \mathcal{J}_{psd}) \tag{2}$$

The PSD is a matrix factorization similar to the decomposition used in the Adaptive Cross Approximation (ACA) [28] and the CURdecomposition [29,30]. Additionally, these references provide algorithms to build the factorization effectively. That decomposition has also been used in the context of model order reduction, for instance in the Empirical Interpolation Method (EIM) proposed in [23,24].

The condition that $A(\mathcal{I}_{psd}, \mathcal{J}_{psd})$ must be non-singular makes it difficult to share sampling points for various matrices $\langle \boldsymbol{\mathcal{A}}^{\chi} \rangle_q$ with $\chi = 1, \ldots, N$ having their own rank.

The gappy POD introduced in [22] aims at relaxing the aforementioned constraint by combining beneficial features of the snapshot POD and the PSD. Indeed, the gappy POD (a) relies on a POD basis that remains computationally affordable, (b) requires only a limited number of rows of the matrix to be approximated, and (c) enables reusing the set of selected rows for different matrices. These properties are key ingredients for an efficient, parsimonious exploration of the reference tensors. The gappy POD approximation T_{gap} of a matrix $A \in \mathbb{R}^{n\times m}$ is given by:

$$A = \underbrace{V[V(\mathcal{I}_{gap}, :)]^{\dagger} A(\mathcal{I}_{gap}, :)}_{= T_{gap}} + E_{gap} \tag{3}$$

where $\dagger$ denotes the Moore–Penrose pseudo-inverse [31] and $\mathcal{I}_{gap}$ is a row selection of s rows and where $V \in \mathbb{R}^{n\times r}$ is a POD basis matrix of rank r such that:

$$A(:, \mathcal{J}_{pod}) = V\, S\, W^T + E_{pod} \tag{4}$$

In the sequel, because the simulation data in $\boldsymbol{\mathcal{A}}^{\chi}$ are outputs of a DAE system, it does not make sense to sample the last index i_d during column sampling of $\langle \boldsymbol{\mathcal{A}}^{\chi} \rangle_q$. Each numerical solution of the DAE system generates all the last components of each tensor $\boldsymbol{\mathcal{A}}^{\chi}$. Hence, the column sampling is restricted to indices $i_{q+1}, \ldots i_{d-1}$, and all the values of i_d in D_d^{χ} are kept. This special column sampling is denoted by $\mathcal{J}_{pod}^{\chi}$. It is performed randomly by using a low-discrepancy Halton sequence [32].

The matrix $V(\mathcal{I}_{gap}, :)$ must have linearly independent columns to ensure that the approximation is meaningful. Since V is a rank-r POD basis, there exists a set of s rows such that this property holds as long as $s \geq r$. Here, $\mathcal{I}_{gap}$ contains at least the interpolation indices related to V. This latter set is denoted by $\mathcal{I}^{\chi}$, such that $V(\mathcal{I}^{\chi}, :)$ is invertible. In the numerical results presented hereafter, $\mathcal{I}^{\chi}$ is obtained using the Q-DEIM algorithm [33] that was shown to be a superior alternative to the better-known DEIM procedure ([34], Algorithm 1).

Unlike the PSD, the gappy POD enables selecting a number of rows that exceeds the rank of the low-rank approximation:

$$\mathcal{I}_{gap} = \mathcal{I}^1 \cup \ldots \cup \mathcal{I}^N \tag{5}$$

This makes it possible to share sampling points between matrices having their own rank. In this case, the interpolation property does not hold as in the PSD case (2).

T_{gap} is the approximation of A by the product of three matrices: V, $\left[V(\mathcal{I}_{gap},:)\right]^{\dagger}$ and $A(\mathcal{I}_{gap},:)$. The TT-cross approximation can be understood as a generalization of such a product of matrices. A tensor $\mathcal{T} \in \mathbb{R}^{n_1\times\cdots\times n_d}$ is said to be in Tensor-Train format (TT format) if its elements are given by the following matrix products:

$$\mathcal{T}(i_1,\ldots,i_d) = G_1(i_1)\ldots G_d(i_d) \in \mathbb{R} \tag{6}$$

where the so-called tensor carriages (or core tensors) are such that for $k = 1,\ldots,d$:

$$G_k(i_k) \in \mathbb{R}^{r_{k-1}\times r_k} \quad \forall i_k \in D_k$$

In the original definition of the tensor-train format [17], the leading and trailing factors (corresponding to $G_1(i_1)$ and $G_d(i_d)$ for any choice of i_1 and i_d) are respectively row and column vectors. Here, the convention $r_0 = r_d = 1$ is adopted so that row matrices $G_1(i_1)$ and column matrices $G_d(i_d)$ can be interpreted as vectors or matrices depending on the context.

The TT format allows significant gains in terms of memory storage and therefore is well-suited to high-order tensors. The storage complexity is $\mathcal{O}(n\bar{r}^2 d)$ where $\bar{r} = \max(r_1,\ldots,r_{d-1})$ and depends linearly on the order d of the tensor. In many applications of practical interest, the small TT-ranks r_k enable alleviating the curse of dimensionality [17].

The sequential computational complexity of the evaluation of a single element of a tensor in TT format is $\mathcal{O}\left(d\bar{r}^2\right)$. Assuming that $\bar{r}$ is small enough, the low computational cost allows a real-time evaluation of the underlying tensor. Therefore, in terms of online exploitation, this representation conforms with the expected requirements of the surrogate model. Figure 1 illustrates the sequence of matrix multiplications required to compute one element of the tensor train.

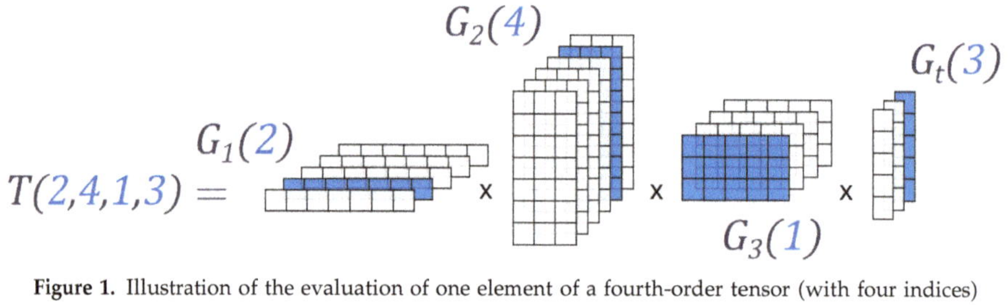

Figure 1. Illustration of the evaluation of one element of a fourth-order tensor (with four indices) having a tensor train decomposition. G_1, G_2, G_3, G_t are the tensor carriages of this decomposition. $\mathcal{T}(2,4,1,3) \in \mathbb{R}$ is obtained by the product of one row vector $G_1(2)$, two matrices $G_2(4)$ and $G_3(1)$, and one column vector $G_t(3)$. The dimensions of these matrices are respectively 1×7, 7×3, 3×5, and 5×1. $G_1(2)$, $G_2(4)$, $G_3(1)$, and $G_t(3)$ are the extraction of one layer (identified by a darker shade) in the tensor G_1, G_2, G_3, G_t, respectively.

The objective of the proposed approach is to build for each physics-based tensor $\mathcal{A}^{\mathcal{X}}$ an approximate tensor $\mathcal{T}^{\mathcal{X}}$ given in TT format by using a nested row sampling of the simulation data. Algorithm 1 provides the set of matrices $\{G_1^{\mathcal{X}},\ldots,G_d^{\mathcal{X}}\}$ that enable defining the tensor-train decompositions and aggregate sets for row sampling. It is a sequential algorithm that navigates from dimension one to dimension $d-1$ of tensors $\mathcal{A}^{\mathcal{X}}$.

The method provided by Algorithm 1 is non-intrusive and relies on the numerical solutions of the DAEs in a black-box fashion.

Algorithm 1: Multiple TT Decomposition.

Input: Tensors $\boldsymbol{\mathcal{A}}^{\chi} \in \mathbb{R}^{n_1 \times \cdots \times n_{d-1} \times n_d^{\chi}}$ for $\chi = 1, \ldots, N$ associated with a DAE system
Output: Sets of matrices $\{G_1^{\chi}, \ldots, G_d^{\chi}\}$ for $\chi = 1, \ldots, N$.
Initialization:
For each χ, define the matrix $A_1^{\chi} \in \mathbb{R}^{(s_0 n_1) \times (n_2 \ldots n_{d-1} n_d^{\chi})}$ with $s_0 = 1$ as the first matricization of the tensor $\boldsymbol{\mathcal{A}}^{\chi}$:

$$A_1^{\chi} = \langle \boldsymbol{\mathcal{A}}^{\chi} \rangle_1 \tag{7}$$

For $k = 1, \ldots, d-1$ **do**

Snapshot POD:
Define consistent sets of sampling columns $\mathcal{J}_k^{\chi}$ and evaluate the DAE to fill the matrices $\tilde{A}_k^{\chi}$ defined as:

$$\tilde{A}_k^{\chi} = A_k^{\chi}\left(:, \mathcal{J}_k^{\chi}\right) \quad \text{for} \quad \chi = 1, \ldots, N$$

Apply the truncated SVD (4) on each $\tilde{A}_k^{\chi}$ with the truncation tolerance ϵ to get the rank-r_k^{χ} matrices:

$$\tilde{A}_k^{\chi} = V_k^{\chi} S_k^{\chi} W_k^{\chi\,T} + E_{pod\,k}^{\chi} \quad \text{with} \quad \left\| E_{pod\,k}^{\chi} \right\| \leq \epsilon \left\| \tilde{A}_k^{\chi} \right\| \tag{8}$$

$$V_k^{\chi} \in \mathbb{R}^{(s_{k-1} n_k) \times r_k^{\chi}} \quad \text{for} \quad \chi = 1, \ldots, N \tag{9}$$

Row Sampling:
From each χ, select a set of rows $\mathcal{I}_k^{\chi}$ applying the Q-DEIM algorithm [33] to the basis V_k^{χ}. Define the union of all selected rows and the corresponding row selection matrix:

$$\mathcal{I}_k = \bigcup_{\chi=1}^{N} \mathcal{I}_k^{\chi} \tag{10}$$

and:

$$s_k = \mathrm{Card}(\mathcal{I}_k) \tag{11}$$

Output Definitions:
Compute the matrices $G_k^{\chi} \in \mathbb{R}^{(s_{k-1} n_k) \times s_k}$ such that:

$$G_k^{\chi} = V_k^{\chi} \left[V_k^{\chi}(\mathcal{I}_k, :) \right]^{\dagger}$$

Tensorization:
Define, formally, the tensors $\boldsymbol{\mathcal{A}}^{\chi,(k+1)} \in \mathbb{R}^{s_k \times n_{k+1} \times \cdots \times n_{d-1} \times n_d^{\chi}}$ such that:

$$\left\langle \boldsymbol{\mathcal{A}}^{\chi,(k+1)} \right\rangle_1 = A_k^{\chi}(\mathcal{I}_k, :) \in \mathbb{R}^{s_k \times \left(n_{k+1} \ldots n_{d-1} n_d^{\chi}\right)} \tag{12}$$

Matricization:
Define, formally, the matrix $A_{k+1}^{\chi} \in \mathbb{R}^{(s_k n_{k+1}) \times \left(n_{k+2} \ldots n_{d-1} n_d^{\chi}\right)}$ as the second matricization of the tensor $\boldsymbol{\mathcal{A}}^{\chi,(k+1)}$:

$$A_{k+1}^{\chi} = \left\langle \boldsymbol{\mathcal{A}}^{\chi,(k+1)} \right\rangle_2 \tag{13}$$

end
Finalization:
For each $\chi = 1, \ldots, N^{\chi}$, define the matrix $G_d^{\chi} \in \mathbb{R}^{\left(s_{d-1} n_d^{\chi}\right) \times s_d}$ with $s_d = 1$ such that:

$$G_d^{\chi} = A_d^{\chi} \tag{14}$$

At each iteration $k = 1, \ldots, d-1$, the snapshot POD method, used to build the POD reduced basis (9), requires sampling a set $\mathcal{J}_k^\chi$. The column sampling amounts to a parsimonious selection of $\tilde{n}_k$ points in the partial discretized parameter domain $\mathcal{D}_{k+1} \times \cdots \times \mathcal{D}_{d-1}$ and an exhaustive sampling of the last dimension for each tensor $\boldsymbol{\mathcal{A}}^\chi$. The considered submatrices $\tilde{A}_k^\chi = A_k^\chi\left(:, \mathcal{J}_k^\chi\right)$ are then constituted of $\tilde{n}_k^\chi = \tilde{n}_k n_d^\chi$ columns (see Figure 2).

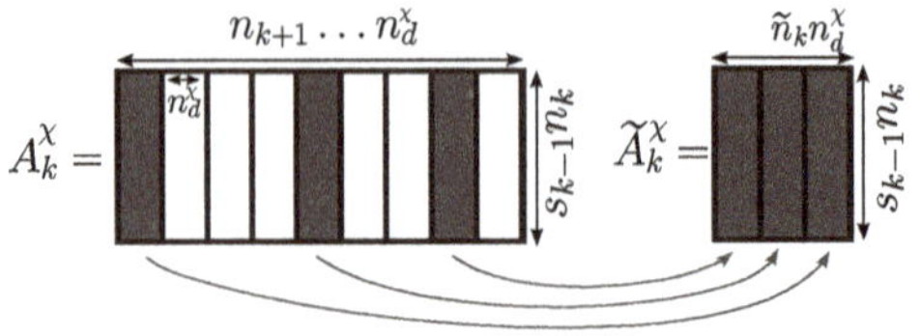

Figure 2. Definition of the submatrix $\tilde{A}_k^\chi$ used to construct the POD reduced basis. In the illustration, the snapshot POD sample size is $\tilde{n}_k = 3$.

In the row sampling step, specific sets of interpolant rows $\mathcal{I}_k^\chi$ are first determined independently for each output χ, but a common, aggregated set $\mathcal{I}_k$ (10) is then used to sample the entries of all outputs. Indeed, computing the elements of all submatrices $A_k^\chi\left(\mathcal{I}_k, :\right)$ requires m_k calls to the DAE system solver with: $m_k = \mathrm{Card}(\mathcal{I}_{k-1})\, \tilde{n}_k$ with $\mathcal{I}_0 = D_1$. Furthermore, the gappy POD naturally accommodates a number of rows larger than the rank r_k^χ for each approximation of A_k^χ, and considering a larger sample size for each individual χ is expected to provide a more accurate approximation.

The tensorization and matricization steps are purely formal. No call to the DAE system solver is done here. They define the way the simulation data must be ordered in matrices to be approximated at the next iteration. The recursive definition of the matrix A_k^χ implies that the latter is equal to the kth matricization of a subtensor extracted from $\boldsymbol{\mathcal{A}}^\chi$. Equivalently, the matrix A_k^χ corresponds to a submatrix of the kth matricization of $\boldsymbol{\mathcal{A}}^\chi$, as illustrated in Figure 3.

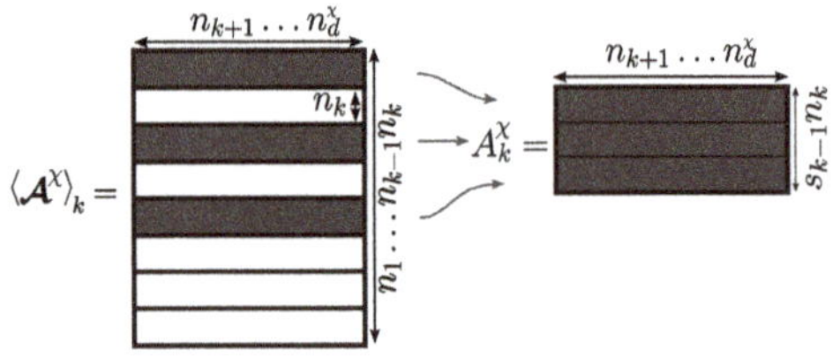

Figure 3. Definition of A_k^χ based on $\boldsymbol{\mathcal{A}}^\chi$. In the illustration, the number of rows selected at the previous iteration $k-1$ is $s_{k-1} = 3$.

To quantify the theoretical accumulation of errors introduced at each iteration, Proposition 1 gives an upper bound for the approximation error associated with a tensor-train decomposition built by the snapshot POD followed by the row sampling steps, when a full column sampling is performed.

Proposition 1. *Consider* $\boldsymbol{\mathcal{A}}^\chi \in \mathbb{R}^{n_1 \times \cdots \times n_{d-1} \times n_d^\chi}$ *and its tensor-train approximation* $\boldsymbol{\mathcal{T}}^\chi$ *constructed by Algorithm 1. Assuming that for all* $k \in [1 : d-1]$*:*

$$\left\| \left(\mathbb{I} - V_k^\chi {V_k^\chi}^T \right) A_k^\chi \right\| \leq \epsilon \left\| A_k^\chi \right\| \tag{15}$$

the following inequality holds:

$$\left\| \boldsymbol{\mathcal{A}}^\chi - \boldsymbol{\mathcal{T}}^\chi \right\| \leq \sum_{k=1}^{d-1} \frac{\epsilon}{\sigma_{\min}\left(V_k^\chi(I_k, :)\right)} \prod_{k'=1}^{k-1} \frac{\min(\sigma_{\max}(V_{k'}^\chi(I_{k'}, :)) + \epsilon,\, 1)}{\sigma_{\min}(V_{k'}^\chi(I_{k'}, :))} \left\| \boldsymbol{\mathcal{A}}^\chi \right\| \tag{16}$$

where $\sigma_{\min}$ *and* $\sigma_{\max}$ *refer to the smallest and the largest singular values of its matrix argument.*

The proof is given in ([35], Proposition 12).

Proposition 1 suggests that the approximation error

$$\|\boldsymbol{\mathcal{A}}^{\chi} - \boldsymbol{\mathcal{T}}^{\chi}\|$$

can be controlled by the truncation tolerances ϵ set by the user. However, the bound (16) tends to be very loose, and the hypothesis (15) may be difficult to verify when the basis V_k^{χ} stems from a column sampling of the matrix A_k^{χ}. Hence, the convergence should be assessed empirically in practical cases.

3. Results

3.1. Outputs' Partitioning as Formal Tensors

The physical model described in Appendix A is represented as the relations between six ($d = 7$) parameters (inputs of the model) and the time-dependent mechanical variables (outputs of the model):

$$(n, K, R_0, Q, b, C) \mapsto \left(\underset{\sim}{\varepsilon}(t), \underset{\sim}{\varepsilon}_{vp}(t), \underset{\sim}{\sigma}(t), p(t)\right)$$

where $\underset{\sim}{\varepsilon}, \underset{\sim}{\varepsilon}_{vp}, \underset{\sim}{\sigma}$ have six components each and pis a scalar. $\underset{\sim}{\varepsilon}, \underset{\sim}{\varepsilon}_{vp}$, and p have the same units, but have different physical meanings.

The surrogate model is defined by introducing $N = 4$ groups of outputs as tensors $\boldsymbol{\mathcal{A}}^{\chi}$. The formal tensors $\boldsymbol{\mathcal{A}}^1$, ... $\boldsymbol{\mathcal{A}}^4$ are related to p, $\underset{\sim}{\varepsilon}, \underset{\sim}{\varepsilon}_{vp}$, and $\underset{\sim}{\sigma}$, respectively.

For each parameter, the interval of definition is discretized by a regular grid with 30 points:

$$n_1 = n_2 = n_3 = n_4 = n_5 = n_6 = 30$$

The time interval discretized is the one used for the numerical solution; it corresponds to a regular grid with $n_t = 537$ points. Then:

$$n_7^1 = n_7^2 = n_7^3 = 6n_t \quad \text{and} \quad n_7^4 = n_t$$

The snapshot POD sample sizes are:

$$\tilde{n}_1 = \tilde{n}_2 = \tilde{n}_3 = \tilde{n}_4 = \tilde{n}_5 = 100 \quad \text{and} \quad \tilde{n}_6 = 30$$

3.2. Performance Indicators

The truncation tolerance is chosen here to be $\epsilon = 10^{-3}$. The construction of the tensor-train decompositions requires solving the system of DAEs $\sum_{k=1}^{d-1} s_k n_k \tilde{n}_k$ times with as many sets of parameter values. In the proposed numerical example, it amounts to 514,050 solutions. Fifteen hours were necessary on a 16-core workstation to carry out the computations. Ninety eight percent of the effort was devoted to the solution of the physical model and the remaining 2% to the decomposition operations.

For a single simulation on a personal laptop computer, the solution of the physical model took 0.7 s, whereas the surrogate model was evaluated in only 1 ms, corresponding to a speed-up of 700.

Storing the multiple TT approximations requires 2,709,405 double-precision floating-point values. For comparison purposes, storing a single solution (constituted by the multiple time-dependent outputs) of the DAE system involves 10,203 values. Therefore, the storage of the tensor-train decompositions is commensurate with the storage of 265 solutions, while it can express the approximation of 30^6 solutions.

For $\chi = 1, \ldots, 4$, the rank r_k^{χ} is bounded from above by the theoretical maximum rank $r_{max,k}^{\chi}$ of the matrix A_k^{χ}. More specifically, $r_{max,k}^{\chi}$ corresponds to the case where A_k^{χ} has full rank and is the kth matricizations of the tensors $\boldsymbol{\mathcal{A}}^{\chi}$. Given the choice of truncation tolerance $\epsilon = 10^{-3}$, the TT-ranks listed in Table 1 show that the resulting tensor trains involve low rank approximations. Table 2 emphasizes that in practice, $r_k^X \ll r_{max,k}^X$ except for $k = 1$ where $r_{max,k}^X$ is already "small".

Table 1. TT-ranks of the outputs of interest and theoretical maximum ranks.

	$k=1$	$k=2$	$k=3$	$k=4$	$k=5$	$k=6$
r_k^1	7	9	10	24	27	30
r_k^2	13	23	29	123	143	134
r_k^3	11	17	20	67	90	100
r_k^4	9	12	14	24	20	21
$r_{max,k}^1 = r_{max,k}^2 = r_{max,k}^3$	30	30^2	30^3	30^4	$6 \times 30 n_t$	$6 \times n_t$
$r_{max,k}^4$	30	30^2	30^3	$30^2 n_t$	$30 n_t$	n_t

Table 2. Ratio between the theoretical maximum ranks and the TT-ranks of the outputs of interest.

	$k=1$	$k=2$	$k=3$	$k=4$	$k=5$	$k=6$
$r_{max,k}^1/r_k^1$	4.3	1.0×10^2	2.7×10^3	3.4×10^4	3.6×10^3	1.1×10^2
$r_{max,k}^2/r_k^2$	2.3	3.9×10^1	9.3×10^2	6.6×10^3	6.8×10^2	2.4×10^1
$r_{max,k}^3/r_k^3$	2.7	5.3×10^1	1.4×10^3	1.2×10^4	1.1×10^3	3.2×10^1
$r_{max,k}^4/r_k^4$	3.3	7.5×10^1	1.9×10^3	2.0×10^4	8.1×10^2	2.6×10^1

3.3. Approximation Error

The accuracy of the surrogate model is estimated a posteriori by measuring the discrepancy between its own outputs and the outputs of the original physical model. The estimation is conducted by comparing solutions associated with 20,000 new samples of parameter set values randomly selected according to a uniform law on each discretized parameter interval. The difference between the surrogate and the physical models is measured based on the following norms:

$$\|x\|_{[0,T]}^2 = \int_0^T x^2 dt \quad \text{et} \quad \left\|\underset{\sim}{X}\right\|_{[0,T]}^2 = \int_0^T \underset{\sim}{X} : \underset{\sim}{X}\, dt$$

where x and $\underset{\sim}{X}$ are respectively scalar and tensor time-dependent functions.

For the mechanical variable Z (where Z can stand for any one of $\underset{\sim}{\varepsilon}$, $\underset{\sim}{\varepsilon}_{vp}$, $\underset{\sim}{\sigma}$ and p), Z^{PM} and Z^{TT} denote the output corresponding respectively to the solution of the DAEs and the surrogate model. A relative error is associated with each mechanical variable, namely:

- Total Strain Tensor: $e_\varepsilon = \dfrac{\left\|\underset{\sim}{\varepsilon}^{PM} - \underset{\sim}{\varepsilon}^{TT}\right\|_{[0,T]}}{\left\|\underset{\sim}{\varepsilon}^{PM}\right\|_{[0,.]}}$;
- Viscoplastic Strain Tensor: $e_{\varepsilon_{vp}} = \dfrac{\left\|\underset{\sim}{\varepsilon}_{vp}^{PM} - \underset{\sim}{\varepsilon}_{vp}^{TT}\right\|_{[0,T]}}{\left\|\underset{\sim}{\varepsilon}^{PM}\right\|_{[0,.]}}$;
- Stress Tensor: $e_\sigma = \dfrac{\left\|\underset{\sim}{\sigma}^{PM} - \underset{\sim}{\sigma}^{TT}\right\|_{[0,T]}}{\left\|\underset{\sim}{\sigma}^{PM}\right\|_{[0,.]}}$;
- Cumulative Viscoplastic Deformation: $e_p = \dfrac{\left\|p^{PM} - p^{TT}\right\|_{[0,T]}}{\left\|\underset{\sim}{\varepsilon}^{PM}\right\|_{[0,.]}}$.

Depending on the parameter values, the viscoplastic part of the behavior may or may not be negligible as measured by the magnitudes of $\|p\|$ and $\left\|\underset{\sim}{\varepsilon}_{vp}\right\|$ relative to $\left\|\underset{\sim}{\varepsilon}\right\|$. Hence, in the proposed application, the focus is on comparing the norm of the approximation error for $\underset{\sim}{\varepsilon}$, $\underset{\sim}{\varepsilon}_{vp}$, and p with respect to the norm of $\underset{\sim}{\varepsilon}$.

The histograms featured in Figure 4a–d present, for each mechanical variables, the empirical distribution of the relative error for all simulation results. The surrogate model given by the tensor-train decompositions features a level of error that is sufficiently low to carry out parametric studies such as the calibration of constitutive laws where errors lower than 2% are typically tolerable.

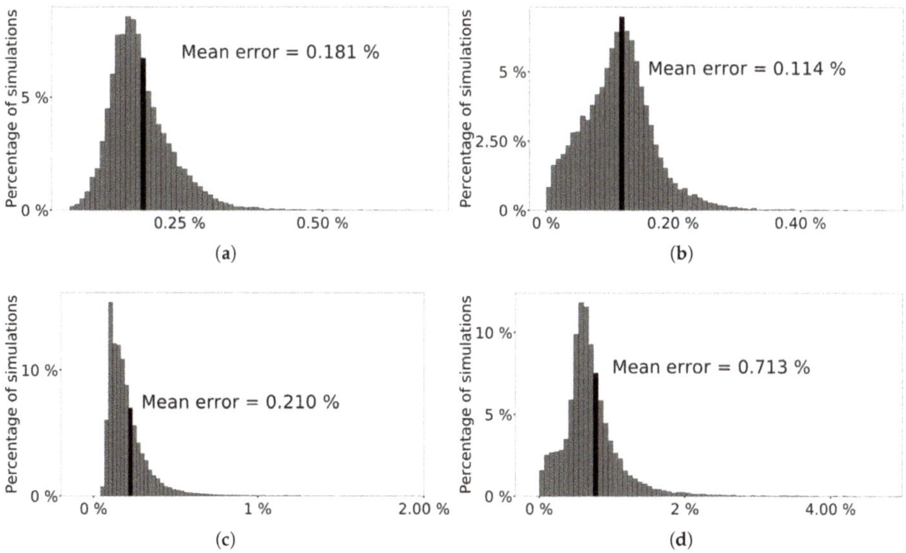

Figure 4. Empirical distribution of the errors for every mechanical variable. (**a**) Empirical distribution for e_{ε}. The size of the histogram bucket is 0.009%; (**b**) Empirical distribution for $e_{\varepsilon_{vp}}$. The size of the histogram bucket is 0.008%; (**c**) Empirical distribution for e_{σ}. The size of the histogram bucket is 0.024%; (**d**) Empirical distribution for e_p. The size of the histogram bucket is 0.066%.

3.4. Convergence with Respect to the Truncation Tolerance

A first surrogate model is constructed from the physical model with the prescribed truncation tolerance $\epsilon = 10^{-3}$. Then, this first surrogate model is used as an input for Algorithm 1. Running the algorithm several times with different truncation tolerances:

$$\epsilon \in \left\{1 \times 10^{-3};\ 2 \times 10^{-3};\ 4.6 \times 10^{-3}; 1 \times 10^{-2};\ 2 \times 10^{-2};\ 4.6 \times 10^{-2};\ 1 \times 10^{-1}\right\}$$

generates as many new surrogate models.

Figure 5a–d present the evolution of the relative error distribution (for the different mechanical variables) with respect to the truncation tolerance based on a random sample of 20,000 parameter set values chosen as in Section 3.3. Figure 6 details the graphical notations. The results empirically show for each mechanical output that the relative error decreases together with ϵ. This is consistent with the expected behavior of the algorithm.

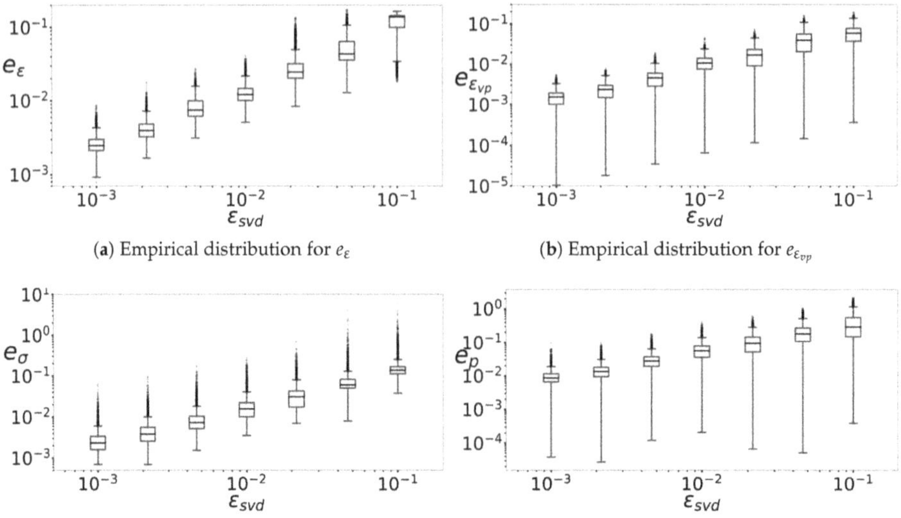

(**a**) Empirical distribution for e_ε (**b**) Empirical distribution for $e_{\varepsilon_{vp}}$

(**c**) Empirical distribution for e_σ (**d**) Empirical distribution for e_p

Figure 5. Empirical distribution of the relative approximation error for every mechanical variable.

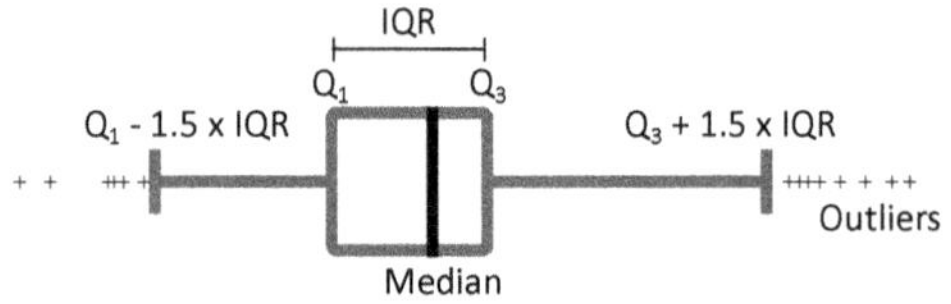

Figure 6. The left and right sides are the first and third quartiles (respectively Q_1 and Q_3). The line inside the box represents the median. The reach of the whiskers past the first and third quartiles is 1.5 times the Interquartile Range (IQR). The crosses represent the outliers lying beyond the whiskers.

The plots in Figure 7a,b show the dependence of the number of stored elements and the number of calls to the physical model on ϵ.

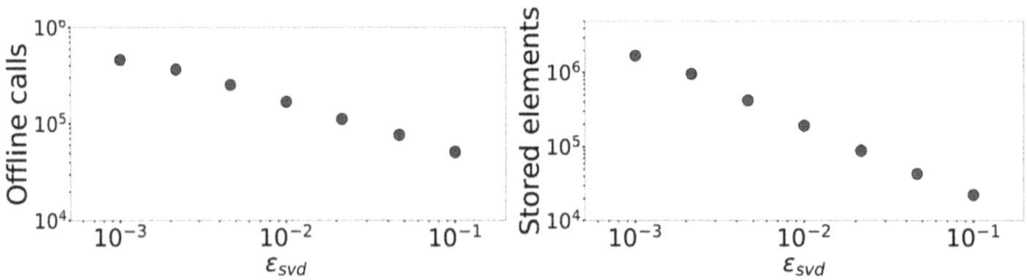

(**a**) (**b**)

Figure 7. Dependence of computational cost and memory storage indicators on ϵ. (**a**) Dependence of the number of calls to the physical model on ϵ; (**b**) Dependence of the number of stored elements on ϵ.

3.5. On Fly Error Estimation

Based on the physical model, the surrogate model gives an approximation of each output of interest. However, the approximate outputs may be inconsistent with the physics in the sense that they may lead to non-zero residuals when introduced into (the discrete version of) the DAE system describing the physical model.

A coherence estimator is an indicator that measures how closely the physical equations are verified by the outputs of the surrogate model. It is reasonable to expect the accuracy of the metamodel to be correlated with the coherence estimator.

Using Equation (A1), let:

$$\underset{\sim}{\sigma}^{eq,TT} = \frac{E}{1+\nu}\left(\underset{\sim e}{\varepsilon}^{TT} + \frac{\nu}{1-2\nu} Tr\left(\underset{\sim e}{\varepsilon}^{TT}\right)\underset{\sim}{\mathbb{I}}\right)$$

and define the associated coherence estimator as follows:

$$\eta_\sigma = \frac{\left\|\underset{\sim}{\sigma}^{TT} - \underset{\sim}{\sigma}^{eq,TT}\right\|_{[0,T]}}{\left\|\underset{\sim}{\sigma}^{TT}\right\|_{[0,T]}} \tag{17}$$

and the effectivity of the estimator as the following ratio:

$$\frac{\eta_\sigma}{e_\sigma}$$

Figure 8 displays the relation between the relative error for $\underset{\sim}{\sigma}$ and the effectivity of the estimator for 20,000 simulation results drawn randomly. The error increases with the final cumulative deformation, that is when the material exhibits a more intense viscoplastic behavior.

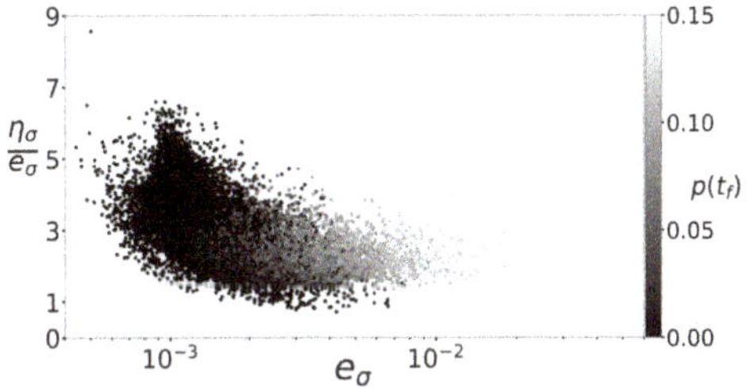

Figure 8. Effectivity of the coherence estimator η_σ (17) associated with σ. The color scale indicates the final cumulative deformation.

Furthermore, the plot shows a correlation between the coherence estimator and the relative error. In particular, the effectivity tends to be larger than one, which indicates that the coherence estimator behaves like an upper bound of the relative error. Excluding a few outliers, the coherence estimator does not overestimate the relative error by more than a factor of seven.

Finally, the effectivity of the coherence estimator empirically converges to one (that is, the estimator becomes sharper) as the magnitude of the relative error increases.

This coherence estimator is very inexpensive to compute and only relies on the outputs of the surrogate model. The results suggest that the coherence estimator could be used as an online error indicator that increases the reliability of the surrogate model at the current point when exploring in real time the parameter domain.

4. Discussion

The TT-cross decomposition enables building an approximation based on a sparse exploration of a reference tensor, by using the gappy POD. Several outputs of a parametric DAE are approximated, assuming they have $d-1$ common indices and one specific index. The present work assesses the performance of tensor-train representations for the approximation of numerical solutions of nonlinear DAE systems. The proposed method enables incorporating a large number of simulation results ($\simeq$500,000 scalar values) to produce a metamodel that is accurate over the entire parameter domain. More specifically, numerical results show that the multiple TT decomposition gives promising results when used as a surrogate model for an elasto-viscoplastic constitutive law. For this particular application, the surrogate model exhibits a satisfying accuracy given the moderate computational effort spent for its construction and the data storage requirements. Moreover, the observed behavior of the proposed empirical coherence estimator indicates that the latter could be exploited to assess the approximation error in real time.

The application to more complex material constitutive laws of industrial interest and involving a larger number of parameters [35] corroborates the aforementioned results in terms of compactness and accuracy of the surrogate models. Surrogate models have the potential to transform the way of carrying out parametric studies in material science. In particular, Reference [35] demonstrates that the exploitation of models based on the multiple TT approach simplifies the process of the calibration of constitutive laws. Future work will investigate the combination of the proposed method with the "usual" model order reduction techniques such as hyper-reduction [36] in order to take into account the space dimension.

Author Contributions: Conceptualization, D.R.; methodology, C.O., D.R., and J.C.; software, C.O.; validation, C.O.; formal analysis, C.O. and J.C.; investigation, C.O.; resources, J.C.; data curation, C.O.; writing, original draft preparation, C.O.; writing, review and editing, D.R. and J.C.; visualization, C.O.; supervision, D.R. and J.C.; project administration, J.C.; funding acquisition, D.R. and J.C.

Funding: This research was funded by the Association Nationale de la Recherche et de la Technologie (ANRT) (Grant Number CIFRE 2014/0923).

Acknowledgments: The authors gratefully acknowledge fruitful discussions with Safran Helicopter Engines.

Conflicts of Interest: The authors declare no conflict of interest. The founding sponsors had no role in the design of the study; in the collection, analyses, or interpretation of data; in the writing of the manuscript; nor in the decision to publish the results.

Abbreviations

The following abbreviations are used in this manuscript:

DAE	Differential Algebraic Equation
DEIM	Discrete Empirical Interpolation Method
EIM	Empirical Interpolation Method
PGD	Proper Generalized Decomposition
POD	Proper Orthogonal Decomposition
PSD	Pseudo-Skeleton Decomposition
SVD	Singular-Value Decomposition
TT	Tensor-Train

Appendix A. Elasto-Viscoplastic Model

The application case consists of a nonlinear constitutive law in elasto-viscoplasticity [37,38] linking the following time-dependent mechanical variables:

- The strain tensor: $\underset{\sim}{\varepsilon} = \underset{\sim}{\varepsilon}_e + \underset{\sim}{\varepsilon}_{vp}$ (dimensionless) (sum of an elastic part and a viscoplastic part);
- The stress tensor: $\underset{\sim}{\sigma}$ (MPa);
- An internal hardening variable: $\underset{\sim}{X}$ (MPa);
- The cumulative viscoplastic deformation: p (dimensionless).

where $\underset{\sim}{\varepsilon}$, $\underset{\sim}{\varepsilon}_e$, $\underset{\sim}{\varepsilon}_{vp}$, $\underset{\sim}{\sigma}$, and $\underset{\sim}{X}$ are second-order tensors in $\mathbb{R}^{3\times3}$.

The hypotheses of the infinitesimal strain theory are assumed to hold.

The model involves eight material coefficients: E, ν, n, K, R_0, Q, b, and C. The Young and Poisson coefficients are set to E = 200,000 MPa and $\nu = 0.3$. Table A1 presents the range of variation of the other material coefficients considered as input parameters of the model.

Table A1. Parameter range of variations considered in the model. When applicable, the unit is indicated between brackets. Note that the dimension of K depends on the parameter n according to Equation (A2).

	n	K (MPa·s^{-n})	R_0 (MPa)	Q (MPa)	b	C (MPa)
Lower Bound	2	100	1	1	0.02	150
Upper Bound	12	10,000	200	2000	2000	150,000

System of Equations

The elastic behavior is governed by:

$$\underset{\sim}{\sigma} = \frac{E}{1+\nu}\left(\underset{\sim}{\varepsilon}_e + \frac{\nu}{1-2\nu}Tr\left(\underset{\sim}{\varepsilon}_e\right)\underset{\sim}{\mathbb{I}}\right) \tag{A1}$$

The viscoplastic behavior is described by the Norton flow rule (A2) formulated with the von Mises criterion (A5). The yield function and the normal to the yield function are given by (A3) and (A4). (A6) gives the definition of the deviatoric stress tensor involved in (A5).

$$\frac{d}{dt}\underset{\sim}{\varepsilon}_{vp} = \underset{\sim}{N}\left(\frac{f}{K}\right)_+^n \tag{A2}$$

$$f = J\left(\underset{\sim}{\sigma}^D - \underset{\sim}{X}\right) - R \tag{A3}$$

$$\underset{\sim}{N} = \frac{3}{2}\frac{\underset{\sim}{\sigma}^D - \underset{\sim}{X}}{J\left(\underset{\sim}{\sigma}^D - \underset{\sim}{X}\right)} \tag{A4}$$

$$J\left(\underset{\sim}{\sigma}^D - \underset{\sim}{X}\right) = \sqrt{\frac{3}{2}\left(\underset{\sim}{\sigma}^D - \underset{\sim}{X}\right):\left(\underset{\sim}{\sigma}^D - \underset{\sim}{X}\right)} \tag{A5}$$

$$\underset{\sim}{\sigma}^D = \underset{\sim}{\sigma} - \frac{1}{3}Tr\left(\underset{\sim}{\sigma}\right)\underset{\sim}{\mathbb{I}} \tag{A6}$$

where $(.)_+$ denotes the positive part function.

The operator $:$ denotes the contracted product defined as:

$$\underset{\sim}{Z}_1 : \underset{\sim}{Z}_2 = \sum_{i=1}^{3}\sum_{j=1}^{3} Z_1^{ij} Z_2^{ij} \quad \text{for} \quad \underset{\sim}{Z}_1, \underset{\sim}{Z}_2 \in \mathbb{R}^{3\times 3}$$

The nonlinear isotropic hardening is modeled by (A7) where (A8) gives the viscoplastic cumulative rate.

$$R = R_0 + Q\left(1 - e^{-bp}\right) \tag{A7}$$

$$\frac{dp}{dt} = \sqrt{\frac{2}{3}\frac{d}{dt}\underset{\sim}{\varepsilon}_{vp}:\frac{d}{dt}\underset{\sim}{\varepsilon}_{vp}} \tag{A8}$$

Finally, the linear kinematic hardening is given by:

$$\underset{\sim}{X} = \frac{2}{3}C\underset{\sim}{\varepsilon}_{vp} \tag{A9}$$

The case of a uniaxial cyclic tensile testing driven by deformation is considered. The loading is applied by imposing $\varepsilon^{11}(t)$ with the pattern shown in Figure A1 and $\sigma^{12}(t) = \sigma^{13}(t) = \sigma^{23}(t) = \sigma^{22}(t) = \sigma^{33}(t) = 0$.

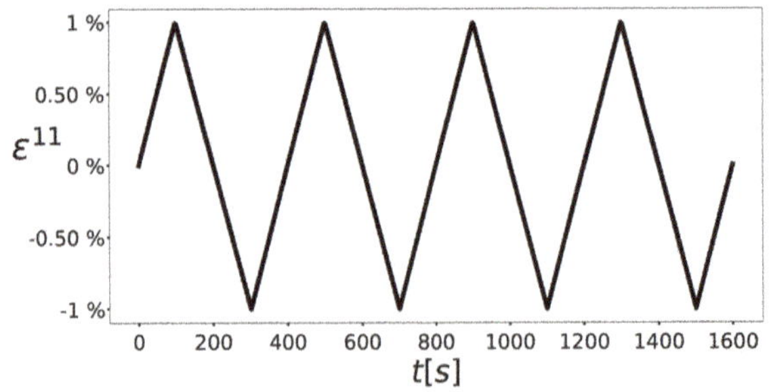

Figure A1. The applied strain component $\varepsilon^{11}(t)$ consists of a triangular pattern of period 400 s with a peak-to-peak amplitude of 2% centered at zero.

The initial conditions for the internal variables are:

$$p(t=0)=0 \quad \text{and} \quad \underset{\sim}{X}(t=0)=\underset{\sim}{0}$$

The model is highly nonlinear. First, the isotropic hardening law introduces an exponential nonlinearity. The most significant nonlinearity arises from the Norton law (A2) featuring the positive part function. Capturing the resulting threshold effect is particularly challenging for surrogate models.

References

1. Ghighi, J.; Cormier, J.; Ostoja-Kuczynski, E.; Mendez, J.; Cailletaud, G.; Azzouz, F. A microstructure sensitive approach for the prediction of the creep behaviour and life under complex loading paths. *Tech. Mech.* **2012**, *32*, 205–220.
2. Le Graverend, J.B.; Cormier, J.; Gallerneau, F.; Villechaise, P.; Kruch, S.; Mendez, J. A microstructure-sensitive constitutive modeling of the inelastic behavior of single crystal nickel-based superalloys at very high temperature. *Int. J. Plast.* **2014**, *59*, 55–83. [CrossRef]
3. Gilchrist, M.D.; Murphy, J.G.; Pierrat, B.; Saccomandi, G. Slight asymmetry in the winding angles of reinforcing collagen can cause large shear stresses in arteries and even induce buckling. *Meccanica* **2017**, *52*, 3417–3429. [CrossRef]
4. Marino, M. Constitutive Modeling of Soft Tissues. In *Encyclopedia of Biomedical Engineering*; Narayan, R., Ed.; Elsevier: Oxford, UK, 2019; pp. 81–110.
5. Marino, M.; von Hoegen, M.; Schröder, J.; Wriggers, P. Direct and inverse identification of constitutive parameters from the structure of soft tissues. Part 1: micro- and nanostructure of collagen fibers. *Biomech. Model. Mechanobiol.* **2018**, *17*, 1011–1036. [CrossRef] [PubMed]
6. Ladevèze, P.; Nouy, A. On a multiscale computational strategy with time and space homogenization for structural mechanics. *Comput. Methods Appl. Mech. Eng.* **2003**, *192*, 3061–3087. [CrossRef]
7. Ammar, A.; Mokdad, B.; Chinesta, F.; Keunings, R. A new family of solvers for some classes of multidimensional partial differential equations encountered in kinetic theory modeling of complex fluids. *J. Non-Newtonian Fluid Mech.* **2006**, *139*, 153–176. [CrossRef]
8. Nouy, A. A priori model reduction through Proper Generalized Decomposition for solving time-dependent partial differential equations. *Comput. Methods Appl. Mech. Eng.* **2010**, *199*, 1603–1626. [CrossRef]
9. Hitchcock, F.L. The Expression of a Tensor or a Polyadic as a Sum of Products. *J. Math. Phys.* **1927**, *6*, 164–189. [CrossRef]
10. Khoromskij, B.N. Tensors-structured numerical methods in scientific computing: Survey on recent advances. *Chemom. Intell. Lab. Syst.* **2012**, *110*, 1–19. [CrossRef]
11. Grasedyck, L.; Kressner, D.; Tobler, C. A literature survey of low-rank tensor approximation techniques. *GAMM-Mitt.* **2013**, *36*, 53–78. [CrossRef]
12. Bigoni, D.; Engsig-Karup, A.P.; Marzouk, Y.M. Spectral Tensor-Train Decomposition. *SIAM J. Sci. Comput.* **2016**, *38*, A2405–A2439. [CrossRef]

13. Harshman, R.A. Foundations of the PARAFAC procedure: Models and conditions for an "explanatory" multi-modal factor analysis. *UCLA Work. Pap. Phon.* **1970**, *16*, 1–84.
14. Kiers, H.A. Towards a standardized notation and terminology in multiway analysis. *J. Chemom.* **2000**, *14*, 105–122. [CrossRef]
15. Tucker, L.R. The extension of factor analysis to three-dimensional matrices. In *Contributions to Mathematical Psychology*; Gulliksen, H., Frederiksen, N., Eds.; Holt, Rinehart and Winston: New York, NY, USA, 1964; pp. 110–127.
16. Hackbusch, W.; Kühn, S. A new scheme for the tensor representation. *J. Fourier Anal. Appl.* **2009**, *15*, 706–722. [CrossRef]
17. Oseledets, I.; Tyrtyshnikov, E. TT-cross approximation for multidimensional arrays. *Linear Algebr. Appl.* **2010**, *432*, 70–88. [CrossRef]
18. Oseledets, I.V. Tensor-train decomposition. *SIAM J. Sci. Comput.* **2011**, *33*, 2295–2317. [CrossRef]
19. Savostyanov, D.; Oseledets, I. Fast adaptive interpolation of multi-dimensional arrays in tensor train format. In Proceedings of the 7th International Workshop on Multidimensional (nD) Systems (nDs), Poitiers, France, 5–7 September 2011; pp. 1–8.
20. Savostyanov, D.V. Quasioptimality of maximum-volume cross interpolation of tensors. *Linear Algebr. Appl.* **2014**, *458*, 217–244. [CrossRef]
21. Sirovich, L. Turbulence and the dynamics of coherent structures. Part 1: Coherent structures. *Q. Appl. Math.* **1987**, *45*, 561–571. [CrossRef]
22. Everson, R.; Sirovich, L. Karhunen-Loève procedure for gappy data. *J. Opt. Soc. Am. A* **1995**, *12*, 1657–1664. [CrossRef]
23. Maday, Y.; Nguyen, N.C.; Patera, A.T.; Pau, S.H. A general multipurpose interpolation procedure: The magic points. *Commun. Pure Appl. Anal.* **2009**, *8*, 383–404. [CrossRef]
24. Barrault, M.; Maday, Y.; Nguyen, N.C.; Patera, A.T. An empirical interpolation method: Application to efficient reduced-basis discretization of partial differential equations. *C.-R. Math.* **2004**, *339*, 667–672. [CrossRef]
25. Ryckelynck, D.; Lampoh, K.; Quilicy, S. Hyper-reduced predictions for lifetime assessment of elasto-plastic structures. *Meccanica* **2016**, *51*, 309–317. [CrossRef]
26. Carlberg, K.; Farhat, C.; Cortial, J.; Amsallem, D. The GNAT method for nonlinear model reduction: Effective implementation and application to computational fluid dynamics and turbulent flows. *J. Comput. Phys.* **2013**, *242*, 623–647. [CrossRef]
27. Tyrtyshnikov, E.; Goreinov, S.; Zamarashkin, N. Pseudo-Skeleton Approximations. Available online: www.inm.ras.ru/library/Tyrtyshnikov/biblio/psa-dan.pdf (accessed on 26 January 2018).
28. Bebendorf, M. Approximation of boundary element matrices. *Numer. Math.* **2000**, *86*, 565–589. [CrossRef]
29. Berry, M.W.; Pulatova, S.A.; Stewart, G. Algorithm 844: Computing sparse reduced-rank approximations to sparse matrices. *ACM Trans. Math. Softw.* **2005**, *31*, 252–269. [CrossRef]
30. Stewart, G. Four algorithms for the the efficient computation of truncated pivoted QR approximations to a sparse matrix. *Numer. Math.* **1999**, *83*, 313–323. [CrossRef]
31. Golub, G.H.; van Loan, C.F. *Matrix Computations*, 4th ed.; The Johns Hopkins University Press: Baltimore, MD, USA, 2013.
32. Halton, J.H. On the efficiency of certain quasi-random sequences of points in evaluating multi-dimensional integrals. *Numer. Math.* **1960**, *2*, 84–90. [CrossRef]
33. Drmac, Z.; Gugercin, S. A new selection operator for the discrete empirical interpolation method—Improved a priori error bound and extensions. *SIAM J. Sci. Comput.* **2016**, *38*, A631–A648. [CrossRef]
34. Chaturantabut, S.; Sorensen, D.C. Nonlinear model reduction via discrete empirical interpolation. *SIAM J. Sci. Comput.* **2010**, *32*, 2737–2764. [CrossRef]
35. Olivier, C. Décompositions Tensorielles et Factorisations de Calculs Intensifs Appliquées à l'Identification de Modèles de Comportement non Linéaire. Ph.D. Thesis, PSL Reasearch University, Evry, France, 2017.
36. Ryckelynck, D.; Vincent, F.; Cantournet, S. Multidimensional a priori hyper-reduction of mechanical models involving internal variables. *Comput. Methods Appl. Mech. Eng.* **2012**, *225*, 28–43. [CrossRef]

37. Besson, J.; Cailletaud, G.; Chaboche, J.L.; Forest, S. *Non-Linear Mechanics of Materials*, 1st ed.; Springer: Dordrecht, The Netherlands, 2010.
38. Lemaitre, J.; Chaboche, J.L. *Mechanics of Solid Materials*; Cambridge University Press: Cambridge, UK, 1994.

Mathematical and Computational Applications

MDPI

Article

Finite Strain Homogenization Using a Reduced Basis and Efficient Sampling

Oliver Kunc and **Felix Fritzen** *

Efficient Methods for Mechanical Analysis, Institute of Applied Mechanics (CE), University of Stuttgart, 70569 Stuttgart, Germany; kunc@mechbau.uni-stuttgart.de

* Correspondence: fritzen@mechbau.uni-stuttgart.de

Received: 31 March 2019; Accepted: 24 May 2019; Published: 27 May 2019

check for updates

Abstract: The computational homogenization of hyperelastic solids in the geometrically nonlinear context has yet to be treated with sufficient efficiency in order to allow for real-world applications in true multiscale settings. This problem is addressed by a problem-specific surrogate model founded on a reduced basis approximation of the deformation gradient on the microscale. The setup phase is based upon a snapshot POD on deformation gradient fluctuations, in contrast to the widespread displacement-based approach. In order to reduce the computational offline costs, the space of relevant macroscopic stretch tensors is sampled efficiently by employing the Hencky strain. Numerical results show speed-up factors in the order of 5–100 and significantly improved robustness while retaining good accuracy. An open-source demonstrator tool with 50 lines of code emphasizes the simplicity and efficiency of the method.

Keywords: computational homogenization; large strain; finite deformation; geometric nonlinearity; reduced basis; reduced-order model; sampling; Hencky strain

MSC: 74Q05; 74B20; 74S30

1. Introduction

1.1. Purpose

The description of solid mechanics under finite strains is of particular interest in both academia and industry. It allows for accurate descriptions of rotations and stretches under mild assumptions. Thus, many geometric effects can be captured. For instance, alignments and rearrangements of the respective structures may trigger pronounced stiffening or softening effects.

In such cases where rotations and deformations are not suitable for linearization, dissipative effects also play a notable role for many materials. Regardless of the kind of dissipation involved in a certain process, hyperelasticity usually persists to a certain extent. Therefore, it is worthwhile investigating this comparatively simple case at first, before introducing history dependence into the description. Prominent examples of materials that require a hyperelastic description at finite strains include carbon black-filled rubber [1] and amorphous glassy polymers [2], to name just two.

The main purpose of this work is the *computationally efficient quasi-static homogenization of hyperelastic solids with full account for geometric nonlinearities*. The employed methodology is twofold. First, a Reduced Basis (RB) model for the microscopic problem is established. The term Reduced Basis, used in this work, is not to be confused with the homonymous method introduced by Barrault, Maday, Nguyen, and Patera [3]. Once set up, it enables more efficient evaluations of the homogenized material response as compared to the Finite Element Method (FEM). Second, an efficient strategy for sampling of the space of macroscopic kinematic states is proposed. This renders the setup phase of the RB model more rational.

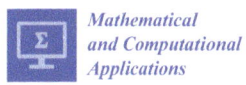

1.2. State of the Art

Efficiently determining the overall solid–mechanical properties of microstructures has been investigated for decades, and a large body of literature is available. Comprehensive review articles, such as [4] and [5], summarize the progress. Here, attention is restrained to few methods most similar or relevant to the present work.

The FE^2 method [6] is theoretically capable of performing realistic two-scale simulations with arbitrary accuracy. Therefore, it serves as a reference method in the context of first-order homogenization based on the assumption of separated length scales. In the FE^2, the evaluations of the unknown macroscopic constitutive law are approximated by microscopic FE simulations. However, this comes along with computational costs that quickly exceed the capabilities of common workstations, both at present and in the foreseeable future. Roughly speaking, the computational effort required on the microscale multiplies with that of the macroscale, hence the method's name. It is thus worthwhile to develop order reduction methods for the microscopic problem.

A common approach within the field of computational homogenization (and well beyond) is to extract essential information from provided in silico data. To this end, schemes based on the Proper Orthogonal Decomposition (POD) compute correlations within snapshot data, [7]. Such methods include the R3M [8] and can be further enhanced by the use of, e.g., the EIM, as in [9]. Numerical comparisons of various schemes were conducted in [10,11]. To the best of the authors' knowledge, all published POD-based methods addressing the finite strain hyperelastic problem choose to reduce the number of degrees of freedom (DOF) of the displacement field. This results in sometimes significant speed-ups. Another important feature is that they allow for reconstruction of the microscopic displacement fields. The application of the snapshot POD to *gradients* of the primal variables has been studied—e.g., for infinitesimal strain hyperelasticity [12] and fluid mechanics [13]—but does not appear to have been investigated for finite strain hyperelasticity yet.

Still, the solution of the reduced equations remains a complex task. It requires evaluations of material laws and numerical integration over the microstructure. Promising progress has been made in the field of efficient integration schemes, see for instance [14,15]. A main reason for the speed-up of these methods is the reduced number of function evaluations.

The highest speed-ups are achievable if the computational effort of the determination of effective microstructural responses can be fully decoupled from underlying microstructural discretizations. Such homogenization methods directly approximate the effective material law by means of a dedicated numerical scheme. Technically, this can be seen as the direct surrogation of unknown functions, e.g., of the effective free energy or stress. For instance, the Material Map [16] interpolates the coefficients of an assumed macroscopic material model. Another example is the NEXP method [17], where the effective stored energy density is approximated using a tensor product of one-dimensional splines. The authors treated the case of small strains by introducing the RNEXP method [12], where the effective stored energy is interpolated by a dedicated kernel scheme.

However, interpolatory and regressional methods suffer the inherent drawback of not providing any explicit information on the microscale. For instance, microscopic displacement or stress fields cannot be reconstructed from the solutions of macroscopic interpolation. Another important open question is how to provide the supporting data points for the interpolation in an efficient manner. The data at these points is usually provided by the solution of a full-order model (FOM) and come along with the corresponding numerical costs. Hence, the positions of data points in the parameter space should be chosen carefully, as unnecessary or redundant solutions of the FOM should be avoided. On the other hand, too sparsely seeded points might not capture the homogenized properties of the microstructure appropriately.

1.3. Main Contributions and Outline

The present work generalizes parts of the previous paper [12] to the finite strain regime. It aims at reducing the computational complexity for the determination of the homogenized microstructural response, which is parametrized by the macroscopic deformation gradient acting as a boundary condition. This is achieved by means of a *Reduced Basis approximation of the microscopic deformation gradient*. The basis is obtained with the aid of a *POD of snapshots of fluctuation fields of the deformation gradient*. Thus, the application of the RB model *does not necessitate the computation of gradients of displacement fields*, and even does not require the displacements to be available at all. In other words, microscopic displacement fields are completely avoided. However, they *can be reconstructed* from the RB approximation of the deformation gradient, uniquely up to rigid body motion.

Another key advantage is the *sleek implementation* of the method. A *demonstration* containing a minimum working example of the RB model with 50 lines of MATLAB/Octave code is provided [18].

As for the setup phase, the snapshot data is created by means of an *efficient sampling procedure* for the microscopic boundary condition. To this end, the set of macroscopic *Hencky strains* is identified as a suitable *linear parameter space*, within which the sampling sites are placed based upon *physical interpretation*. This allows for *control*of the resolution of certain key characteristics of the effective material response while keeping the total number of samples within bounds.

The Reduced Basis method is presented in Section 2. The basis identification is based on the sampling strategy developed in Section 3. Numerical examples are presented in Section 4. Both the numerical and the theoretical findings are summarized and discussed in Section 5.

1.4. Notation

The set of real numbers and the subset of positive numbers greater than zero are denoted by $\mathbb{R}$ and $\mathbb{R}_+$, respectively. Matrices are marked by two underlines and vectors by one underline, e.g., $\underline{\underline{A}}$, $\underline{a}$. Vectors are assumed to be columns, and the dot product of two vectors of the same size is understood as the Euclidean scalar product, $\underline{x} \cdot \underline{y} = \underline{x}^{\mathsf{T}} \underline{y}$. First order and second order tensors in coordinate-free description are denoted by bold letters, e.g., $\boldsymbol{A}$, $\boldsymbol{a}$. No conclusion can be drawn on the order of a tensor based on its capitalization. Here, the underlying space is always the Euclidean space $\mathbb{R}^3$ with its standard basis. First order and second order tensors can be represented as vectors and matrices, e.g., $\boldsymbol{A} \leftrightarrow \underline{A} \in \mathbb{R}^3$ and $\boldsymbol{B} \leftrightarrow \underline{\underline{B}} \in \mathbb{R}^{3\times 3}$, respectively. Norms of vectors and matrices respectively denote the Euclidean and the Frobenius norm. The norm of a tensor of second order equals the norm of its matrix representation for the chosen basis. Fourth order tensors are denoted by blackboard bold symbols other than $\mathbb{R}$, e.g., $\mathbb{C}$ and $\mathbb{I}$. Components of tensors of order M are with respect to the Euclidean tensorial basis $\boldsymbol{e}^{(1)} \otimes \boldsymbol{e}^{(2)} \otimes \cdots \otimes \boldsymbol{e}^{(M)}$, e.g., A_{ij}, B_{ij} for second order tensors $\boldsymbol{A}$, $\boldsymbol{B}$ and C_{ijkl},C'_{ijkl} for $\mathbb{C}$, $\mathbb{C}'$. The following contractions are defined:

$$\boldsymbol{A} \cdot \boldsymbol{B} = \sum_{i,j=1}^{3} A_{ij} B_{ij}\,, \qquad \mathbb{C} \cdot \boldsymbol{B} = \sum_{i,j,k,l=1}^{3} C_{ijkl} B_{kl}\, \boldsymbol{e}^{(i)} \otimes \boldsymbol{e}^{(j)}\,,$$

$$\boldsymbol{A} \cdot \mathbb{C} = \sum_{i,j,k,l=1}^{3} A_{ij} C_{ijkl}\, \boldsymbol{e}^{(k)} \otimes \boldsymbol{e}^{(l)}\,, \qquad \mathbb{C} \cdot \mathbb{C}' = \sum_{i,j,k,l,m,n=1}^{3} C_{ijmn} C'_{mnkl}\, \boldsymbol{e}^{(i)} \otimes \boldsymbol{e}^{(j)} \otimes \boldsymbol{e}^{(k)} \otimes \boldsymbol{e}^{(l)}\,.$$

Let $\Omega \subset \mathbb{R}^3$ be the domain occupied by a physical body undergoing elastic deformations, and let Ω_0 be its initial configuration. Then, $\boldsymbol{x}$ and $\boldsymbol{X}$ describe the coordinates of material points within the current configuration Ω and within the reference state Ω_0, respectively. Their difference is the displacement $\boldsymbol{u} = \boldsymbol{x} - \boldsymbol{X}$, see Figure 1. The gradient of a vector field $\boldsymbol{v} = \boldsymbol{v}(\boldsymbol{X})$ is defined as a *right gradient* and denoted by $\dfrac{\partial \boldsymbol{v}}{\partial \boldsymbol{X}} = \boldsymbol{v} \otimes \nabla_{\mathrm{X}}$. The divergence of a second order tensor field is the vector field resulting from *row-wise* divergence. The boundaries of the respective configurations are denoted by $\partial\Omega$ and $\partial\Omega_0$. The set of square-integrable Lebesgue functions on the reference domain is tagged $L^2(\Omega_0)$.

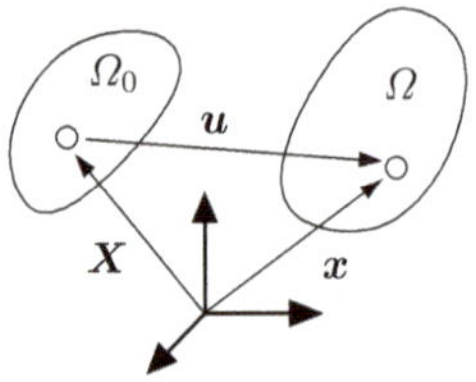

Figure 1. Initial (Ω_0) and current (Ω) configuration, with elementary kinematic quantities.

The displacement gradient $H = u \otimes \nabla_X$ and the deformation gradient $F = x \otimes \nabla_X$ are related through $F = H + I$, where I is the second order identity tensor in three dimensions. The determinant $J = \det F$ measures the relative volumetric change due to the present deformation.

Unimodular quantities, i.e., second order tensors with determinant ones, may be emphasized by a hat, e.g., $\widehat{F} = J^{-1/3}F$. This multiplicative decomposition is sometimes attributed to Flory [19] and also goes by the name Dilatational-Deviatoric Multiplicative Split (DDMS).

In the two-scale context, overlined symbols represent quantities on the macroscopic scale, e.g., $\overline{A}$, $\overline{a}$, while symbols without overline correspond to their microscopic counterpart, e.g., A, a. Equivalently, macroscopic quantities are called *global* and microscopic ones are called *local*. The volume average of a general local field φ

$$\langle \varphi \rangle = \langle \varphi(\bullet) \rangle = \frac{1}{|\Omega_0|} \int_{\Omega_0} \varphi(\bullet) \, \mathrm{d}V \tag{1}$$

is essential to the theory. The dependence of a microscopic quantity A on both the microscopic coordinates X and a macroscopic quantity $\overline{B}$ is denoted by $A = A(X; \overline{B})$. In such a case, the components of the macroscopic quantity $\overline{B}$ are called *parameters* of the microscopic function $A(\bullet; \overline{B})$. The application of the volume averaging operator is abbreviated by $\langle A \rangle = \langle A(\bullet; \overline{B}) \rangle$. The case of a concatenated function $f(A) = f(A(X; \overline{B}))$ is analogous, i.e. $\langle f \rangle = \langle f(A) \rangle = \langle f(A(\bullet; \overline{B})) \rangle$, regardless of the tensorial order of the image of the function f.

1.5. Material Models

In this work, hyperelastic materials are investigated. They are characterized by *stored energy density functions* $W = W(F)$. The first Piola–Kirchhoff stress

$$P(F) = \frac{\partial W}{\partial F}(F) \tag{2}$$

and the corresponding fourth-order stiffness tensor

$$\mathbb{C}(F) = \frac{\partial^2 W}{\partial F^2}(F) \tag{3}$$

characterize the material response.

Henceforth, for reasons of readability, the stored energy density function W will be spoken of as an energy, and the terms *stored* and *density* will not always be mentioned explicitly. In the infinitesimal strain framework, hyperelastic energies have been formulated to model deformation plasticity (e.g., [12,17,20]). Although these models are only valid for purely proportional loading conditions, they provide means to simulate highly nonlinear material behavior in certain scenarios comparably easily within the context of hyperelasticity. Note that genuine dissipative processes require additional state describing variables with corresponding evolution laws.

The proposed method is *suitable for any type of hyperelastic constitutive law*. As the modeling of complex material behavior is not the main focus of this study, the Neo–Hookean law

$$W(\boldsymbol{F}) = W_{\mathrm{DDMS}}(J, \widehat{\boldsymbol{F}}) = \frac{K}{4}\left[(J-1)^2 + (\ln J)^2\right] + \frac{G}{2}\left(\mathrm{tr}(\widehat{\boldsymbol{F}}^{\mathsf{T}}\widehat{\boldsymbol{F}}) - 3\right) \tag{4}$$

is used, with K the bulk modulus and G the shear modulus. The volumetric part of the energy is taken from [21]. Using the DDMS, a decoupled dependence on the volumetric and isochoric part of the deformation is assumed, which is a common way to model the distinct material behavior with respect to these two contributions, see e.g., [22].

1.6. Problem Setting of First Order Homogenization

Assuming stationarity and separability of scales, the following coupled and deformation-driven problems can be derived by means of asymptotic expansion of the displacement $\boldsymbol{u}$ and subsequent first order approximation. This procedure is carried out in [23] with much detail. Here, the technical process is omitted and only the resulting equations are stated.

1.6.1. Macroscopic Problem

Balance of linear momentum

$$\mathrm{Div}_{\overline{\mathrm{X}}}(\overline{\boldsymbol{P}}) + \overline{\boldsymbol{b}} = \boldsymbol{0}, \tag{5}$$

where $\overline{\boldsymbol{b}}$ denote bulk forces, and balance of angular momentum

$$\overline{\boldsymbol{F}}^{-1}\overline{\boldsymbol{P}} = \overline{\boldsymbol{P}}^{\mathsf{T}}\overline{\boldsymbol{F}}^{-\mathsf{T}}, \tag{6}$$

along with well-posed boundary conditions that constitute the *macroscopic boundary value problem*. This system of equations is closed by means of the hyperelasticity law, cf. (2),

$$\overline{\boldsymbol{P}}(\overline{\boldsymbol{F}}) = \frac{\partial \overline{W}}{\partial \overline{\boldsymbol{F}}}. \tag{7}$$

Note that, in general, $\overline{W}$ is a priori not available and (7) is thus a purely formal relation. For reasons of readability, the dependence of any quantity on the macroscopic material coordinate $\overline{\boldsymbol{X}}$ is usually spared, e.g., $\overline{\boldsymbol{F}} = \overline{\boldsymbol{F}}(\overline{\boldsymbol{X}})$, $\overline{\boldsymbol{H}} = \overline{\boldsymbol{H}}(\overline{\boldsymbol{X}})$, or $\overline{\boldsymbol{u}} = \overline{\boldsymbol{u}}(\overline{\boldsymbol{X}})$.

1.6.2. Microscopic Problem

The *microscopic boundary value problem* is given by the balance equations

$$\mathrm{Div}_{\mathrm{X}}(\boldsymbol{P}) = \boldsymbol{0} \tag{8}$$

$$\boldsymbol{F}^{-1}\boldsymbol{P} = \boldsymbol{P}^{\mathsf{T}}\boldsymbol{F}^{-\mathsf{T}} \tag{9}$$

and suitable boundary conditions, e.g., as discussed in [24]. In this work, *periodic displacement fluctuation boundary conditions* are employed. The microscopic displacements take the form

$$\boldsymbol{u}(\boldsymbol{X};\overline{\boldsymbol{F}}) = \overline{\boldsymbol{u}} + \overline{\boldsymbol{H}}\boldsymbol{X} + \boldsymbol{w}(\boldsymbol{X};\overline{\boldsymbol{F}}). \tag{10}$$

Therein, the macroscopic displacement is independent of microscopic quantities. The second term, $\overline{\boldsymbol{H}}\boldsymbol{X}$, corresponds to a homogeneous deformation of the microstructure. The third term, $\boldsymbol{w}(\boldsymbol{X})$, is a *displacement fluctuation* with the *zero mean property* $\langle \boldsymbol{w} \rangle = \boldsymbol{0}$. Hence, the deformation gradient reads

$$\boldsymbol{F}(\boldsymbol{X};\overline{\boldsymbol{F}}) = \overline{\boldsymbol{F}} + \widetilde{\boldsymbol{H}}(\boldsymbol{X};\overline{\boldsymbol{F}}) = \overline{\boldsymbol{F}} + \widetilde{\boldsymbol{F}}(\boldsymbol{X};\overline{\boldsymbol{F}}), \tag{11}$$

where the fluctuation part $\widetilde{H} = w \otimes \nabla_X = \widetilde{F}$, has the zero mean property

$$\left\langle \widetilde{F} \right\rangle = 0. \tag{12}$$

Thus, the volume average of the local deformation gradient equals its macroscopic counterpart,

$$\overline{F} = \langle F \rangle. \tag{13}$$

This motivates the identification of $\overline{F}$ as *the boundary condition* to the microscopic problem (8). As for the material response, the following relationships can be deduced:

$$\overline{W} = \langle W \rangle, \tag{14}$$

$$\overline{P} = \langle P \rangle, \tag{15}$$

$$\overline{\mathbb{C}} \neq \langle \mathbb{C} \rangle. \tag{16}$$

Equations (13) and (15) are called kinematic and static *coupling relations*, respectively. The inequality (16) generally holds for heterogeneous microstructures, even in the most simple case of infinitesimal strains and linear elasticity. More precisely, the volume average overestimates the effective stiffness in the spectral sense,

$$x \cdot \langle \mathbb{C} \rangle \cdot x \geq x \cdot \overline{\mathbb{C}} \cdot x \qquad \forall \text{ second order tensors } x. \tag{17}$$

2. Reduced Basis Homogenization for Hyperelasticity

2.1. Formulation

The *Reduced Basis (RB)* scheme is based on a direct approximation of the microscopic deformation gradient F from Equation (11) without the need to explicitly have the corresponding displacement at hand. The initial approximation is given by

$$F_{\xi}(X;\overline{F},\underline{\xi}) = \overline{F} + \sum_{i=1}^{N} B^{(i)}(X)\xi_i. \tag{18}$$

The arguments $\overline{F}$ and $\underline{\xi} \in \mathbb{R}^N$ are the boundary condition to (8) and the reduced coefficient vector, respectively. Note that the macroscopic coordinate $\overline{X}$ is not assumed to influence the RB approximation, i.e., the same approximation is made throughout the macrostructure. The set $\{B^{(i)}\}_{i=1}^{N}$ is linearly independent within the space $L^2(\Omega_0)$ and is called *RB of* F. In a later context, it will also be referred to as the set of *ansatz functions*. In order to enforce the relationship

$$\left\langle F_{\xi} \right\rangle = \overline{F} \tag{19}$$

regardless of the reduced coefficients $\underline{\xi}$, the basis functions are asserted to have the *fluctuation* property, i.e., for $i = 1, \ldots, N$

$$\left\langle B^{(i)} \right\rangle = 0. \tag{20}$$

For now, the RB is assumed to be given.

The ansatz (18) allows for evaluation of the energy at the macroscale as a function of the macroscopic kinematic variable $\overline{F}$ and of the reduced coefficients $\underline{\xi}$,

$$\overline{W}_{\xi}(\overline{F},\underline{\xi}) = \left\langle W(F_{\xi}) \right\rangle. \tag{21}$$

By the principle of minimum energy, the optimal coefficients

$$\underline{\xi}^*(\overline{\boldsymbol{F}}) = \arg\min_{\underline{\xi}\in\mathbb{R}^N} \overline{W}_\xi(\overline{\boldsymbol{F}}, \underline{\xi}) \tag{22}$$

are sought after. The unconstrained and unique solvability of this task is assumed for the moment and will be addressed in Section 2.4. The solution of (22) defines the *RB approximation* of the deformation gradient

$$\boldsymbol{F}^{\text{RB}}(\boldsymbol{X};\overline{\boldsymbol{F}}) = \overline{\boldsymbol{F}} + \sum_{i=1}^{N} \boldsymbol{B}^{(i)}(\boldsymbol{X})\xi_i^*(\overline{\boldsymbol{F}}). \tag{23}$$

The microscopic energy, stress, and stiffness within the microstructure are then approximated by

$$W^{\text{RB}}(\boldsymbol{X};\overline{\boldsymbol{F}}) = W(\boldsymbol{F}^{\text{RB}}(\boldsymbol{X};\overline{\boldsymbol{F}})), \tag{24}$$

$$\boldsymbol{P}^{\text{RB}}(\boldsymbol{X};\overline{\boldsymbol{F}}) = \frac{\partial W}{\partial \boldsymbol{F}}(\boldsymbol{F}^{\text{RB}}(\boldsymbol{X};\overline{\boldsymbol{F}})), \tag{25}$$

$$\mathbb{C}^{\text{RB}}(\boldsymbol{X};\overline{\boldsymbol{F}}) = \frac{\partial^2 W}{\partial \boldsymbol{F}^2}(\boldsymbol{F}^{\text{RB}}(\boldsymbol{X};\overline{\boldsymbol{F}})), \tag{26}$$

respectively. The resulting effective energy is readily given by

$$\overline{W}^{\text{RB}}(\overline{\boldsymbol{F}}) = \left\langle W^{\text{RB}} \right\rangle. \tag{27}$$

Also, the effective responses $\overline{\boldsymbol{P}}^{\text{RB}}(\overline{\boldsymbol{F}})$ and $\overline{\mathbb{C}}^{\text{RB}}(\overline{\boldsymbol{F}})$ may now be calculated. However, before going into detail on that, it is advantageous to first elaborate on the solution of the minimization problem (22). This short survey will reveal essential properties of some occurring quantities that are important for the determination of the effective material response.

The necessary, first order optimality conditions to (22) define the components of the residual vector $\underline{r} \in \mathbb{R}^N$,

$$r_i(\overline{\boldsymbol{F}}, \underline{\xi}) = \frac{\partial \overline{W}_\xi}{\partial \xi_i}(\overline{\boldsymbol{F}}, \underline{\xi}) = \left\langle \frac{\partial W}{\partial \boldsymbol{F}}(\boldsymbol{F}_\xi) \cdot \frac{\partial \boldsymbol{F}_\xi}{\partial \xi_i} \right\rangle = \left\langle \boldsymbol{P}(\boldsymbol{F}_\xi) \cdot \boldsymbol{B}^{(i)} \right\rangle = 0 \qquad (i = 1, \ldots, N). \tag{28}$$

Note 1. *The solution stress field $\boldsymbol{P}^{\text{RB}}$ is $L^2(\Omega_0)$-orthogonal to the RB ansatz functions $\{\boldsymbol{B}^{(i)}\}_{i=1}^N$.*

A viable choice for the solution of the minimization problem (22) is the Newton–Raphson scheme, which necessitates the Jacobian $\underline{\underline{D}} \in \text{Sym}(\mathbb{R}^{N\times N})$ with the components

$$D_{ij} = \left\langle \boldsymbol{B}^{(i)} \cdot \frac{\partial \boldsymbol{P}}{\partial \xi_j}(\boldsymbol{F}_\xi) \right\rangle = \left\langle \boldsymbol{B}^{(i)} \cdot \mathbb{C}(\boldsymbol{F}_\xi) \cdot \boldsymbol{B}^{(j)} \right\rangle = D_{ji} \qquad (i, j = 1, \ldots, N). \tag{29}$$

Then, the kth iteration to the solution $\underline{\xi}^*(\overline{\boldsymbol{F}})$ reads

$$\underline{\xi}^{(k)} = \underline{\xi}^{(k-1)} - \left(\underline{\underline{D}}^{(k-1)}\right)^{-1} \underline{r}^{(k-1)} \qquad (k \geq 1). \tag{30}$$

The initial guess $\underline{\xi}^{(0)}$ can be zero or a more sophisticated alternative.

The deduction of the effective material response by means of Note 1 and the Jacobian $\underline{\underline{D}}$ is given in Appendix B. The following list summarizes the homogenized quantities arising from the $\boldsymbol{F}$-RB:

$$\overline{\boldsymbol{F}} = \left\langle \boldsymbol{F}^{\mathrm{RB}} \right\rangle \qquad \text{see (19)}$$
$$\overline{W}^{\mathrm{RB}} = \left\langle W^{\mathrm{RB}} \right\rangle \qquad \text{see (27)}$$
$$\overline{\boldsymbol{P}}^{\mathrm{RB}} = \left\langle \boldsymbol{P}^{\mathrm{RB}} \right\rangle \qquad \text{see Appendix B.1}$$
$$\overline{\mathbb{C}}^{\mathrm{RB}} = \left\langle \mathbb{C}^{\mathrm{RB}} \right\rangle - \sum_{i,j=1}^{N} D_{ij}^{-1} \left\langle \mathbb{C}^{\mathrm{RB}} \cdot \boldsymbol{B}^{(i)} \right\rangle \otimes \left\langle \boldsymbol{B}^{(j)} \cdot \mathbb{C}^{\mathrm{RB}} \right\rangle \qquad \text{see Appendix B.2}$$

In total, the problem of determining both the local field $\boldsymbol{F}$ and the homogenized material properties depends only on N degrees of freedom, namely on the N coefficients ξ_i. Usually, the number of DOF N is in the range of 10 to 150, which compares to the full order model's complexity that can easily reach more than 10^5 DOF.

Remark 1. *Despite this impressive reduction of the number of DOF, the computational costs associated with the assembly of the residual $\underline{r}$ and of the Jacobian $\underline{\underline{D}}$ still relate to the number of quadrature points of the microstructural discretization.*

This method extends corresponding methods for the geometrically linear case, where the infinitesimal strain tensor $\boldsymbol{\varepsilon} = \mathrm{sym}(\boldsymbol{H})$ is considered. For more information on these topics, the reader is referred to the authors' previous work [12] or standard literature, such as ([25], part II.C).

2.2. Identification of the Reduced Basis

The basis $\{\boldsymbol{B}^{(i)}\}_{i=1}^{N}$ is computed by means of a classical snapshot POD. In contrast to many other POD based reduction methods, it is important to point out that here, the primal variable is *not* taken to be the displacement field , $\boldsymbol{u}$. Instead, the POD is performed on deformation gradient fluctuations, $\widetilde{\boldsymbol{F}}$.

During the snapshot POD, *snapshots* are first created by means of high-fidelity solutions to the nonlinear microscopic problem (8) with different *snapshot boundary conditions* $\overline{\boldsymbol{F}}^{(j)}$, $j = 1, \ldots, N_{\mathrm{s}}$, which are also called *training boundary conditions*. Each of these boundary conditions leads to a solution field $\boldsymbol{F}^{(j)}(\boldsymbol{X}; \overline{\boldsymbol{F}})$. Typically, the Finite Element Method (FEM) or solvers making use of the Fast Fourier Transform (e.g., [26]) are used to this end. The RB method presented here is independent of the discretization method utilized to obtain full field solutions. It is merely necessary to know the quadrature weights and the related discrete values of the solutions $\boldsymbol{F}^{(j)}(\boldsymbol{X}; \overline{\boldsymbol{F}}^{(j)})$. For better readability, the continuous notation is maintained for the moment. The corresponding fluctuation fields are computed by means of local subtraction of the macroscopic deformation gradient

$$\widetilde{\boldsymbol{F}}^{(j)}(\boldsymbol{X}; \overline{\boldsymbol{F}}^{(j)}) = \boldsymbol{F}^{(j)}(\boldsymbol{X}; \overline{\boldsymbol{F}}^{(j)}) - \overline{\boldsymbol{F}}^{(j)} \qquad (j = 1, \ldots, N_{\mathrm{s}}). \tag{31}$$

Each of these N_{s} fluctuation fields $\widetilde{\boldsymbol{F}}^{(j)}$ represent one snapshot.

Second, the most dominant features of the snapshots are extracted. This is done by means of the eigendecomposition of the correlation matrix. It consists of the mutual $L^2(\Omega_0)$ scalar products of the snapshots, $\langle \widetilde{\boldsymbol{F}}^{(i)} \cdot \widetilde{\boldsymbol{F}}^{(j)} \rangle$ $(i, j = 1, \ldots, N_{\mathrm{s}})$, cf. (1). The remaining procedure is standard, see for instance [7] or [27]: the N_{s} eigenvalues λ_j, corresponding to the eigenvectors $\underline{E}^{(j)} \in \mathbb{R}^{N_{\mathrm{s}}}$, are sorted in descending order and truncated by only considering the first N values, $\lambda_1 \geq \ldots \geq \lambda_N$. The decision on a particular threshold index N is based on consideration of the Schmidt–Eckhard–Young–Mirsky theorem. Finally, the RB is constructed as

$$B^{(i)}(X) = \sum_{j=1}^{N_s} \frac{1}{\sqrt{\lambda_i}} \left(\underline{E}^{(i)}\right)_j \widetilde{F}^{(j)}(X) \quad (i = 1, \ldots, N) \tag{32}$$

where the factor $1/\sqrt{\lambda_i}$ accounts for $L^2(\Omega_0)$ normalization of the base elements.We conclude that the RB is a collection of $L^2(\Omega_0)$ orthonormal $\widetilde{H}$-like quantities.

2.3. Mathematical Motivation of the Reduced Basis Model

Next, the obtained deformation gradient field F^{RB} and the associated stress field P^{RB} are shown to weakly solve the original problem (8), $\mathrm{Div}_X(P) = 0$.

Let $\delta w \in H_0^1(\Omega_0)$ be an admissible test function, i.e., a once weakly differentiable periodic displacement fluctuation field, and let $\delta H = \delta w \otimes \nabla_X$ denote its gradient. Then, the well-known weak form of (8) is equivalent to the principle of virtual work,

$$\delta \overline{W} = \int_{\Omega_0} P \cdot \delta H \, \mathrm{d}V = 0. \tag{33}$$

The residual $\underline{r}$ from (28) coincides with the integral of the weak form, if the test function δw is chosen suitably. As the basis elements $B^{(i)}$ are linear combinations of deformation gradient fluctuations $w^{(j)} \otimes \nabla_X$, cf. (32), the equivalence of (28) and (33) is obvious.

It follows that the Reduced Basis scheme is a *Galerkin* procedure, taking the displacement fields corresponding to the RB elements $B^{(i)}$ as both ansatz and test functions. Hence, the *RB model is equivalent to the FEM*, but with basis functions that are globally supported in $\Omega_0 \backslash \partial\Omega_0$. In other words, the basis functions of the RB method span a subspace of dimension N within the high-dimensional space of FE basis functions. Although this property is shared with RB schemes based on POD of displacement snapshots, a notable difference is that this novel approach directly operates on fields entering the constitutive equations.

2.4. Details on the Coefficient Optimization

The coefficient optimization task (22) leads to a weak solution of the microscopic boundary value problem, as was just shown. Hence, the well-established theories on which the FEM is built, e.g., the calculus of variations, are applicable to the presented method just as much. This implies that the well-known issues with suitable convexity conditions and with existence and uniqueness of minimizers apply to the RB method, too. We focus on ad hoc numerical treatments of these issues. For more details on the theoretical part, the reader is referred to standard literature, such as [28], and recent developments in this matter, e.g., [29].

A constraint to the optimization problem is the physical condition

$$J = \det(F_{\underline{\xi}}(X)) > 0 \quad \forall X \in \Omega_0, \tag{34}$$

which means that no singular ($J = 0$) or self-penetrating ($J < 0$) deformations shall occur. This reduces the set of admissible coefficients $\underline{\xi}$ to a subset of $\mathbb{R}^N$ that is nontrivial to access. The positiveness of the microscopic determinant of the deformation gradient introduces a constraint to the, thus far, unconstrained minimum problem (22), representing the weak form of (8) in the RB setting.

In case of a violation of the inequality (34), the implementation of the RB method is prone to failure as soon as the constitutive law (4) is evaluated. This occurs only in the context of volume averaging, i.e., for the assembly of the residual, the Jacobian, or the effective energy, stress, or stiffness. The numerical quadrature approximating the volume averaging operation is

$$\langle \bullet \rangle = \frac{1}{|\Omega_0|} \int_{\Omega_0} \bullet(\boldsymbol{X}) \, \mathrm{d}V \approx \frac{1}{|\Omega_0|} \sum_{p=1}^{N_{\mathrm{qp}}} \bullet(\boldsymbol{X}_p) \, w_p. \tag{35}$$

Here, N_{qp}, $\boldsymbol{X}_p$, and w_p respectively denote the number of quadrature points, their positions, and the corresponding positive weights. Moreover, even if the inequality (34) is satisfied everywhere, the local field $\boldsymbol{F}_{\xi}$ might possibly have some positive but overly small values of the determinant, $0 < \det(\boldsymbol{F}_{\xi}(\boldsymbol{X})) \ll 1$, that are unphysical. In such a case, the energy optimization, cf. (22), would be dominated by these nearly singular points. Instead of allowing the optimization to focus almost exclusively on these exceptional points, we interpret unphysically small values of the determinant as limitations to the reliability of the RB method. On the other hand, large values $\det(\boldsymbol{F}_{\xi}(\boldsymbol{X})) \gg 1$ are not too detrimental to the functionality of the scheme, although such values are just as questionable.

Thus, the following *weighted numerical quadrature rule* is introduced:

$$\langle \bullet \rangle \approx \left(\sum_{p=1}^{N_{\mathrm{qp}}} \bullet(\boldsymbol{X}_p) \, \phi(J_p) \, w_p \right) \Big/ \left(\sum_{p=1}^{N_{\mathrm{qp}}} \phi(J_p) w_p \right). \tag{36}$$

Therein, the almost smooth *cutoff function* $\phi : \mathbb{R} \to \mathbb{R}_{\geq 0}$ is empirically defined by

$$\phi(J) = \begin{cases} 1 & \text{if } J > 0.6 \\ 0.5 \, \mathrm{erf}(30\,J - 15) + 0.5 & \text{if } 0.4 < J \leq 0.6 \\ 0 & \text{if } J \leq 0.4 \end{cases} . \tag{37}$$

which makes use of the well-known error function. The cutoff function is depicted in Figure 2. This reliability indicator could, in principle, be modified, e.g., the steepness, the smoothness, and its center could be adjusted. Thus, it should be regarded as an example only.

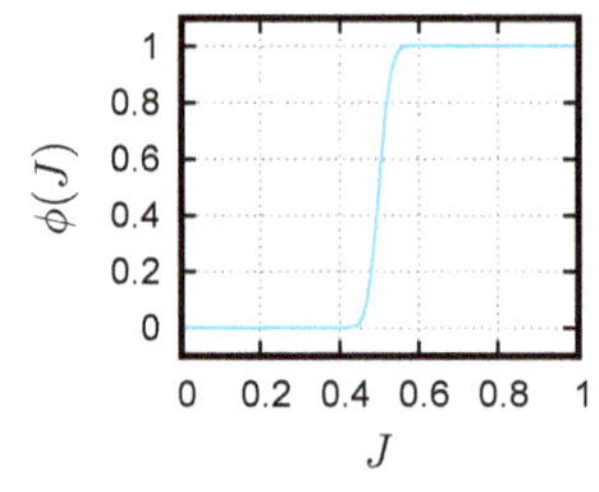

Figure 2. Cutoff function ϕ. Its value is used as a reliability factor in the numerical quadrature.

This weighted numerical quadrature rule is used henceforward for all *numerical* volume averaging operations. Its application will not be noted explicitly. However, the *theoretical* derivation of the RB method, as described in Section 2.1, is not affected by this, i.e., volume averaging operations remain exact as far as the theory is concerned. The two most important consequences of this numerical tweak are:

- The RB method is robust with respect to outlier values of the determinant. The modified quadrature rule extends the set of coefficient vectors ξ for which effective quantities can be computed, albeit approximately, to the whole space $\mathbb{R}^N$.
- The significance of local fields varies with the value of the cutoff function. When ϕ attains values less than one, information is considered accordingly less reliable. In this sense, microscopic information is filtered based on a trust region for J defined by ϕ can be seen as a reliability indicator.

3. Sampling

3.1. General Considerations

The proposed sampling strategy builds on the previous work [12], in which the authors proposed an analogous sampling procedure for the small strain setting. However, substantial modifications are required in order to account for the finite rotational part, $\overline{\boldsymbol{R}}$, of the macroscopic deformation gradient, $\overline{\boldsymbol{F}}$, and the nonlinearity of the volumetric part of the deformation, J, with respect to the local displacements, $\boldsymbol{u}$. For the setup of the Reduced Basis model as described in Section 2.2, the space of macroscopic deformation gradients,

$$\overline{\mathcal{F}} = \{\text{second order tensors } \overline{\boldsymbol{F}} \mid \det \overline{\boldsymbol{F}} > 0\}, \tag{38}$$

needs to be sampled, i.e., the discrete sampling set $\overline{\mathcal{F}}_{\mathrm{s}} = \{\overline{\boldsymbol{F}}^{(m)}\}_{m=1}^{N_{\mathrm{s}}} \subset \overline{\mathcal{F}}$ is to be defined. Two contradictory requirements need to be satisfied when constructing $\overline{\mathcal{F}}_{\mathrm{s}}$:

1. The samples should be *densely and homogeneously distributed* within the space of all admissible macroscopic kinematic configurations. This is owing to the desire that the POD may extract correlation information from a holistic and unbiased set. In other words, the samples should be as *uniformly random* as possible *within the anticipated query domain of the surrogate.*
2. The sample number N_{s} should not exceed a certain limit. Only with this property may the RB be identified within the bounds of available computational resources (e.g., memory and CPU time).

3.2. Large Strain Sampling Strategy

The set of admissible macroscopic deformation gradients $\overline{\mathcal{F}}$ is a subset of a nine-dimensional space ($\underline{\underline{F}} \in \mathbb{R}^{3\times 3} \sim \mathbb{R}^9$) restricted by the inequality

$$\det \overline{\boldsymbol{F}} > 0. \tag{39}$$

Therefore, a regular grid in the components of $\overline{\boldsymbol{F}}$ might lead to a prohibitively large amount of samples, and even to a violation of (39). For instance, such a grid with a rather moderate resolution of just 10 samples of each component would require 1 billion solutions of the FOM. Also, the subsequent POD would involve a snapshot matrix and/or correlation matrix of accordingly huge dimensionality.

In order to decrease the dimension of the sampling space, recall the polar decomposition of the deformation gradient, $\overline{\boldsymbol{F}} = \overline{\boldsymbol{R}}\,\overline{\boldsymbol{U}}$, where $\overline{\boldsymbol{R}}$ is a rotation and $\overline{\boldsymbol{U}}$ is the symmetric positive definite (s.p.d.) stretch tensor. Material objectivity implies the energy function to be independent of the frame of reference,

$$\overline{W}(\overline{\boldsymbol{R}}\,\overline{\boldsymbol{U}}) = \overline{W}(\overline{\boldsymbol{U}}), \tag{40}$$

and the transformation behavior

$$\overline{\boldsymbol{P}}(\overline{\boldsymbol{R}}\,\overline{\boldsymbol{U}}) = \overline{\boldsymbol{R}}\,\overline{\boldsymbol{P}}(\overline{\boldsymbol{U}}) \tag{41}$$

for the first Piola–Kirchhoff stress and

$$\overline{C}_{ijkl}(\overline{\boldsymbol{R}}\,\overline{\boldsymbol{U}}) = \sum_{m,n=1}^{3} \delta_{ik}\overline{U}_{lm}^{-1}\overline{P}_{mj}(\overline{\boldsymbol{U}}) + \overline{R}_{im}\overline{C}_{mjnl}(\overline{\boldsymbol{U}})\overline{R}_{kn} \qquad (i,j,k,l = 1,2,3) \tag{42}$$

for the components of the corresponding stiffness tensor $\mathbb{C}$, see Appendix A. These known facts lead to

Note 2. *In order to collect representative samples of the hyperelastic response functions* $\overline{W}$, $\overline{\mathbf{P}}$, *and* $\overline{\mathbb{C}}$, *it suffices to evaluate them on samples of the stretch tensor* $\overline{\boldsymbol{U}} \sim \mathbb{R}^6$ *instead of evaluating them on samples of the deformation gradient* $\overline{\mathbf{F}} \sim \mathbb{R}^9$.

This effectively reduces the number of dimensions of $\overline{\mathcal{F}}$ from nine to six. The same dimensionality is attained when considering the response functions with respect to a symmetric measure of strain, e.g., as is done in [30] where the tangent stiffness is efficiently computed using a perturbation technique. However, such measures are unsuitable for reduction by means of the proposed RB method.

The remaining six-dimensional space of s.p.d. tensors is not linear but a convex cone (which does not include the zero element). Moreover, linearly combining elements within this space quickly leads to values of $\overline{J} = \det \overline{\boldsymbol{U}}$ describing unphysically large changes in volume. For instance, $\overline{\boldsymbol{U}} = 1.3\,\boldsymbol{I}$ equates to more than 100% volumetric extension, which is well beyond the regime of usual hyperelastic materials that are often nearly incompressible. On the other hand, 100% deviatoric strain is within the range of many standard materials, such as rubber. Hence, *in order to describe the space of practically relevant stretch tensors*, we propose to apply the DDMS to the macroscopic stretch tensor,

$$\overline{\boldsymbol{U}} = \overline{J}^{1/3}\,\widehat{\overline{\boldsymbol{U}}}. \tag{43}$$

Let $\widehat{\overline{\mathcal{U}}}$ denote the manifold of unimodular macroscopic stretch tensors $\widehat{\overline{\boldsymbol{U}}} = (\overline{J})^{-1/3}\overline{\boldsymbol{U}}$. The multiplicative split (43) is the basis for:

Proposition 1. *The set of practically relevant macroscopic stretch tensors* $\overline{\mathcal{U}}$ *can be sampled via sampling of both the macroscopic determinants,*

$$\left\{\overline{J}^{(m)}\right\}_{m=1}^{N_{\mathrm{dil}}} \subset \mathbb{R}_+,$$

and the macroscopic unimodular stretch tensors,

$$\left\{\widehat{\overline{\boldsymbol{U}}}^{(j)}\right\}_{j=1}^{N_{\mathrm{dev}}} \subset \widehat{\overline{\mathcal{U}}},$$

where N_{dil} *and* N_{dev} *are the numbers of the samples. The sampling set is determined by the product set*

$$\left\{\left(\overline{J}^{(m)}\right)^{1/3}\widehat{\overline{\boldsymbol{U}}}^{(j)}\right\}_{m,j=1}^{m=N_{\mathrm{dil}},\,j=N_{\mathrm{dev}}} \subset \overline{\mathcal{U}}. \tag{44}$$

The choice of the dilatational samples is relatively straightforward. For many common materials, the expected range of $\overline{J}$ is rather narrow, so a few equisized or adaptive sub-intervals around $\overline{J} = 1$ deliver sufficient resolution.

For the space of unimodular s.p.d. matrices representing $\widehat{\overline{\boldsymbol{U}}} \in \widehat{\overline{\mathcal{U}}}$, basic results of Lie group theory can be exploited. We restrict to stating well-known facts that are necessary at this point. For more details, the interested reader is referred to standard text books, such as [31].

Corollary 1. *Let*

$$\mathrm{symsl} = \left\{\underline{\underline{Y}} \in \mathbb{R}^{3\times 3} \,|\, \underline{\underline{Y}} = \underline{\underline{Y}}^{\mathsf{T}}, \mathrm{tr}(\underline{\underline{Y}}) = 0\right\}$$

be the tangent space *and*

$$\mathrm{SymSL}_+ = \left\{\underline{\underline{U}} \in \mathbb{R}^{3\times 3} \,|\, \underline{\underline{U}} = \underline{\underline{U}}^{\mathsf{T}}, \det \underline{\underline{U}} = 1, \underline{x}^{\mathsf{T}}\,\underline{\underline{U}}\,\underline{x} > 0\,\forall \underline{x} \in \mathbb{R}^3\right\}$$

be the manifold *of unimodular s.p.d. matrices. The matrix exponential maps the tangent space bijectively onto the manifold,*

$$\exp: \quad \mathrm{symsl} \to \mathrm{SymSL}_+ .$$

The proof of this statement is standard, e.g., by means of the eigenvalue decomposition, and does not necessitate the reference to the abstract setting of Lie groups. In fact, all of the following arguments could be given without the notion of tangent spaces and manifolds. However, this notion is a fundamental concept in nonlinear mechanics. For a very descriptive and comprehensive work on this topic, the reader is referred to [32]. We choose to build upon this concept, as it comes along with vivid interpretations of the function spaces $\overline{\mathcal{U}}$ and $\hat{\overline{\mathcal{U}}}$.

The set SymSL_+ is the set of matrix representations of the stretch tensors $\hat{\overline{\mathbb{U}}} \in \hat{\overline{\mathcal{U}}}$. The tangent space symsl is the set of *Hencky strains*, which is linear. Hence, by virtue of the matrix exponential, the sampling of the nonlinear manifold $\hat{\overline{\mathcal{U}}}$ can be reduced to the sampling of a linear space. Moreover, the restrictions of symmetry and zero trace render the tangent space five-dimensional. This property is, by definition, shared with the manifold SymSL_+.

The two-dimensional case is now addressed for the sake of visualization. In this setting, the nonlinearity of the manifold and the structure of the sampling set $\overline{\mathcal{U}}$ can be illustrated figuratively. With the subscript (2) denoting two-dimensional quantities, a basis of the tangent space is given by

$$\underline{\underline{Y}}^{(1)}_{(2)} = \sqrt{\frac{1}{2}}\begin{bmatrix} 1 & 0 \\ 0 & -1 \end{bmatrix} \quad \text{and} \quad \underline{\underline{Y}}^{(2)}_{(2)} = \sqrt{\frac{1}{2}}\begin{bmatrix} 0 & 1 \\ 1 & 0 \end{bmatrix}.$$

The stretch tensors are obtained through

$$\underline{\underline{\overline{U}}}_{(2)} = \overline{J}^{\frac{1}{2}} \exp\left(t\left(\alpha \underline{\underline{Y}}^{(1)}_{(2)} + \beta \underline{\underline{Y}}^{(2)}_{(2)}\right)\right) = \begin{bmatrix} a & b \\ b & d \end{bmatrix}.$$

A visualization of such samples is depicted in Figure 3. There, for the sake of visual clarity, the determinant $\overline{J}$ is sampled by four equidistant (and rather unrealistic) values between 0.1 and 4. The value $t \in [-2, 2]$ is called *deviatoric amplitude*. A densely uniform sampling $\varphi \in [0, 2\pi)$ yields the coefficients $\alpha = \cos\varphi$ and $\beta = \sin\varphi$.

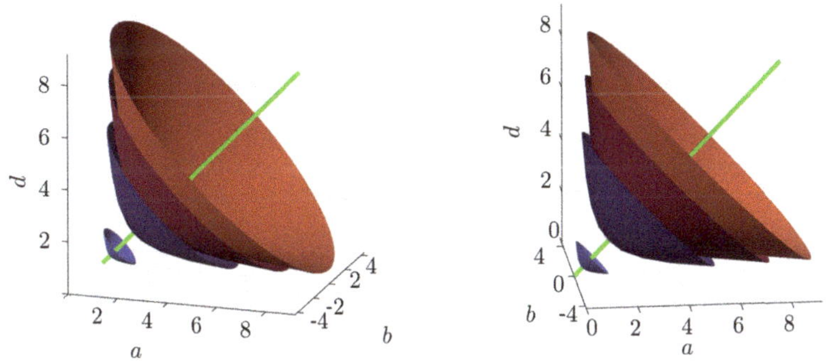

Figure 3. Visualizations of the family of $\overline{\mathbb{U}}$-manifolds with constant determinant $\overline{J} \in \{0.1, \ldots, 4.0\}$ for the two-dimensional case from two different perspectives. The green line represents the set $\{\lambda \boldsymbol{I} \,|\, \lambda > 0\}$.

This emphasizes the important role of the DDMS in the context of sampling, as utilized in (44)—it allows for the identification of a physically meaningful sampling domain that is much smaller than the surrounding space of all admissible stretch tensors. On a side note, the isodet surfaces are perpendicular to the line representing the dilatational stretch tensors.

The proposed sampling procedure for $\overline{\mathcal{U}}$ in three dimensions is given in Algorithm 1. For this purpose, an orthonormal basis $\underline{\underline{Y}}^{(1)}, \ldots, \underline{\underline{Y}}^{(5)}$ of the tangent space symsl is fixed, cf. Appendix C. The numbers of different kind of samples N_{det}, N_{dir}, and N_{amp} relate to the quantities N_{dil} and N_{dev} introduced in (44) by $N_{\text{det}} = N_{\text{dil}}$ and $N_{\text{dir}} N_{\text{amp}} = N_{\text{dev}}$.

Algorithm 1: Sampling of the macroscopic stretch tensor.

Input : $\overline{J}_{\min}, \overline{J}_{\max}$ minimum and maximum determinant with $\overline{J}_{\min} < 1 < \overline{J}_{\max}$
$t_{\max} > 0$ maximum deviatoric amplitude
N_{det} number of macroscopic determinants
N_{dir} number of deviatoric directions
N_{amp} number of deviatoric amplitudes

Output: $N_{\text{det}} N_{\text{dir}} N_{\text{amp}}$ samples of $\overline{\boldsymbol{U}}$

1 Place N_{det} determinants regularly between the extremal values,

$$\overline{J}_{\min} = \overline{J}^{(1)} < \ldots < 1 < \ldots < \overline{J}^{(N_{\text{det}})} = \overline{J}_{\max} .$$

2 Generate any approximately uniform distribution of N_{dir} directions in $\mathbb{R}^5$, e.g., as in [12],

$$\left\{ \underline{N}^{(n)} \right\}_{n=1}^{N_{\text{dir}}} \subset \left\{ \underline{N} \in \mathbb{R}^5 \, : \, \|\underline{N}\| = 1 \right\} .$$

3 Place N_{amp} deviatoric amplitudes regularly between 0 and the expected maximum value,

$$0 \leq t^{(1)} < \ldots < t^{(N_{\text{amp}})} = t_{\max} .$$

4 Return the set of samples of $\overline{\boldsymbol{U}}$:

$$\left\{ \left(\overline{J}^{(m)} \right)^{1/3} \exp \left(t^{(p)} \left[\sum_{k=1}^{5} \left(\underline{N}^{(n)} \right)_k \underline{\underline{Y}}^{(k)} \right] \right) \right\}_{m,n,p=1}^{m=N_{\text{det}}, n=N_{\text{dir}}, p=N_{\text{amp}}} \subset \overline{\mathcal{U}} \tag{45}$$

The order of Steps 1 to 3 is interchangeable. Details on these parts are now presented:

Step 1. Uniform seeding of the determinants is actually not required, but any pattern implying the sampling determinants $\{\overline{J}^{(k)}\}_{k=1}^{N_{\text{det}}}$ to be dense in $[\overline{J}_{\min}, \overline{J}_{\max}]$ as $N_{\text{det}} \to \infty$ works without loss of generality. In this way, the dilatational response may be resolved adaptively.

Step 2. The generation of uniform point distributions on spheres is a research topic on its own, see [33] for an overview. The method described in [12] is based on energy minimization, which is also used in the present work. Some point sets of various sizes are included in the example program [18]. More detailed investigations on this topic and an open-source code of a point generation program are part of another work, [34]. Alternatively, Equal Area Points [35] may be used as a rough but quickly computable approximation of such point sets.

Step 3. As in Step 1, the uniform placement of the deviatoric amplitudes, $t^{(p)}$, may be substituted by adaptive alternatives. In [12], we have suggested to use an exponential distance function.

The result of Steps 2 and 3, i.e., the sampling of the tangent space, is exemplified in Figure 4 for the two-dimensional case ($\boldsymbol{u} \in \mathbb{R}^2$) and for $N_{\text{dir}} = 5$ and $N_{\text{amp}} = 3$. There, an adaptive spacing of the deviatoric amplitudes is applied. This might be beneficial for capturing strongly changing material behavior near the relaxed state.

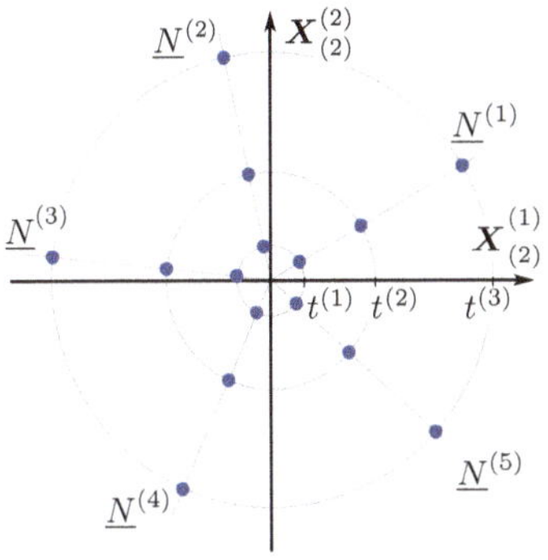

Figure 4. Example of a discretization of the two-dimensional tangent space. The samples are placed along the equidistant (in higher dimensions—uniformly distributed) directions with nonuniformly increasing amplitude.

In general, the vector $\underline{N} \in \mathbb{R}^5$, $\|\underline{N}\| = 1$, corresponding to a macroscopic Hencky strain characterizes the *direction of the applied stretch* $\overline{U}$, which we also coin the *load case*.

3.3. Application of the Stretch Tensor Trained Reduced Basis Model

Since the RB model is trained on only the rotationally invariant part of $\overline{F}$ but should be applied to general deformation gradients, the transformation rules (41) and (42) are incorporated during the evaluation of the surrogate model. Details on the procedure are given in Algorithm 2.

Algorithm 2: Online phase of the stretch tensor trained Reduced Basis method

Input : $\overline{F}$ macroscopic deformation gradient
Output: $\overline{P}^{\mathrm{RB}}(\overline{F})$, $\overline{\mathbb{C}}^{\mathrm{RB}}(\overline{F})$ effective material response
1 Compute polar decomposition $\overline{F} = \overline{R}\,\overline{U}$.
2 Compute approximations of effective stress $\overline{P}^{\mathrm{RB}}(\overline{U})$ and effective stiffness $\overline{\mathbb{C}}^{\mathrm{RB}}(\overline{U})$.
3 Transform effective responses to correct frame $\overline{P}^{\mathrm{RB}}(\overline{F})$, $\overline{\mathbb{C}}^{\mathrm{RB}}(\overline{F})$, using $\overline{R}$, cf. (41), (42).

4. Numerical Examples

4.1. Reduced Basis for a Fibrous Microstructure

The applicability of the proposed RB method in combination with the sampling scheme is now numerically studied for a fibrous microstructure roughly resembling polymers with woven reinforcements. The goal is to prove the efficiency of the F-RB scheme in principle and under "worst-case" conditions, the latter meaning that the phase contrast is chosen to be rather large. Yet, at this stage, it is explicitly not aspired to provide fully realistic examples. These would require investigations on the proper size of the microstructure and should employ dissipative material laws, both of which are not within the scope of this work.

A cubical microstructure with two fibrous inclusions is considered, see Figure 5a and cf. [36] for a related example. The inclusions are periodic and occupy approximately 0.7% of the volume. The mesh contains 35,516 nodes in 25,633 quadratic tetrahedron elements (C3D10).

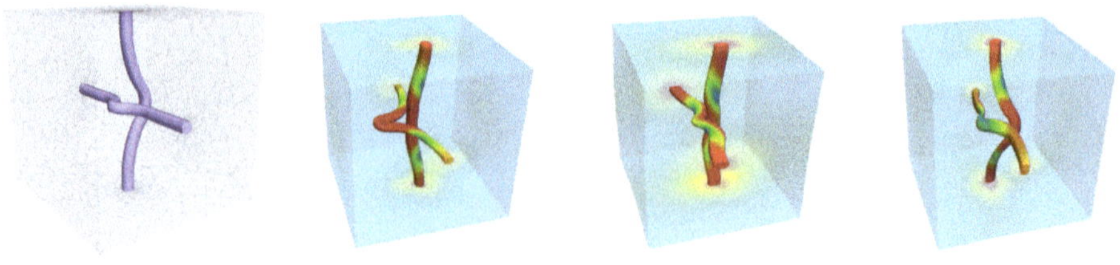

(a) (b) (c) (d)

Figure 5. (**a**) fibrous microstructure. (**b–d**) random $\boldsymbol{F}$-reduced basis (RB) elements.

For the matrix, the material constants are chosen to be $K_m = 400$ MPa and $G_m = 0.4$ MPa, resembling rubber-like material properties. For the fibers, the values are $K_f = 800$ MPa and $G_f = 240$ MPa. The latter parameters approximate the behavior of stiffer polymers, such as polyethylene. The phase contrast is 600 with respect to the shear moduli, and the Poisson ratios are $\nu_m = 0.4995$ and $\nu_f = 0.3636$.

The training boundary conditions are defined with $N_{dir} = 128$ deviatoric directions, $\underline{N}^{(n)}$, and $N_{amp} = 10$ deviatoric amplitudes, $t^{(p)}$, which are regularly distributed in the interval $[0.05, 0.5]$. In order to also consider response to volumetric extension in the training data, an additional set of N_{det}=10 boundary conditions of the form $\left(\overline{J}^{(m)}\right)^{1/3}\boldsymbol{I}$ is included in the training set, with the determinant $\overline{J}^{(m)}$ being linearly increased from 1 to 1.02.

The validation load cases are 640 mixed dilatational-deviatoric boundary conditions. Along $N_{dir} = 64$ new deviatoric directions, both the deviatoric amplitudes ($t^{(p)} = 0.05, \ldots, 0.5$) and the dilatational amplitudes ($\overline{J}^{(m)} = 0.9995, \ldots, 0.995$) are applied in 10 equidistant increments.

The results for various values N of the RB-size are compared with the results of FE simulations with the same boundary conditions. To this end, the error measures

$$\mathrm{err}_W = \frac{\|\overline{W}^{RB} - \overline{W}^{FEM}\|}{\|\overline{W}^{FEM}\|} \qquad \text{and} \qquad \mathrm{err}_P = \frac{\|\overline{\boldsymbol{P}}^{RB} - \overline{\boldsymbol{P}}^{FEM}\|}{\|\overline{\boldsymbol{P}}^{FEM}\|} \tag{46}$$

are employed. Figures 6 and 7 visualize the results.

The distribution of the energy error, err_W, improves monotonically as the RB is enriched from $N = 8$ to $N = 128$ elements. This enrichment corresponds to the inclusion of additional subtrahends in the computation of $\overline{\mathbb{C}}^{RB}$, improving the spectral over-estimation by the volume average of the stiffness, cf. (17). It is also noteworthy that the error tends to be higher for larger magnitudes of the applied kinematic boundary condition, although that is not always the case.

In contrast to this, the stress error err_P distribution monotonically improves only up to a certain threshold value of the number of basis elements, which is $N = 64$ in this example. For the greater bases with $N = 96$ and $N = 128$, the quality of the results deteriorates as far as the stress error is concerned. This is most likely due to an excessively oscillatory nature of the higher order modes—at some critical level $1 \ll i = N_{crit} < N$, the POD constructs eigenvectors $\underline{E}^{(i)}$ with the $L^2(\Omega_0)$-norm $\sqrt{\lambda_i} \ll 1$. Therefore, the POD would construct basis vectors out of *numerical fluctuations*, which would be unphysical. While the enrichment of the optimization space with unphysical information cannot increase the minimum energy error err_W, it might lead to fluctuations in the displacement field that have significant impact on the overall stress response. This is especially the case for numerical fluctuations within the stiff inclusion phase where low overall displacement errors still could lead to notable impact on the effective stress.

Nonetheless, all observed errors are less than 20% and stay below 3% for the optimal sampled size $N = 64$. For half the basis size, $N = 32$, the errors max at approximately 5%, which is still acceptable considering the uncertainties involved in realistic two-scale simulations. Note that the maximum errors strongly depend on the maximum load amplitude, which is chosen to be very large in this example (50% deviatoric strain and 0.5% compression).

The runtimes of the RB model for different sizes N are depicted in Figure 8. A nearly linear growth of the runtime with respect to the basis size can be asserted. It is noteworthy that the online time of the RB method is strongly dominated by the assembly of the Jacobian $\underline{\underline{D}}$. Therefore, a *Quasi-Newton implementation* was chosen, resulting in only two assemblies per load increment.

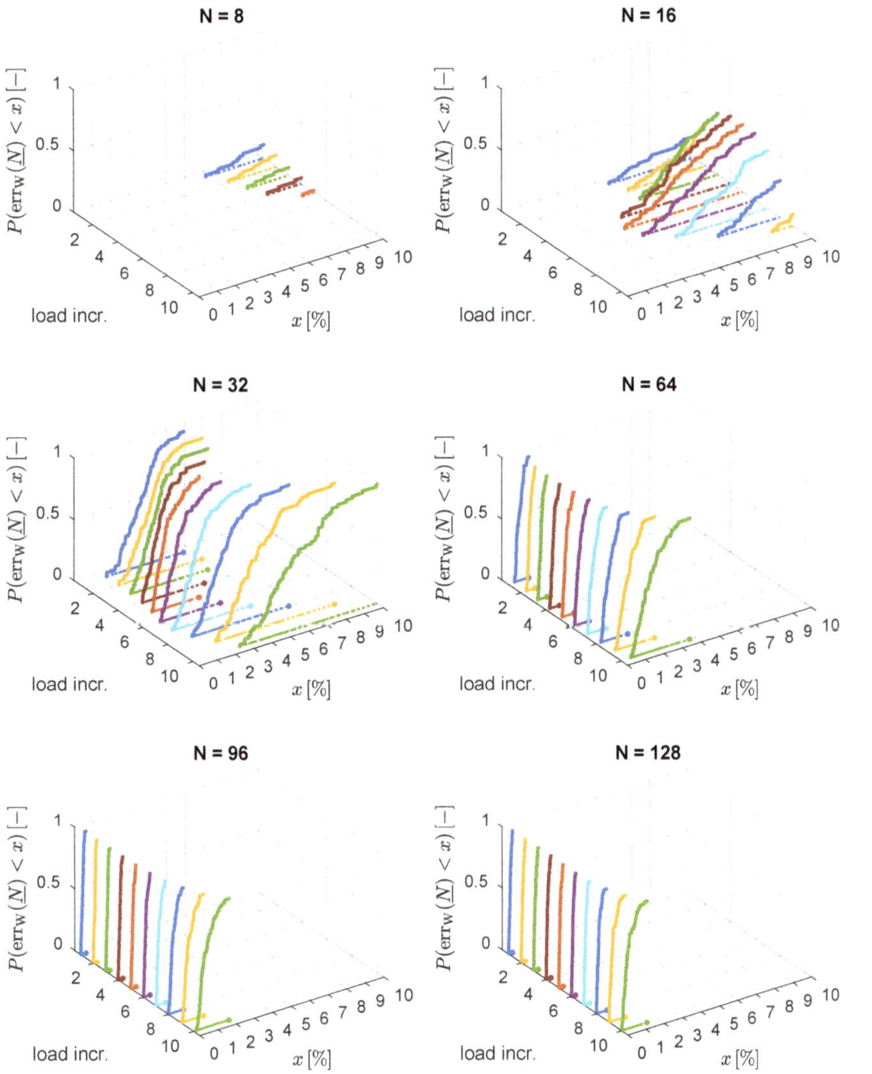

Figure 6. Cumulative energy error distribution per direction for the *RB of the fibrous microstructure* under validation boundary conditions.

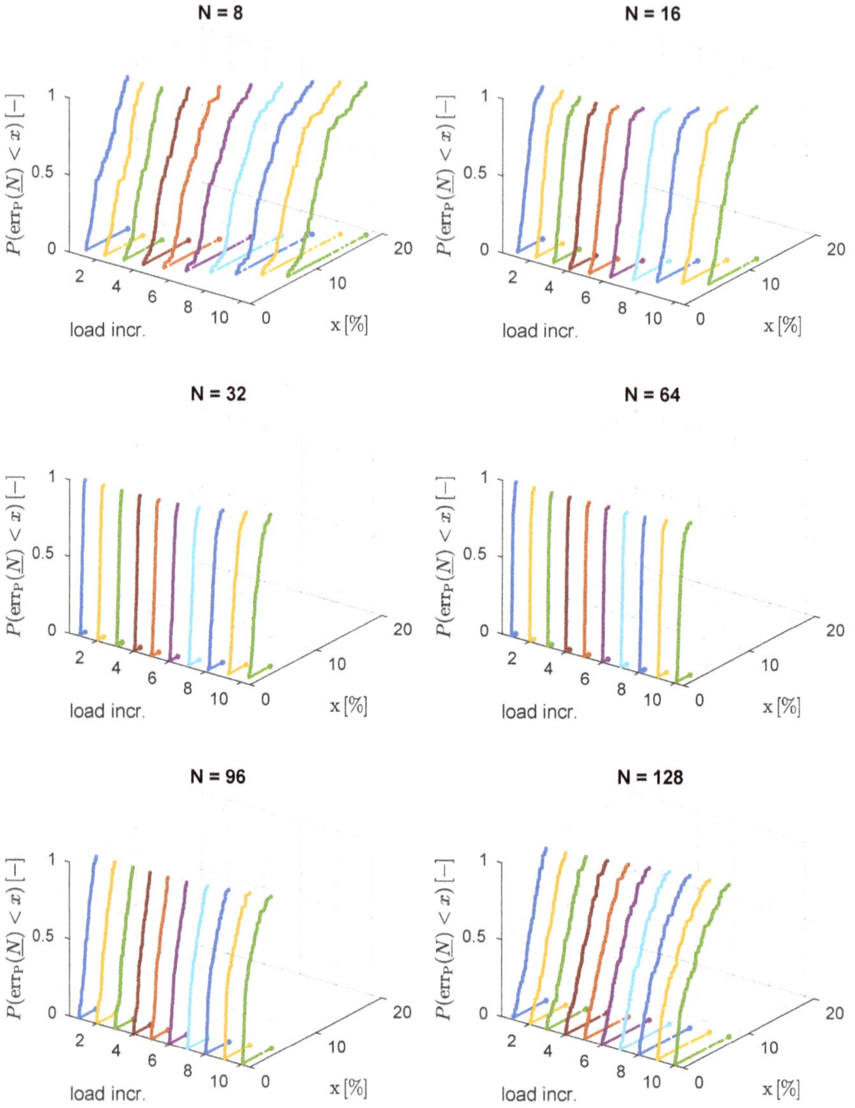

Figure 7. Cumulative stress error distribution per direction for the *RB of the fibrous microstructure* under validation boundary conditions.

Speed-ups become impressive when very large load increments are considered. In all examples observed thus far, *the RB converges to the final load amplitude in a single increment, requiring 7–13 Quasi–Newton iterations, with only 2–4 assemblies of the Jacobi matrix* $\underline{\underline{D}}$ *and a runtime of 10–50 seconds.*This is in strong contrast to the FEM which is very sensitive to large load increments as they come along with a high probability of a violation of the condition $\det(\boldsymbol{F}) > 0$. By means of standard implementation, such occurrences are usually treated with cutbacks of the load increment, which is detrimental to the runtime of the FEM.

No rigorous speed-up analysis is intended at this point. Both the codes of the FEM and of the RB method are fairly efficient in-house C/C++ developments and perform exact line searches. While the FEM has not yet been equipped with a Quasi–Newton procedure, the linear solver makes use of parallelization. This is in contrary to the current implementation of the RB method. Depending on the microstructure (especially the geometry, material nonlinearities, and phase contrast), the loading conditions, and the size N of the RB, *speed-up factors are in the order of 5–100.*

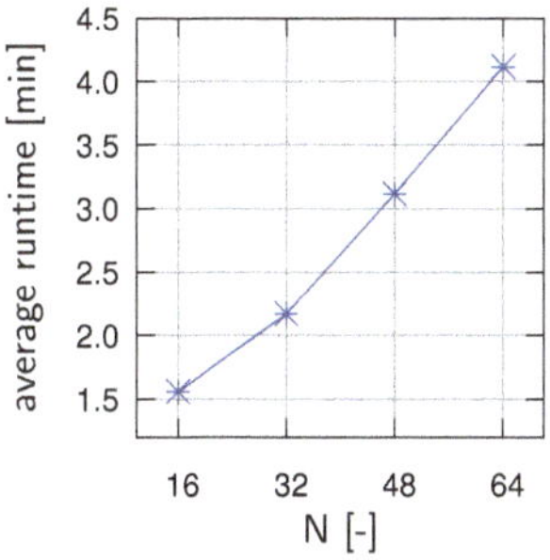

Figure 8. Laptop computer runtimes of the RB model for the fibrous microstructure for various sizes N of the basis. Each data point represents the time needed for all 10 load increments. The spread of the individual times of the 64 validation cases around these average values is negligible.

4.2. Reduced Basis for a Stiffening Microstructure

The second example takes the "worst-case" approach further by considering a noncubical microstructure with even higher phase contrast and significant topological nonlinearity. To this end, a cuboid microstructure occupying the domain $[-0.5, 0.5] \times [-0.3, 0.3] \times [-0.05, 0.05] \subset \mathbb{R}^3$ and containing a hash-like inclusion is investigated, see Figure 9a. The mesh is periodic and contains 33,923 nodes in 21,726 quadratic tetrahedron elements (C3D10). The reinforcement makes up approximately 13.3% of the volume. Due to this large volume fraction, a pronounced geometry-induced nonlinearity of the effective response is expected under uniaxial loading conditions along the x-axis. As it is elongated, the hashlike part is straightened and thus increasingly aligned with the external loading, see Figure 9b. Such effects might be desirable when designing microstructures for functional materials.

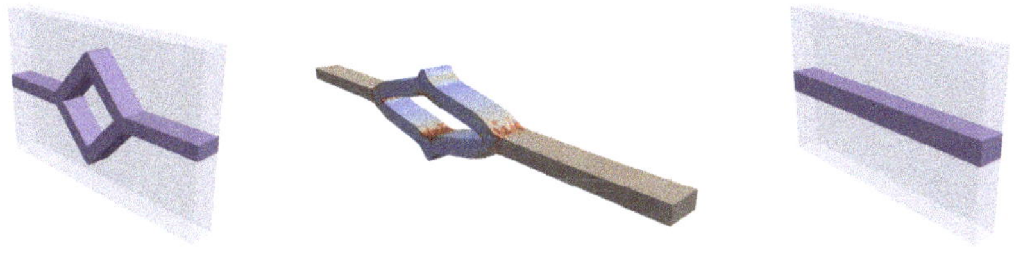

(a) (b) (c)

Figure 9. (**a**) Cuboidal microstructure with hashlike inclusion phase. (**b**) Deformed state under uniaxial tension loading. Only inclusion is shown, coloring indicates $\overline{P}_{xx}$. (**c**) Straight inclusion substitute microstructure, leading to a comparable effective stress.

The material parameters are $K_{\rm m} = 19.867$ MPa, $G_{\rm m} = 0.4$ MPa, $K_{\rm f} = 19,867$ MPa, and $G_{\rm f} = 400$ MPa, implying a Poisson ratio of 0.49 in both materials and a phase contrast of 1000.

The training boundary conditions are the deviatoric ones of the set considered in Section 4.1, i.e., $N_{\rm dir} = 128$ deviatoric directions and $N_{\rm amp} = 10$ regularly spaced deviatoric amplitudes from the interval $[0.05, 0.5]$. No dilatational training cases are considered, i.e., only points from a five-dimensional submanifold of the space $\overline{\mathcal{U}}$ are sampled.

Uniaxial tension boundary conditions are applied for the validation. More precisely, the stretch component $\overline{\boldsymbol{U}}_{xx}$ is increased from 1.0 to 1.5 in 10 increments of equal size. The other components are chosen such that all but the xx-component of the effective stress $\overline{\boldsymbol{P}}$ vanish.

Figure 10 depicts the results for different sizes N of the RB. The influence of the stiffening effect on the stress curve is emphasized by the black dashed line corresponding to a similar microstructure with a straight, cuboid inclusion that leads to the same final stress value under these boundary conditions, see Figure 9c.

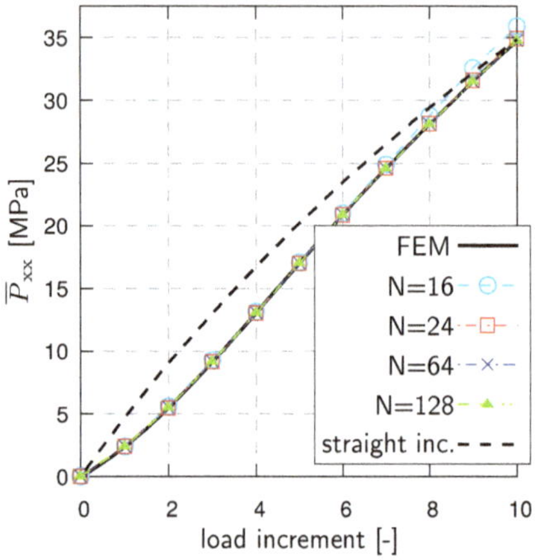

Figure 10. Stress curves for a microstructure with geometric stiffening (cf. Figure 9a), comparing the FEM result to the RB for various number of basis elements N. A similar microstructure without geometric stiffening but with the same final stress value (cf. Figure 9c) leads to the black dashed curve.

In this example, the geometric stiffening effect is captured by the RB with high accuracy, with as few as $N = 24$ basis elements. For moderate stretches, even an RB size of $N = 16$ suffices. These results are somewhat more impressive when noticing that the applied boundary condition contains more than 1.2% volumetric compression, i.e. the validation loading actually lies outside the submanifold covered during training.

In order to examine the action of the cutoff function ϕ, the following two indicators are introduced:

$$c_{qp} = \#\,\text{quadrature points with } (\phi(J) < 1)\,, \tag{47}$$

$$V_{excl} = \left(|\Omega_0| - \sum_{p=1}^{N_{qp}} \phi(J_p) w_p \right) / |\Omega_0|. \tag{48}$$

The first quantity, c_{qp}, counts the number of quadrature points at which the cutoff function has an influence. The second one, V_{excl}, measures the relative excluded volume, interpreting the value of ϕ as a scaling of the corresponding quadrature weight. The values of these indicators are depicted in Figure 11.

Most notably, the cutoff function does not have any effect before the eighth load increment in this example. Only for large load amplitudes does this numerical stability tweak become necessary. Even then, the number of points at which it has an influence is small, considering the total number of quadrature points, $N_{qp} = 86,904$. This example is representative for all conducted numerical experiments.

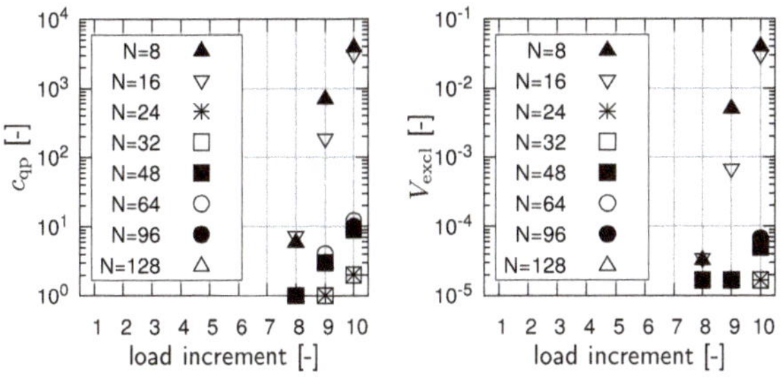

Figure 11. (Left) Number of quadrature points at which the cutoff function ϕ attains a value less than one. (Right) Relative excluded volume.

5. Discussion

5.1. Discussion of the Reduced Basis Method

5.1.1. Relation of the RB Homogenization to Analytical Estimates

Zero coefficients, $\underline{\xi} = \underline{0}$, correspond to the Taylor homogenization, i.e., to the nonlinear counterpart of the Voigt estimate [37], which provides an upper bound for the material response, cf. (17). Starting with the initial guess $\underline{\xi}^{(0)} = \underline{0}$, the evolution of the coefficients corresponds to a (possibly not monotonous) relaxation of this overly stiff response into a more natural state. In view of improved computational efficiency, a nonzero initial guess $\underline{\xi}^{(0)}$ combined with an exact line search has proven reasonable and easy to realize. For instance, such a guess might stem from previous load steps or an interpolation/extrapolation of available coefficient data.

5.1.2. Reconstruction of Displacement Fields

Given an RB approximation of the deformation gradient, $\boldsymbol{F}^{\mathrm{RB}}$, one can reconstruct the corresponding displacement field uniquely up to rigid body motion. This is possible due to the linear dependence of the deformation gradient fluctuations on the displacement fluctuations. Recall the definition of the RB in (32),

$$\boldsymbol{B}^{(i)} = \sum_{j=1}^{N_s} \frac{1}{\sqrt{\lambda_i}} \left(\underline{E}^{(i)}\right)_j \widetilde{\boldsymbol{F}}^{(j)} \quad (i = 1, \dots, N).$$

The corresponding displacement fluctuations are

$$\widetilde{\boldsymbol{u}}_{\mathrm{B}}^{(i)} = \sum_{j=1}^{N_s} \frac{1}{\sqrt{\lambda_i}} \left(\underline{E}^{(i)}\right)_j \widetilde{\boldsymbol{u}}^{(j)} \quad (i = 1, \dots, N). \tag{49}$$

The displacement fluctuation fields $\widetilde{\boldsymbol{u}}^{(j)}$ are defined by $\widetilde{\boldsymbol{u}}^{(j)}(\boldsymbol{X}) = \boldsymbol{u}^{(j)}(\boldsymbol{X}) - \overline{\boldsymbol{H}}^{(j)}\boldsymbol{X}$, where the displacement fields $\boldsymbol{u}^{(j)}$ are the solutions computed during training, and $\overline{\boldsymbol{H}}^{(j)} = \overline{\boldsymbol{U}}^{(j)} - \boldsymbol{I}$. Thus, a displacement field compatible to the RB result $\boldsymbol{F}^{\mathrm{RB}}(\boldsymbol{X}; \overline{\boldsymbol{F}})$ is given by

$$\boldsymbol{u}^{\mathrm{RB}}(\boldsymbol{X}; \overline{\boldsymbol{F}}) = \overline{\boldsymbol{H}}\boldsymbol{X} + \sum_{i=1}^{N} \xi_i^*(\overline{\boldsymbol{F}}) \widetilde{\boldsymbol{u}}_{\mathrm{B}}^{(i)}(\boldsymbol{X}). \tag{50}$$

The missing term $\overline{\boldsymbol{u}}(\overline{\boldsymbol{X}})$, cf. (10), cannot be reconstructed.

5.1.3. Relation to Classical Displacement-Based POD Methods

In a certain sense, the entries of the correlation matrix used in the offline phase, cf. Section 2.2, are *"weighted" scalar products* of the displacement fluctuation fields $\widetilde{w}^{(i)}$ within the Sobolev space $H_0^1(\Omega_0)$. "Weighted" is to be understood in that the zeroth order derivative is multiplied by zero. Classical displacement-based POD methods compute correlations of the fluctuations $\widetilde{w}^{(i)}$ within the Lebesgue space $L^2(\Omega_0)$. The change to $H_0^1(\Omega_0)$-like scalar products is physically motivated by the fact that the local energy $W = W(\boldsymbol{F})$ explicitly depends on the gradient of the displacement, $\boldsymbol{F} = \boldsymbol{u} \otimes \nabla_X + \boldsymbol{I}$, but *not* on the displacement, $\boldsymbol{u}$.

5.1.4. Advantages Compared to General Displacement-Based Schemes

The proposed method is advantageous compared to both displacement-based POD methods and the classical FEM for the following reasons:

- *No gradients need to be computed from displacement fields*, which displacement-based schemes always require prior to the evaluation of the material law.
- The residual $\underline{r}$ and the Jacobian $\underline{\underline{D}}$ are *algorithmically sleek and trivial to implement*.
- The *absence of element formulations* in the assembly of the reduced residual $\underline{r}$ and of the Jacobi matrix $\underline{\underline{D}}$ contributes to both the simplicity and the efficiency of the method—no incidence matrices occur, allowing for *linear memory access*. Moreover, the algebraic operations associated with reference element formulations are bypassed. This is also in favor of *parallel computations*. Such an implementation is still outstanding for the problem at hand, but has been conducted for related problems in the small strain setting in [38].

Although the storage of the basis $\{\boldsymbol{B}^{(i)}\}_{i=1}^N$ requires $9N_{\mathrm{qp}}N$ double precision values, the basis is compact enough to be completely loaded into random access memory of standard computers. For instance, the bases considered in Section 4 occupy only ~200 Mb of memory for $N = 32$.

We now briefly address the *algorithmic complexity* associated with the proposed $\boldsymbol{F}$-RB method and with the $\boldsymbol{u}$-RB method that was employed in previous works, such as [9] and [8]. To this end, the fully discretized versions of the residual $\underline{r}$ and of the Jacobian $\underline{\underline{D}}$ as well as discrete quantities associated with the $\boldsymbol{u}$-RB method are introduced in the following listing.

- $\underline{P}(\boldsymbol{X}_p) \in \mathbb{R}^9$: Nine values of the stress $\boldsymbol{P}(\boldsymbol{F}_{\xi}(\boldsymbol{X}_p))$ at the quadrature point $\boldsymbol{X}_p \in \Omega_0$
- $\underline{\underline{C}}(\boldsymbol{X}_p) \in \mathbb{R}^{9\times 9}$: Symmetric stiffness tensor
- $\underline{\underline{B}}(\boldsymbol{X}_p) \in \mathbb{R}^{9\times N}$: $\boldsymbol{F}$-RB matrix containing the nine values of each basis elements $\boldsymbol{B}^{(i)}$ as columns
- w_p: The quadrature weight at $\boldsymbol{X}_p$
- N_{DOF}: Three times the number of nodes, $\sim N_{\mathrm{qp}}$
- $\underline{r}^{\mathrm{FE}} \in \mathbb{R}^{N_{\mathrm{DOF}}}$: Global FE residual vector
- $\underline{\underline{B}}^{\mathrm{u}} \in \mathbb{R}^{N_{\mathrm{DOF}}\times N}$: $\boldsymbol{u}$-RB matrix of which the columns contain the nodal displacement values
- $\underline{\underline{K}}^{\mathrm{FE}} \in \mathbb{R}^{N_{\mathrm{DOF}}\times N_{\mathrm{DOF}}}$: Global FE stiffness matrix

Table 1 compares the *algorithmic complexity* of the presented $\boldsymbol{F}$-RB method with that of standard $\boldsymbol{u}$-RB schemes. First of all, both methods share a *quadratic* dependence of their Jacobi matrices on the number of modes, N. Therefore, the assembly of the Jacobian is usually the most costly operation. Secondly, both approaches' complexities suffer a linear dependency on the number of quadrature points, N_{qp}. In the displacement-based approach, this is included in the assembly of the residual and of the stiffness, which relate to the factor 9 and 9^2, respectively. Thirdly, the novel $\boldsymbol{F}$-RB scheme spares the computational overhead associated with FE formulations $\underline{r}^{\mathrm{FE}}$ and $\underline{\underline{K}}^{\mathrm{FE}}$. More details on this matter are currently being investigated.

Table 1. Algorithmic complexities of the well-established $\boldsymbol{u}$-based RB method and the novel $\boldsymbol{F}$-based RB method. The assembly of the FE residual and of the FE stiffness depend on N_{qp}.

RB Method	Quantity	Complexity
$\boldsymbol{F}$-based	$\underline{r} = \sum_{p=1}^{N_{\mathrm{qp}}} \underline{\underline{B}}(\boldsymbol{X}_p)^{\mathsf{T}} \underline{P}(\boldsymbol{X}_p) w_p$	$\mathcal{O}(9NN_{\mathrm{qp}})$
	$\underline{\underline{D}} = \sum_{p=1}^{N_{\mathrm{qp}}} \underline{\underline{B}}(\boldsymbol{X}_p)^{\mathsf{T}} \underline{\underline{C}}(\boldsymbol{X}_p) \underline{\underline{B}}(\boldsymbol{X}_p) w_p$	$\mathcal{O}([9^2N + 9N^2]N_{\mathrm{qp}})$
$\boldsymbol{u}$-based	$\underline{r} = \left(\underline{\underline{B}}^{\mathrm{u}}\right)^{\mathsf{T}} \underline{r}^{\mathrm{FE}}$	$\mathcal{O}(NN_{\mathrm{DOF}})$ + assembly of $\underline{r}^{\mathrm{FE}}$
	$\underline{\underline{D}} = \left(\underline{\underline{B}}^{\mathrm{u}}\right)^{\mathsf{T}} \underline{\underline{K}}^{\mathrm{FE}} \underline{\underline{B}}^{\mathrm{u}}$	$\mathcal{O}(NN_{\mathrm{DOF}}^2 + N^2 N_{\mathrm{DOF}})$ + assembly of $\underline{\underline{K}}^{\mathrm{FE}}$

5.1.5. Outlook

Future research should aim at an application of the introduced Reduced Basis method within realistic two-scale simulations, in analog to [12,38–40]. Hyperreduction methods, cf. [41], might give rise to additional speed-ups in the online phase. Further, modifications of the cutoff function, ϕ, should be investigated—a function with compact support might be more appropriate. The construction of the RB from large sets of snapshots is computationally intense, as much data needs to be processed. In the above examples, the POD consumed multiple hours of time. Hierarchical approximations, such as [42], might mitigate the effects by enabling parallel computations. Overall, the long-term perspective is to extend this RB framework efficiently to the context of dissipative materials.

5.2. Discussion of the Sampling Strategy

The proposed sampling strategy is meant to serve as a template. As exemplified in Section 4.1, the samplings can be modified and still lead to a coverage of the set of macroscopic boundary conditions that is sufficient for the problem at hand. The example of Section 4.2 took this idea further and showed that it might not even be necessary to sample the macroscopic determinant. Hence, the sampling can sometimes be reduced to the five-dimensional subspace of isochoric macroscopic stretch tensors.

In any case, the exact choice of both the inputs to Algorithm 1 and the distributions of the deviatoric amplitudes and the macroscopic determinants remains to be based on knowledge and sophisticated guesses, at least at the current state of the art. Further research on this matter might lead to a refined alternative to Algorithm 1, possibly involving the evaluation of error estimators.

Author Contributions: Conceptualization, O.K. and F.F.; Funding acquisition, F.F.; Methodology, O.K. and F.F.; Project administration, F.F.; Software, O.K.; Supervision, F.F.; Writing—original draft, O.K. and F.F.

Funding: This research was funded by Deutsche Forschungsgemeinschaft (DFG) within the Emmy–Noether programm under grant DFG-FR2702/6.

Conflicts of Interest: The authors declare no conflict of interest.

Abbreviations

The following abbreviations are used in this manuscript:

RB	Reduced Basis
FE(M)	Finite Element (Method)
POD	Proper Orthogonal Decomposition
DOF	degree(s) of freedom
FOM	full-order model
s.p.d.	symmetric positive definite
DDMS	Dilatational-Deviatoric Multiplicative Split

Appendix A. Material Objectivity

The components of the stiffness tensor $\mathbb{C}$ show the following transformation behavior.

$$C_{ijkl}(\boldsymbol{F}) = \frac{\partial P_{ij}(\boldsymbol{F})}{\partial F_{kl}} \overset{(41)}{=} \sum_{m=1}^{3} \frac{\partial R_{im} P_{mj}(\boldsymbol{U})}{\partial F_{kl}} = \sum_{m=1}^{3} \left(\frac{\partial R_{im}}{\partial F_{kl}} P_{mj}(\boldsymbol{U}) + R_{im} \frac{\partial P_{mj}(\boldsymbol{U})}{\partial F_{kl}} \right) \tag{A1}$$

$$= \sum_{m,n,o=1}^{3} \left(\delta_{ik} U_{lm}^{-1} P_{mj}(\boldsymbol{U}) + R_{im} \frac{\partial P_{mj}(\boldsymbol{U})}{\partial U_{no}} \frac{\partial U_{no}}{\partial F_{kl}} \right) \tag{A2}$$

$$= \sum_{m,n,o=1}^{3} \left(\delta_{ik} U_{lm}^{-1} P_{mj}(\boldsymbol{U}) + R_{im} C_{mjno}(\boldsymbol{U}) R_{kn} \delta_{ol} \right) \tag{A3}$$

$$= \sum_{m,n=1}^{3} \left(\delta_{ik} U_{lm}^{-1} P_{mj}(\boldsymbol{U}) + R_{im} C_{mjnl}(\boldsymbol{U}) R_{kn} \right) \tag{A4}$$

Here, δ_{ij} is the Kronecker symbol, and $i, j, k, l = 1, 2, 3$.

Appendix B. Effective Material Responses of the RB

Let $\mathbb{I}$ denote the fourth order identity tensor and let the arguments of the $\boldsymbol{F}$-RB approximation (23) be omitted, i.e. here $\boldsymbol{F}^{\mathrm{RB}} = \boldsymbol{F}^{\mathrm{RB}}(\boldsymbol{X}; \overline{\boldsymbol{F}})$. Its derivative with respect to the boundary condition $\overline{\boldsymbol{F}}$ is

$$\frac{\partial \boldsymbol{F}^{\mathrm{RB}}}{\partial \overline{\boldsymbol{F}}} = \mathbb{I} + \sum_{i=1}^{N} \boldsymbol{B}^{(i)} \otimes \frac{\partial \xi_i^*}{\partial \overline{\boldsymbol{F}}}(\overline{\boldsymbol{F}}). \tag{A5}$$

Appendix B.1. Effective Stress

$$\begin{aligned} \overline{\boldsymbol{P}}^{\mathrm{RB}} &= \frac{\partial \overline{W}^{\mathrm{RB}}}{\partial \overline{\boldsymbol{F}}} = \frac{\partial}{\partial \overline{\boldsymbol{F}}} \left\langle W^{\mathrm{RB}} \right\rangle = \left\langle \frac{\partial W^{\mathrm{RB}}}{\partial \boldsymbol{F}} \cdot \frac{\partial \boldsymbol{F}^{\mathrm{RB}}}{\partial \overline{\boldsymbol{F}}} \right\rangle \\ &\overset{(A5)}{=} \left\langle \boldsymbol{P}^{\mathrm{RB}} \right\rangle + \sum_{i=1}^{N} \left\langle \boldsymbol{P}^{\mathrm{RB}} \cdot \left(\boldsymbol{B}^{(i)} \otimes \frac{\partial \xi_i^*}{\partial \overline{\boldsymbol{F}}}(\overline{\boldsymbol{F}}) \right) \right\rangle \\ &= \left\langle \boldsymbol{P}^{\mathrm{RB}} \right\rangle + \sum_{i=1}^{N} \left\langle \boldsymbol{P}^{\mathrm{RB}} \cdot \boldsymbol{B}^{(i)} \right\rangle \otimes \frac{\partial \xi_i^*}{\partial \overline{\boldsymbol{F}}}(\overline{\boldsymbol{F}}) \qquad \overset{(28)}{=} \left\langle \boldsymbol{P}^{\mathrm{RB}} \right\rangle \end{aligned} \tag{A6}$$

Appendix B.2. Effective Stiffness

$$\begin{aligned} \overline{\mathbb{C}}^{\mathrm{RB}} &= \frac{\partial \overline{\boldsymbol{P}}^{\mathrm{RB}}}{\partial \overline{\boldsymbol{F}}} \overset{(A6)}{=} \frac{\partial}{\partial \overline{\boldsymbol{F}}} \left\langle \boldsymbol{P}^{\mathrm{RB}} \right\rangle = \left\langle \frac{\partial^2 W^{\mathrm{RB}}}{\partial \overline{\boldsymbol{F}} \partial \boldsymbol{F}} \right\rangle = \left\langle \frac{\partial^2 W^{\mathrm{RB}}}{\partial \boldsymbol{F}^2} \cdot \frac{\partial \boldsymbol{F}^{\mathrm{RB}}}{\partial \overline{\boldsymbol{F}}} \right\rangle \\ &\overset{(A5)}{=} \left\langle \mathbb{C}^{\mathrm{RB}} \right\rangle + \sum_{i=1}^{N} \left\langle \mathbb{C}^{\mathrm{RB}} \cdot \left(\boldsymbol{B}^{(i)} \otimes \frac{\partial \xi_i^*}{\partial \overline{\boldsymbol{F}}}(\overline{\boldsymbol{F}}) \right) \right\rangle \\ &= \left\langle \mathbb{C}^{\mathrm{RB}} \right\rangle + \sum_{i=1}^{N} \left\langle \mathbb{C}^{\mathrm{RB}} \cdot \boldsymbol{B}^{(i)} \right\rangle \otimes \frac{\partial \xi_i^*}{\partial \overline{\boldsymbol{F}}}(\overline{\boldsymbol{F}}) \end{aligned} \tag{A7}$$

For $\frac{\partial \xi_i^*}{\partial \overline{\boldsymbol{F}}}(\overline{\boldsymbol{F}})$, we demand that the residual $r_i(\overline{\boldsymbol{F}}, \underline{\xi})$ from (28) is stable with respect to the boundary condition $\overline{\boldsymbol{F}}$ when converged to $r_i(\overline{\boldsymbol{F}}, \underline{\xi}^*(\overline{\boldsymbol{F}})) = 0$,

$$\begin{aligned}
\frac{\partial r_i}{\partial \overline{\boldsymbol{F}}}(\overline{\boldsymbol{F}}, \underline{\xi}^*(\overline{\boldsymbol{F}})) &= \left\langle \boldsymbol{B}^{(i)} \cdot \frac{\partial \boldsymbol{P}^{\mathrm{RB}}}{\partial \overline{\boldsymbol{F}}} \right\rangle \\
&= \left\langle \boldsymbol{B}^{(i)} \cdot \left(\frac{\partial \boldsymbol{P}^{\mathrm{RB}}}{\partial \boldsymbol{F}} \frac{\partial \boldsymbol{F}^{\mathrm{RB}}}{\partial \overline{\boldsymbol{F}}} \right) \right\rangle \\
&= \left\langle \boldsymbol{B}^{(i)} \cdot \mathbb{C}^{\mathrm{RB}} \right\rangle + \sum_{j=1}^{N} \left\langle \boldsymbol{B}^{(i)} \cdot \mathbb{C}^{\mathrm{RB}} \cdot \frac{\partial \boldsymbol{F}^{\mathrm{RB}}}{\partial \xi_j^*} \frac{\partial \xi_j^*}{\partial \overline{\boldsymbol{F}}} \right\rangle \\
&= \left\langle \boldsymbol{B}^{(i)} \cdot \mathbb{C}^{\mathrm{RB}} \right\rangle + \sum_{j=1}^{N} \underbrace{\left\langle \boldsymbol{B}^{(i)} \cdot \mathbb{C}^{\mathrm{RB}} \cdot \boldsymbol{B}^{(j)} \right\rangle}_{D_{ij}} \frac{\partial \xi_j^*}{\partial \overline{\boldsymbol{F}}} \quad = \boldsymbol{0}.
\end{aligned} \tag{A8}$$

It follows that

$$\frac{\partial \xi_j^*}{\partial \overline{\boldsymbol{F}}}(\overline{\boldsymbol{F}}) = -\sum_{i=1}^{N} \left(\underline{\underline{D}}^{-1} \right)_{ij} \left\langle \boldsymbol{B}^{(i)} \cdot \mathbb{C}^{\mathrm{RB}} \right\rangle. \tag{A9}$$

Appendix C. Basis for Symmetric Traceless Second Order Tensors

$$\begin{aligned}
&\underline{\underline{Y}}^{(1)} = \sqrt{\frac{1}{6}} \begin{bmatrix} 2 & 0 & 0 \\ 0 & -1 & 0 \\ 0 & 0 & -1 \end{bmatrix} \qquad \underline{\underline{Y}}^{(2)} = \sqrt{\frac{1}{2}} \begin{bmatrix} 0 & 0 & 0 \\ 0 & 1 & 0 \\ 0 & 0 & -1 \end{bmatrix} \\
&\underline{\underline{Y}}^{(3)} = \sqrt{\frac{1}{2}} \begin{bmatrix} 0 & 1 & 0 \\ 1 & 0 & 0 \\ 0 & 0 & 0 \end{bmatrix} \qquad \underline{\underline{Y}}^{(4)} = \sqrt{\frac{1}{2}} \begin{bmatrix} 0 & 0 & 1 \\ 0 & 0 & 0 \\ 1 & 0 & 0 \end{bmatrix} \qquad \underline{\underline{Y}}^{(5)} = \sqrt{\frac{1}{2}} \begin{bmatrix} 0 & 0 & 0 \\ 0 & 0 & 1 \\ 0 & 1 & 0 \end{bmatrix}
\end{aligned} \tag{A10}$$

References

1. Rendek, M.; Lion, A. Amplitude dependence of filler-reinforced rubber: Experiments, constitutive modelling and FEM—Implementation. *Int. J. Solids Struct.* **2010**, *47*, 2918–2936. [CrossRef]
2. Nguyen, V.D.; Lani, F.; Pardoen, T.; Morelle, X.; Noels, L. A large strain hyperelastic viscoelastic-viscoplastic-damage constitutive model based on a multi-mechanism non-local damage continuum for amorphous glassy polymers. *Int. J. Solids Struct.* **2016**, *96*, 192–216. [CrossRef]
3. Barrault, M.; Maday, Y.; Nguyen, N.C.; Patera, A.T. An 'empirical interpolation' method: Application to efficient reduced-basis discretization of partial differential equations. *C. R. Math.* **2004**, *339*, 667–672. [CrossRef]
4. Geers, M.; Yvonnet, J. Multiscale modeling of microstructure-property relations. *MRS Bull.* **2016**, *41*, 610–616. [CrossRef]
5. Saeb, S.; Steinmann, P.; Javili, A. Aspects of Computational Homogenization at Finite Deformations: A Unifying Review From Reuss' to Voigt's Bound. *Appl. Mech. Rev.* **2016**, *68*, 050801. [CrossRef]
6. Feyel, F. Multiscale FE^2 elastoviscoplastic analysis of composite structures. *Comput. Mater. Sci.* **1999**, *16*, 344–354. [CrossRef]
7. Sirovich, L. Turbulence and the Dynamics of Coherent Structures. Part 1: Coherent Structures. *Q. Appl. Math.* **1987**, *45*, 561–571. [CrossRef]
8. Yvonnet, J.; He, Q.C. The reduced model multiscale method (R3M) for the non-linear homogenization of hyperelastic media at finite strains. *J. Comput. Phys.* **2007**, *223*, 341–368. [CrossRef]

9. Radermacher, A.; Reese, S. POD-based model reduction with empirical interpolation applied to nonlinear elasticity. *Int. J. Numer. Methods Eng.* **2016**, *107*, 477–495. [CrossRef]
10. Radermacher, A.; Reese, S. Proper orthogonal decomposition-based model reduction for non-linear biomechanical analysis. *Int. J. Mater. Eng. Innov.* **2013**, *4*, 149–165. [CrossRef]
11. Soldner, D.; Brands, B.; Zabihyan, R.; Steinmann, P.; Mergheim, J. A numerical study of different projection-based model reduction techniques applied to computational homogenisation. *Comput. Mech.* **2017**, *60*, 613–625. [CrossRef]
12. Fritzen, F.; Kunc, O. Two-stage data-driven homogenization for nonlinear solids using a reduced order model. *Eur. J. Mech. A Solids* **2018**, *69*, 201–220. [CrossRef]
13. Akkari, N.; Casenave, F.; Moureau, V. Time Stable Reduced Order Modeling by an Enhanced Reduced Order Basis of the Turbulent and Incompressible 3D Navier–Stokes Equations. *Math. Comput. Appl.* **2019**, *24*, 45. [CrossRef]
14. An, S.; Kim, T.; James, D.L. Optimizing cubature for efficient integration of subspace deformations. *ACM Trans. Graph.* **2009**, *27*, 165:1–165:10. [CrossRef]
15. Hernández, J.; Caicedo, M.; Ferrer, A. Dimensional hyper-reduction of nonlinear finite element models via empirical cubature. *Comput. Methods Appl. Mech. Eng.* **2017**, *313*, 687–722. [CrossRef]
16. Temizer, I.; Zohdi, T. A numerical method for homogenization in non-linear elasticity. *Comput. Mech.* **2007**, *40*, 281–298. [CrossRef]
17. Yvonnet, J.; Monteiro, E.; He, Q.C. Computational homogenization method and reduced database model for hyperelastic heterogeneous structures. *J. Multiscale Comput. Eng.* **2013**, *11*, 201–225. [CrossRef]
18. Kunc, O. GitHub repository ReducedBasisDemonstrator. Available online: https://github.com/EMMA-Group/ReducedBasisDemonstrator (accessed on 27 May 2019).
19. Flory, P. Thermodynamic relations for high elastic materials. *Trans. Faraday Soc.* **1961**, *57*, 829–838. [CrossRef]
20. Bilger, N.; Auslender, F.; Bornert, M.; Michel, J.C.; Moulinec, H.; Suquet, P.; Zaoui, A. Effect of a nonuniform distribution of voids on the plastic response of voided materials: a computational and statistical analysis. *Int. J. Solids Struct.* **2005**, *42*, 517–538. [CrossRef]
21. Doll, S.; Schweizerhof, K. On the Development of Volumetric Strain Energy Functions. *J. Appl. Mech.* **1999**, *67*, 17–21. [CrossRef]
22. Simo, J. A framework for finite strain elastoplasticity based on maximum plastic dissipation and the multiplicative decomposition: Part I. Continuum formulation. *Comput. Methods Appl. Mech. Eng.* **1988**, *66*, 199–219. [CrossRef]
23. Pruchnicki, E. Hyperelastic homogenized law for reinforced elastomer at finite strain with edge effects. *Acta Mech.* **1998**, *129*, 139–162. [CrossRef]
24. Miehe, C. Computational micro-to-macro transitions for discretized micro-structures of heterogeneous materials at finite strains based on the minimization of averaged incremental energy. *Comput. Methods Appl. Mech. Eng.* **2003**, *192*, 559–591. [CrossRef]
25. Castañeda, P.P.; Suquet, P. Nonlinear Composites. *Adv. Appl. Mech.* **1998**, *34*, 172–302. [CrossRef]
26. Kabel, M.; Böhlke, T.; Schneider, M. Efficient fixed point and Newton–Krylov solvers for FFT-based homogenization of elasticity at large deformations. *Comput. Mech.* **2014**, *54*, 1497–1514. [CrossRef]
27. Quarteroni, A.; Manzoni, A.; Negri, F. *Reduced Basis Methods for Partial Differential Equations: An Introduction*; Springer: Cham, Switzerland, 2016.
28. Ball, J.M. Convexity conditions and existence theorems in nonlinear elasticity. *Arch. Ration. Mech. Anal.* **1976**, *63*, 337–403. [CrossRef]
29. Schneider, M. Beyond polyconvexity: An existence result for a class of quasiconvex hyperelastic materials. *Math. Methods Appl. Sci.* **2017**, *40*, 2084–2089. [CrossRef]
30. Miehe, C. Numerical computation of algorithmic (consistent) tangent moduli in large-strain computational inelasticity. *Comput. Methods Appl. Mech. Eng.* **1996**, *134*, 223–240. [CrossRef]
31. Faraut, J. *Analysis on Lie Groups: An Introduction*; Cambridge University Press: Cambridge, UK, 2008.
32. Neff, P.; Eidel, B.; Martin, R.J. Geometry of Logarithmic Strain Measures in Solid Mechanics. *Arch. Ration. Mech. Anal.* **2016**, *222*, 507–572. [CrossRef]
33. Brauchart, J.S.; Grabner, P.J. Distributing many points on spheres: Minimal energy and designs. *J. Complex.* **2015**, *31*, 293–326. [CrossRef]

34. Kunc, O.; Fritzen, F. Generation of energy-minimizing point sets on spheres and their application in mesh-free interpolation and differentiation. *Adv. Comput. Math.* **2018**. Under review.
35. Leopardi, P. A partition of the unit sphere into regions of equal area and small diameter. *Electron. Trans. Numer. Anal.* **2006**, *25*, 309–327.
36. Kim, H.J.; Swan, C.C. Algorithms for automated meshing and unit cell analysis of periodic composites with hierarchical tri-quadratic tetrahedral elements. *Int. J. Numer. Methods Eng.* **2003**, *58*, 1683–1711. [CrossRef]
37. Voigt, W. *Lehrbuch der Kristallphysik*; Vieweg+Teubner Verlag: Wiesbaden, Germany, 1966.
38. Fritzen, F.; Hodapp, M. The finite element square reduced (FE^{2R}) method with GPU acceleration: Towards three-dimensional two-scale simulations. *Int. J. Numer. Methods Eng.* **2016**, *107*, 853–881. [CrossRef]
39. Rambausek, M.; Göküzüm, F.S.; Nguyen, L.T.K.; Keip, M.A. A two-scale FE-FFT approach to nonlinear magneto-elasticity. *Int. J. Numer. Methods Eng.* **2019**, *117*, 1117–1142. [CrossRef]
40. Kochmann, J.; Wulfinghoff, S.; Ehle, L.; Mayer, J.; Svendsen, B.; Reese, S. Efficient and accurate two-scale FE-FFT-based prediction of the effective material behavior of elasto-viscoplastic polycrystals. *Comput. Mech.* **2017**. [CrossRef]
41. Ryckelynck, D. A priori hyperreduction method: an adaptive approach. *J. Comput. Phys.* **2005**, *202*, 346–366. [CrossRef]
42. Himpe, C.; Leibner, T.; Rave, S. Hierarchical Approximate Proper Orthogonal Decomposition. *SIAM J. Sci. Comput.* **2018**, *40*, A3267–A3292. [CrossRef]

Mathematical and Computational Applications

Article

Data Pruning of Tomographic Data for the Calibration of Strain Localization Models

William Hilth [1], David Ryckelynck [1,*] and Claire Menet [2]

[1] MAT—Centre des Matériaux, Mines ParisTech, PSL Research University, CNRS UMR 7633, 10 rue Desbruères, 91003 Evry, France; william.hilth@mines-paristech.fr

[2] University of Lyon, INSA de Lyon, UMR CNRS 5510, 20 Avenue Albert Einstein, 69100 Villeurbanne, France; claire.menet@montupet-group.com

* Correspondence: david.ryckelynck@mines-paristech.fr

Received: 12 November 2018; Accepted: 22 January 2019; Published: 28 January 2019

Abstract: The development and generalization of Digital Volume Correlation (DVC) on X-ray computed tomography data highlight the issue of long-term storage. The present paper proposes a new model-free method for pruning experimental data related to DVC, while preserving the ability to identify constitutive equations (i.e., closure equations in solid mechanics) reflecting strain localizations. The size of the remaining sampled data can be user-defined, depending on the needs concerning storage space. The proposed data pruning procedure is deeply linked to hyper-reduction techniques. The DVC data of a resin-bonded sand tested in uniaxial compression is used as an illustrating example. The relevance of the pruned data was tested afterwards for model calibration. A Finite Element Model Updating (FEMU) technique coupled with an hybrid hyper-reduction method aws used to successfully calibrate a constitutive model of the resin bonded sand with the pruned data only.

Keywords: archive; model reduction; 3D reconstruction; inverse problem plasticity; data science

1. Introduction

With the development and the generalization of digital image correlation (DIC) (see Chu et al. [1]) or digital volume correlation (DVC) (see Bay et al. [2]) techniques on Computed Tomography (CT) data, the volume of data acquired has drastically increased. This raises new challenges, such as data storage, data mining or the development of relevant experiments-simulations dialog methods such as model validation and model calibration.

In experimental mechanics, the access to full 3D fields such as displacement or strain fields is far richer than 1D load–displacement curves. These data can drive finite element simulations for model calibration. Although extremely convincing, the increasing resolution of the full-field measurement tools, such as X-ray Computed Tomography, leads to an explosion of the volume of data to store. The long term storage of CT datasets is nowadays an issue (see Ooijen et al. [3]).

This paper proposes a numerical method for pruning 3D dataset related to DVC when it becomes necessary to free up storage capacity. Often, when new experimental results need to be saved, storage memory must be released. The pruned data contain information similar to the original data, but with less memory required. The proposed approach aims to prune experimental data while preserving the ability to identify constitutive equations (i.e., closure equations in solid mechanics) reflecting strain localizations. It is a mechanical based approach to prune DVC data. Outside a reduced experimental domain (RED), the experimental data are deleted. Original experimental data are preserved solely in the RED. We also propose a calibration procedure whose computational complexity is consistent with the pruning of the experimental data.

Compression of data is known to be a convenient approach to restore storage capacity. For instance, MP3 files are a fairly common way to reduce the size of audio files for daily use (see Pan [4]). However, a non-negligible loss of information is needed, but controlled. The MP3 compression roughly consists in filtering certain components of the non-reduced audio file that are actually non-audible for most people. In other words, the MP3 algorithm was made to prune the audio data that are not absolutely necessary. Usually, the compression rate is around 12. In the same philosophy, there can be a way to massively compress the experimental data taken from experiments with a controlled loss of information based on an algorithm that detects the pertinent information. This has been proposed in [5] by using a sensitivity analysis with respect to variations of calibration parameters. These parameters are the coefficients of a given model that should reflect the experimental observations. The result is that the pruned data are dedicated to a given model. In this paper, a model-free approach is proposed. It aims to make possible various calibrations with different models after data pruning. Here, the relevant information are local but situated in regions submitted to strain localization. The data submitted to the pruning procedure are the outputs of a Digital Volume Correlation that reconstructs the displacement field $\mathbf{u}(\mathbf{x}, t)$ from observations at time instants $(t_j)_{j=1,\ldots,N_t}$, over a spatial domain Ω, where $\mathbf{x}$ is a position vector. The geometry of the experimental sample is approximated by a mesh and the determined displacement is decomposed on finite element (FE) shape functions [6].

The proposed method can be linked to data pruning or data cleaning methods described in the literature for machine learning [7]. The aim of these procedures are not to reduce data storage but to improve the data quality by accurate outliers detection for instance [8]. In [9], a data pruning method is employed to filter the noise in the dataset.

Using the FE approximation of the experimental fields paves the way to further simulations. In the calibration procedure, the full-field measurements are used as inputs of an inverse problem that aims to determine a given set of parameters $\boldsymbol{\mu} = \{\mu_1, \ldots, \mu_m\}$. These parameters are the coefficients of given constitutive equations. Their values are unknown or not known precisely. The most straightforward method is called Finite Element Model Updating (FEMU) (see Kavanagh and Clough [10] and Kavanagh [11]). It is a rather common way to optimize a set of parameters taking into account the experimental data and balance equations in mechanics. It consists in computing the discrepancy between the FE approximation of the experimental fields and the FE simulations. Thus, an optimization loop is done on $\boldsymbol{\mu}$ where the FE method is used as a tool for assessing the relevance of the parameter set. The objective function, or cost function, of the optimization can focus on the difference between the computed and experimental displacement fields (FEMU-U), forces (FEMU-F, or force balance method), or the strain fields (FEMU-ε) or a mix between all these sub-methods. A review of FEMU applications can be found in [12]. The method is particularly suitable for:

- Non-isotropic materials (e.g., materials having mechanical properties that depend on their orientation [13,14], such as the human skin [15]);
- Heterogeneous materials such as composites [16];
- Heterogeneous tests such as open-hole tests (e.g., [13,14]) or CT-samples [17];
- Special cases of local phenomena such as strain localization or necking (e.g., Forestier et al. [18], Giton et al. [19]) or the illustrating case of the present paper;
- Multi-materials configurations (e.g., solder joints studied in [20] or heterogeneous material identification done in [21]); and
- Determination of the boundary conditions [22].

One of the recent developments concerning FEMU is to couple this method with reduced order models (ROMs) to cut down the computation time in the parameters optimization loop. An example of such recent developments can be found in [23] where a method called FEMU-b is highlighted, or in [24]. The FEMU-b consists in determining an intermediate space of predominant empirical modes associated to a reduction procedure, such as the Proper Orthogonal Decomposition (see Aubry et al. [25]) or the Proper Generalized Decomposition (PGD) [26]. The discrepancy is computed between the experimental and simulated reduced variables, where the reduced variables are solutions of reduced equations. In this paper, we show that the proposed data pruning method is consistent with a reduced order modeling of the equations to be calibrated. A FEMU-b is introduced, so we take into account the lack of experimental data due to the pruning procedure.

In [27], it has been shown that ROMs can be supplemented by a reduced integration domain (RID), by following a hyper-reduction method. In this method, a RID is a subdomain of a body, where the reduced equations are set up. In the proposed approach, we do not modify the cubature scheme involved in mechanical equations, as proposed by Hernandez et al. [28], but we restrict the cubature to a subdomain. This leads the way for data pruning methods that preserve calibration capabilities. Here, the dimensionality reduction of experimental data enables the restriction of experimental data to a RED. This RED is a subdomain of the specimen where the experimental data are sampled. It is not necessarily a connected domain. The flowchart of the proposed approach to data pruning is shown in Figure 1. After pruning, the data related to the domain occupied by a specimen, denoted by Ω, are restricted to a RED denoted by Ω_R. The way the model calibration is done, depends on the nature of the data available in a storage system. If the data are not pruned, then a conventional calibration by the FEMU method is possible. Otherwise, calibration by a FEMU-b method is recommended. In this paper, the calibration capabilities after data pruning are assessed by using the FEMU with an hybrid hyper-reduction method (H^{2}ROM) [29]. Hence, the FEMU-b is not done on the complete domain but on the RED determined by the data pruning. The result is a fast calibration procedure, with low memory requirement and a validated data pruning protocol. Contrary to usual hyper-reduction methods, the domain where the equations to be calibrated are setup is not generated by using simulation data. It derives from the data pruning procedure applied to experimental data.

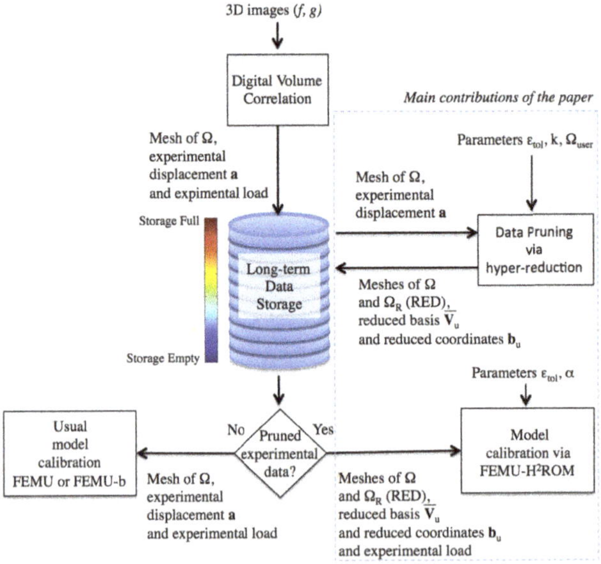

Figure 1. Pruning of experimental data related to DVC, via hyper-reduction. Calibration capabilities of constitutive equations are preserved after data pruning. The experimental data related to the domain occupied by a specimen, denoted by Ω, are restricted to a reduced experimental domain denoted by Ω_R. The way the model calibration is done, depends on the nature of the data available in the storage system.

The remaining part of the paper is structured as follows. In Section 3, the proposed method for data pruning is described. The DVC is recalled. A dimensionality reduction then hyper-reduction are performed to compute the pruned data. The pruning procedure is applied in Section 4 on a resin-bonded sand tested in in situ uniaxial compression with X-ray tomography. In Section 5, the calibration of an elastoplastic model enables validating the pruning protocol. Details on the experimental data are available in the form of supplementary files. These data allow the proposed data pruning to be reproduced.

2. Notations

Second-order tensors are denoted by $\underset{\sim}{a}$. Matrices are denoted by capital bold letters $\mathbf{A}$ and vectors are denoted by bold lowercase characters $\mathbf{a}$. The colon notation is used to denote the extraction of a submatrix or a vector (at column i for example): $\mathbf{a} = \mathbf{A}[:, i]$. Sets of indices are denoted by calligraphic characters $\mathcal{A}$. The element of a matrix $\mathbf{A}$ at row i and column j is denoted A_{ij} or $A_\alpha[i, j]$ when the matrix notation $\mathbf{A}_\alpha$ has a subscript. $\overline{\mathbf{a}}$ is the restriction of $\mathbf{a}$ to the reduced experimental domain.

3. Data Pruning by Following an Hyper-Reduction Scheme

In the proposed approach, the experimental displacements observed on the domain occupied by a specimen are restricted to the RED Ω_R. The smaller is the extent of Ω_R, the smaller is the memory requirement to store the pruned data. Without any constraint, the best memory saving is obtained by saving the parameters $\boldsymbol{\mu}$ that best replicate the experimental data. In that situation, usual FEMU methods are sufficient. Here, the following constraint is taken into consideration. The data pruning should not prevent changes in the way constitutive equations are set up, as these equations may evolve in the future. Knowledge in mechanics is evolving and so are models. Thus, after the data have been pruned, the experimental data saved in the storage system must allow the calibration of constitutive equations. To ensure consistency between the computational complexity of the calibration procedure and the accuracy of the pruning data, we propose hyper-reduced equations for this calibration. In our opinion, it does not make sense to perform complex simulations during such a calibration with a poor representation of the experimental data.

3.1. Digital Volume Correlation

Let us consider a specimen occupying the domain Ω undergoing a certain mechanical test. With image acquisition techniques, grayscale images are obtained in 3D. The Digital Volume Correlation aims to determine the displacement field $\mathbf{u}$ at every position $\mathbf{x}$ in Ω at a given deformed state at time t. f and g are the gray levels at the reference and deformed states. They are related by the equation:

$$g(\mathbf{x}) = f(\mathbf{x} + \mathbf{u}(\mathbf{x}, t)) \tag{1}$$

The best displacement field is estimated via the minimization of the following residual:

$$\phi^2(\mathbf{u}, t) = \int_\Omega [\mathbf{u}(\mathbf{x}, t).\nabla f(\mathbf{x}) + f(\mathbf{x}) - g(\mathbf{x})]^2 \, d\mathbf{x} \tag{2}$$

where ∇f is the gradient of f. This is an ill-posed problem. To get a well-posed problem, the displacement field can be restricted to a kinematic subspace. Here, the displacement field is assumed to be decomposed over a set of vector functions $\boldsymbol{\psi}_j(\mathbf{x})$ that corresponds to the shape functions of a FE model defined on Ω.

$$\mathbf{u}(\mathbf{x}, t) = \sum_{i=1}^{N_d} a_i(t) \boldsymbol{\psi}_i(\mathbf{x}) \tag{3}$$

where N_d is the number of degrees of freedom of the mesh, a_i the ith nodal degree of freedom in the FE model. $\mathbf{a}$ denotes the vector of degrees of freedom to be determined. With this restriction to the kinematic subspace, the function ϕ is now a quadratic form of the a_i, and its minimization is a linear system, set up for each observation of a deformed state:

$$\mathbf{M}\mathbf{a} = \mathbf{h} \tag{4}$$

where the matrix $\mathbf{M}$ and the vector $\mathbf{h}$ are:

$$M_{ij} = \int_{\Omega} (\boldsymbol{\psi}_i(\mathbf{x}).\nabla f(\mathbf{x})) \left(\boldsymbol{\psi}_j(\mathbf{x}).\nabla f(\mathbf{x})\right) \mathrm{d}x \tag{5}$$

$$h_i = \int_{\Omega} [g(\mathbf{x}) - f(\mathbf{x})]\, \boldsymbol{\psi}_i(\mathbf{x}).\nabla f(\mathbf{x}) \mathrm{d}x \tag{6}$$

In the sequel, N_t observations of the specimen deformation at time instants t_j, $j = 1, \ldots, N_t$, are considered. The DVC gives access to the final correlated displacement field $\mathbf{u}(\mathbf{x}, t_j)$ for each observations, through the coefficient vector $\mathbf{a}(t_j)$. From the displacement field, a strain field $\underset{\sim}{\varepsilon}$ is extracted assuming small strains:

$$\underset{\sim}{\varepsilon} = \frac{1}{2}\left(\underset{\sim}{\nabla}\mathbf{u} + \underset{\sim}{\nabla}\mathbf{u}^T\right) \tag{7}$$

This strain is thus calculated at each Gauss point of the mesh used for the DVC. For pressure dependent or plastic materials, it can be convenient to subdivide the strain field in its deviatoric part and its hydrostatic part:

$$\underset{\sim}{\varepsilon} = \underset{\sim}{\varepsilon}^s + \underset{\sim}{\varepsilon}^v, \quad \text{with } \underset{\sim}{\varepsilon}^v = \mathrm{tr}\left(\underset{\sim}{\varepsilon}\right)\underset{\sim}{\boldsymbol{I}} \tag{8}$$

where $\underset{\sim}{\boldsymbol{I}}$ is the unit tensor.

It is worth noting that the pruning procedure only focuses on the displacement and not on the strain. It is considered that the strain can be computed in post-processing (thanks to Equation (7)) and are not worth saving. The strain tensor is actually considered as temporary data used to compute a reduced experimental domain.

3.2. Dimensionality Reduction

The first step of the pruning procedure consists in performing a dimensionality reduction of the experimental data. It is based on singular value decomposition. This approach is similar to the Principal Component Analysis (PCA). However, here, a reduced basis of empirical modes is obtained without centering the data.

The experimental data from DVC are saved into two matrices, $\mathbf{Q}_u$ and $\mathbf{Q}_\varepsilon$ defined as:

$$Q_u[i, j] = a_i(t_j),\ i = 1, \ldots, N_d,\ j = 1, \ldots, N_t \tag{9}$$

and

$$Q_\varepsilon[i, j] = \varepsilon^s_{\alpha\beta}(e_\gamma, t_j) \tag{10}$$

$$Q_\varepsilon[i, j + N_t] = \varepsilon^v_{\alpha\beta}(e_\gamma, t_j) \tag{11}$$

where e_γ is the γth Gauss point, and:

$$\begin{aligned} &i = \beta + 3(\alpha - 1) + 9(\gamma - 1) \\ &\alpha = 1, \ldots, 3,\ \beta = 1, \ldots, 3,\ \gamma = 1, \ldots, N_g \\ &j = 1, \ldots, N_t \end{aligned}$$

with N_g being the number of integration points in the mesh. $\mathbf{Q}_u$ is a $N_d \times N_t$ matrix and $\mathbf{Q}_\varepsilon$ is a $(9N_g) \times (2N_t)$ matrix. For the sake of simplicity, we do not account for the symmetry of the strain tensor.

The first step of the pruning procedure consists in performing a first dimensionality reduction of the DVC data. Only the reduced basis and coordinate are kept instead of the snapshot matrix $\mathbf{Q}_u$. The procedure is also done on the snapshot matrix of the stain $\mathbf{Q}_\varepsilon$ but not in order to reduce storage (as the stain data are not saved). The corresponding reduced basis is used as a temporary tool to compute afterwards the reduced domain. The determination of the empirical modes is performed thanks to a Singular-Value Decomposition (SVD):

$$\mathbf{Q}_u = \mathbf{V}_u \mathbf{S}_u (\mathbf{W}_u)^T + \mathbf{R}_u \tag{12}$$

$$\mathbf{Q}_\varepsilon = \mathbf{V}_\varepsilon \mathbf{S}_\varepsilon (\mathbf{W}_\varepsilon)^T + \mathbf{R}_\varepsilon \tag{13}$$

where $\mathbf{V}_x \in \mathbb{R}^{N_d \times N_x}$, with $x = u$ or ε, is an empirical reduced basis for displacement or strain, respectively, $N_x \leq \text{rank}(\mathbf{Q}_x)$, $\mathbf{S}_x = \text{diag}(\sigma_{x1}, \ldots, \sigma_{xN_x}) \in \mathbb{R}^{N_x \times N_x}$, $\sigma_{x1} \geq \sigma_{x2} \geq, \ldots, \geq \sigma_{xN_x}$ and $\mathbf{W}_x \in \mathbb{R}^{N_t \times N_x}$. Both $\mathbf{V}_x$ and $\mathbf{W}_x$ are orthogonal. The residual $\mathbf{R}_x$ has a 2-norm such as:

$$\|\mathbf{R}_x\|_2 = \sigma_{x\,N_x+1} < \epsilon_{tol}\, \sigma_{x1}, \quad x = u \text{ or } \varepsilon \tag{14}$$

where ϵ_{tol} is a numerical parameter (typically, 10^{-3}). According to the Eckart–Young theorem, the matrix $\mathbf{V}_x\,(\mathbf{V}_x)^T\,\mathbf{Q}_x$ is the best approximation of rank N_x for $\mathbf{Q}_x$ by using the reduced basis $\mathbf{V}_x$.

The relevance of the dimensionality reduction of the displacement data appears to be conditioned by the difference between the number of time steps N_t and the order of the approximation N_u, as $\mathbf{Q}_u \in \mathbb{R}^{N_d \times N_t}$ and $\mathbf{V}_u \in \mathbb{R}^{N_d \times N_u}$. In situ tests observed in X-ray CT tend to have few time steps so the first dimensionality reduction may not be efficient. Moreover, due to the resolution of the Computed Tomography, data have generally an important number of degrees of freedom. In other words, the snapshot matrix $\mathbf{Q}_u$ has a lot of lines (N_d) but few columns (N_t). The memory cost is mostly due to the number of dof of the problem. That is why the proposed pruning protocol is based on a hyper-reduction method in order to reduce significantly this number of dof.

3.3. Reduced Experimental Domain

The proposed pruning method has its roots in the hyper-reduction method [30]. We are not able to prove that the proposed approach has a strong physical basis for pruning data according to an appropriate metric. The proposed approach is heuristic, but it fulfills some mathematical properties. A hyper-reduced order model is a set of FE equations restricted to a RID when seeking an approximate solution of FE equations with a given reduced basis. In other words, this approach accounts for the low rank of the reduced approximation to set up the reduced equations of a given FE model. Let us denote by $\mathbf{a}^{FE} \in \mathbb{R}^{N_d}$ the solution of FE equations that aims to replicate the experimental vector $\mathbf{a}$, by using the same mesh. For a given reduced basis of rank N_R $\mathbf{V} \in \mathbb{R}^{N_d \times N_R}$, the approximate reduced solution of the FE equations is denoted by $\mathbf{a}^R$ such that:

$$\mathbf{a}^R = \mathbf{V}\,\mathbf{b}^R \tag{15}$$

where $\mathbf{b}^R \in \mathbb{R}^{N_R}$ are the variables of the reduced order model. It turns out that the rank of the reduced FE equations must be N_R in order to find a unique solution $\mathbf{b}^R$. Since N_d is usually larger than N_R, few FE equations that preserve the rank of the reduced system exist. By following the hyper-reduction method proposed in [30], this selection is achieved by considering balance equations set up on a RID. In former works on hyper-reduction, the RID were generated by using simulation data.

Here, the RED is similar to a RID, but its construction uses solely experimental data, that is to say that the reduced basis used to perform this row selection comes from Equations (12) and (13). That is why the pruning method is called a model-free approach. One of the advantages of such method is that the data pruning does not have to be performed again if the constitutive model is changed. The RED is denoted by $\Omega_R \subset \Omega$.

In the hyper-reduction method, the RID is generated by the assembly of elements containing interpolation points related to various reduced bases. These reduced bases are usually extracted from simulation data generated by a given mechanical model for various parameter variations [30]. Here, the RED construction is based exclusively on the reduced bases related to $\mathbf{Q}_u$ and $\mathbf{Q}_\varepsilon$. The RED is the union of several subdomains: Ω_u and Ω_ε generated from the reduced matrices $\mathbf{V}_u$ and $\mathbf{V}_\varepsilon$, a domain denoted by Ω_+ corresponding to a set of neighboring elements to the previous subdomains, and a zone of interest (ZOI) denoted by Ω_{user}. In the sequel, Ω_{user} is set up to evaluate the force applied by the experimental setup on the specimen.

Ω_u is designed as if we would like to reconstruct experimental displacements outside Ω_u by using $\mathbf{V}_u$ and given experimental displacement in Ω_u. On a restricted subdomain Ω_u, we only have access to a restricted set of nodal displacements. The set of their indices is denoted by $\mathcal{P}_u$. The set of remaining displacement indices is denoted by $\mathcal{H}_u$ such that $\mathbf{a}[\mathcal{H}_u]$ is the vector to be reconstructed by knowing $\mathbf{a}[\mathcal{P}_u]$. Various approaches have been proposed in the literature to perform this kind of reconstruction. They are related to data completion [31] or data imputation [32] for instance. Here, we have the opportunity to choose the set $\mathcal{P}_u$, because the reconstruction issue is only formal. By using the DEIM method proposed in [33], we can obtain the set $\mathcal{P}_u$ such that $\mathbf{V}_u[\mathcal{P}_u, :]$ is a square and invertible matrix. Then, in that situation, the number of selected degrees of freedom in $\mathcal{P}_u$ is the number of empirical modes in $\mathbf{V}_u$. However, in the present application, this set could be too small to get robust calibrations after data pruning. Then, we propose a modification of the DEIM algorithm in order to multiply the number of selected indices by a given factor k. We name this algorithm k-Selection with empIrical Modes (k-SWIM). The modified algorithm is shown in Algorithm 1. When $k = 1$, this algorithm is exactly the same as the usual DEIM algorithm in [34]. The issue here is not to replicate experimental data via an interpolation scheme, but via calibrated FE simulations (by using $k > 1$). In the sequel, the set of selected indices by using k-SWIM is denoted by $\mathcal{P}_u^{(k)}$. The same reasoning is applied to the reconstruction of the experimental strain tensors. The k-SWIM algorithm applied to $\mathbf{V}_\varepsilon$ defines $\mathcal{P}_\varepsilon^{(k)}$.

For given sets of indices $\mathcal{P}_u^{(k)}$ and $\mathcal{P}_\varepsilon^{(k)}$, the RED is:

$$\Omega_R := \Omega_u \cup \Omega_\varepsilon \cup \Omega_+ \cup \Omega_{user}, \qquad \Omega_u := \cup_{j \in \mathcal{P}_u^{(k)}} \mathrm{supp}(\boldsymbol{\psi}_j) \qquad \Omega_\varepsilon := \cup_{j \in \mathcal{P}_\varepsilon^{(k)}} \mathrm{supp}(\boldsymbol{\psi}_j^\varepsilon). \tag{16}$$

where supp is the support of the function and $\boldsymbol{\psi}_j^\varepsilon$ are the shape functions related to the strain tensor in the FE model used to compute $\mathbf{a}$.

Algorithm 1: k-SWIM Selection of Variables with EmpIrical Modes

Input : integer k, linearly independent empirical modes $\mathbf{v}_l \in \mathbb{R}^d, l = 1, \ldots, M$
Output: variables index set $\mathcal{P}^{(k)}$

1 set $\mathcal{P}_0 := \emptyset, j = 0, \mathbf{U}_1 = [\,]$; `// initialization`
2 **for** $l = 1, \ldots, M$ **do**
3 $\quad \mathbf{r}_l \leftarrow \mathbf{v}_l - \mathbf{U}_l\left((\mathbf{U}_l[\mathcal{P}_j, :])^T \mathbf{U}_l[\mathcal{P}_j, :]\right)^{-1} (\mathbf{U}_l[\mathcal{P}_j, :])^T \mathbf{v}_l[\mathcal{P}_j]$; `// residual vector`
4 $\quad$ **for** $n = 1, \ldots, k$ **do**
5 $\quad\quad j \leftarrow j + 1$; `// add the k largest value of the residual`
6 $\quad\quad i_j \leftarrow \arg\max_{i \in \{1,\ldots,d\} \setminus \mathcal{P}_{j-1}} |\mathbf{r}_l[i]|$; `// index selection`
7 $\quad\quad \mathbf{r}_l[i_j] \leftarrow 0$; `// variable already selected`
8 $\quad\quad \mathcal{P}_j \leftarrow \mathcal{P}_{j-1} \cup \{i_j\}$; `// extend index set`
9 $\quad \mathbf{U}_{l+1} \leftarrow [\mathbf{v}_1, \ldots, \mathbf{v}_l]$; `// truncated reduced matrix`
10 set $\mathcal{P}^{(k)} := \mathcal{P}_j$.

Algorithm 1 is properly defined if in Line 3 the matrix $(\mathbf{U}_l[\mathcal{P}_j, :])^T \mathbf{U}_l[\mathcal{P}_j, :]$ is invertible, for $l > 1$ with $j = (l-1)\,k$, or equivalently if the following property is fulfilled.

Theorem 1. $\mathbf{U}_{l+1}[\mathcal{P}_{j+k},:]^T\,\mathbf{U}_{l+1}[\mathcal{P}_{j+k},:]$ *is invertible for* $l > 0$ *and* $j = (l-1)\,k$.

Proof. Let us assume that $\mathbf{U}_l[\mathcal{P}_j,:]^T\,\mathbf{U}_l[\mathcal{P}_j,:]$ is invertible for $l > 1$ and $j = (l-1)\,k$. Then, we compute $\mathbf{r}_l$. $(\mathbf{v}_l)_{l=1}^M$ is a set of linearly independent vectors. Thus, $\max_{i\in\{1,\ldots,d\}}|\mathbf{r}_l[i]| > 0$. Let us introduce the first additional index, $j^\star = (l-1)\,k+1$, $\mathcal{P}_{j^\star} = \mathcal{P}_j \cup \{\arg\max_{i\in\{1,\ldots,d\}}|\mathbf{r}_l[i]|\}$ and the following residual vector:

$$\mathbf{r}_l^\star = \mathbf{v}_l[\mathcal{P}_{j^\star}] - \mathbf{U}_l[\mathcal{P}_{j^\star},:]\,(\,(\mathbf{U}_l[\mathcal{P}_j,:])^T\,\mathbf{U}_l[\mathcal{P}_j,:])^{-1}\,(\mathbf{U}_l[\mathcal{P}_j,:])^T\,\mathbf{v}_l[\mathcal{P}_j]$$

Then, $\mathbf{r}_l^\star = \mathbf{r}_l[\mathcal{P}_{j^\star}]$ and $\|\mathbf{r}_l^\star\|_2 > |\mathbf{r}_l[j^\star]| > 0$. Thus, $\mathbf{U}_{l+1}[\mathcal{P}_{j^\star},:]$ is full column rank. Since $\mathcal{P}_{j^\star} \subset \mathcal{P}_{j+k}$, then $\mathbf{U}_{l+1}[\mathcal{P}_{j+k},:]$ is full column rank and $\mathbf{U}_{l+1}[\mathcal{P}_{j+k},:]^T\,\mathbf{U}_{l+1}[\mathcal{P}_{j+k},:]$ is invertible. In addition, $\mathbf{U}_2[\mathcal{P}_k,:] = \mathbf{v}_1[\mathcal{P}_k]$ is a non-zero vector. Then, $\mathbf{U}_2[\mathcal{P}_k,:]^T\,\mathbf{U}_2[\mathcal{P}_k,:] > 0$ is invertible. □

Another interesting property is the possible cancellation of the data pruning by using a large value of the parameter k in the input of Algorithm 1. The following property holds.

Theorem 2. *If* $k = N_d$ *and if* $|\mathbf{V}_u[i,1]| > 0\,\forall i = 1,\ldots,N_d$, *then* $\Omega_R = \Omega$. *The RED covers the full domain and all the data are preserved.*

Proof. By following Algorithm 1, for $l = 1$ with $k = N_d$ and $\mathbf{V}_u$ as inputs (in the algorithm, $d = N_d$), we obtain $\mathbf{q}_l = \mathbf{V}_u[:,1]$. If $|\mathbf{V}_u[i,1]| > 0\,\forall i = 1,\ldots,N_d$, then $\mathcal{P}_k = \{1,\ldots,N_d\}$. Hence, $\mathcal{P}_u^{(N_d)} = \{1,\ldots,N_d\}$ and $\Omega_u = \Omega$ and $\Omega_R = \Omega$. □

The second theorem is quite restrictive. In practice, large values of k, with $k < N_d$, enable preserving all the data. The value of k has to be chosen according to the size of the memory that we would like the free up.

3.4. Experimental Data Restricted to the RED

For a given RED, Ω_R, a set of selected degrees of freedom indices can be defined as:

$$\mathcal{F} = \left\{i \in \{1,\ldots,N_d\} \mid \int_{\Omega\backslash\Omega_R} \boldsymbol{\psi}_i^2(\mathbf{x})\mathrm{d}x = 0\right\} \tag{17}$$

The degrees of freedom in $\mathcal{F}$ are not connected to elements outside Ω_R. We denote by $\mathcal{I}$ the degrees of freedom on the interface between Ω_R and $\Omega\backslash\Omega_R$:

$$\mathcal{I} = \left\{i \in \{1,\ldots,N_d\} \mid \; i \notin \mathcal{F},\; \int_{\Omega_R} \boldsymbol{\psi}_i^2(\mathbf{x})\mathrm{d}x > 0\right\} \tag{18}$$

The union of these two set is denoted by $\overline{\mathcal{F}}$:

$$\overline{\mathcal{F}} = \mathcal{I} \cup \mathcal{F} \tag{19}$$

It contains all the indices of the degree of freedom in Ω_R.
We denote by $\overline{\mathbf{Q}}_u \in \mathbb{R}^{\mathrm{card}(\overline{\mathcal{F}})\times N_t}$ the experimental data restricted to the RED, such that:

$$\overline{\mathbf{Q}}_u = \mathbf{Q}_u[\overline{\mathcal{F}},:] \tag{20}$$

An additional SVD is performed on these experimental data such that:

$$\overline{\mathbf{Q}}_u = \overline{\mathbf{V}}_u\mathbf{S}'_u(\mathbf{W}'_u)^T + \mathbf{R}'_u \tag{21}$$

$$\mathbf{b}_u(t_j) = (\overline{\mathbf{V}}_u)^T\mathbf{a}(t_j)[\overline{\mathcal{F}}],\; j = 1,\ldots,N_t \tag{22}$$

When the RED is available, the experimental data are restricted to Ω_R and the data to be stored are:

1. The pruned reduced basis $\overline{\mathbf{V}}_u$, and the consecutive reduced coordinates $(\mathbf{b}_u(t_j))_{j=1,\dots,N_t}$.
2. The full mesh of Ω and the mesh of Ω_R ($\mathcal{F}$ and $\mathcal{I}$).
3. The load history applied to the specimen on the subdomain Ω_{user}.
4. Usual metadata related to the experiment (temperature, material parameters, etc.).

It is also advised to store the statistical distribution of a value of interest in the full domain and in the reduced domain. These data can be saved as histograms, for example. In this present paper, the shear strain distribution was saved, as this variable is extremely interesting in the case of strain localization. The additional memory cost is actually negligible as it consists in storing a few hundred floats.

The data concerning the strains are not stored as they can be computed with the displacement data thanks to Equation (7).

Generally, in-situ experiments observed in X-ray CT do not have numerous time steps, hence the above dimensionality reduction via SVD does not reduce drastically the size of the data to store. This is illustrated with the following example in Section 4. The hyper-reduction of the domain is actually the predominant step for data pruning.

3.5. Reduced Mechanical Equations Set Up on the Reduced Experimental Domain

Let us denote by $\mathbf{r}^{FE}$ the residual of the FE equations that have to be calibrated such that:

$$\mathbf{r}^{FE}(\mathbf{a}^{FE}) = 0 \tag{23}$$

For the sake of simplicity, we do not introduce the parameters $\boldsymbol{\mu}$ in the FE residual. Since the experimental data are restricted to the RED by following a hyper-reduced setting, the mechanical equations submitted to the calibration procedure are also hyper-reduced. We denote by $\overline{\mathbf{r}}^{FE}$ the partial computation of the FE residual restricted to the RED. $\overline{\mathbf{r}}^{FE}$ is the FE residual computed solely on a mesh of the RED. This mesh is termed reduced mesh. To account for the reduced mesh, a renumbering of the set $\mathcal{F}$, denoted by $\mathcal{F}^\star$, is defined such that:

$$\mathcal{F} = \overline{\mathcal{F}}[\mathcal{F}^\star] \tag{24}$$

where $\mathcal{F}^\star$ is the set of degrees of freedom related to the reduced mesh, that corresponds to $\mathcal{F}$ in the full mesh. They are located in blue squares in Figure 2b. Similarly, we define $\mathcal{I}^\star$ such that:

$$\mathcal{I} = \overline{\mathcal{F}}[\mathcal{I}^\star] \tag{25}$$

where $\mathcal{I}^\star$ is the set of degrees of freedom related to the reduced mesh that belongs to the interface between the RED and the remaining part of the full domain. The various sets of degrees of freedom are shown in Figure 2.

We assume that:

$$\overline{\mathbf{r}}^{FE}(\mathbf{a}'[\overline{\mathcal{F}}])[\mathcal{F}^\star] = \mathbf{r}^{FE}(\mathbf{a}')[\mathcal{F}] \quad \forall \mathbf{a}' \in \mathbb{R}^{N_d} \tag{26}$$

This assumption means that the FE residuals at lines selected by $\mathcal{F}$, for any prediction $\mathbf{a}'$, can be computed on the reduced mesh, where the residuals depend only on degrees of freedom in $\overline{\mathcal{F}}$. It is relevant in mechanical problems without contact condition, in the framework of first strain-gradient theory. We refer the reader to [35] for the extension of the hyper-reduction method to contact problems.

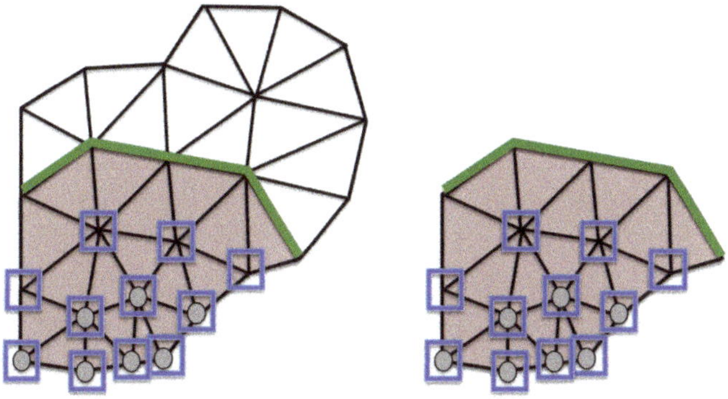

Figure 2. Schematic view of the reduced experimental domain, with linear triangular elements. In both figures, Ω_R is red. On the left, there is the mesh of Ω. On the right, there is the reduced mesh (i.e., the mesh of Ω_R only). In (**a**), the nodes having their degrees of freedom in $\mathcal{I}$ are on the green line, the nodes having their degrees of freedom in $\mathcal{F}$ are in blue squares, and the grey nodes have their degrees of freedom in $\mathcal{R}$. In (**b**), the green line, the blue squares and the grey nodes are related to $\mathcal{I}^\star$, $\mathcal{F}^\star$ and $\mathcal{R}^\star$, respectively.

Both simulation data and experimental data are incorporated in a reduced basis dedicated to the calibration procedure, after the pruning of the experimental data. In the sequel, this reduced basis is extracted from data restrained to the RED, by using the SVD. Let us denote by $\mathbf{X}$ all the data available on the full mesh. Then, after the restriction of data to the reduced mesh, the empirical reduced basis is related to $\overline{\mathbf{X}} = \mathbf{X}[\overline{\mathcal{F}},:]$:

$$\overline{\mathbf{X}} = \overline{\mathbf{V}}\,\overline{\mathbf{S}}\,\overline{\mathbf{W}}^T + \overline{\mathbf{R}},\ \overline{\mathbf{V}} \in \mathbb{R}^{\mathrm{card}(\overline{\mathcal{F}})\times N_R} \tag{27}$$

with $\|\overline{\mathbf{R}}\| < \epsilon_{tol}\ \max(\mathrm{diag}(\overline{\mathbf{S}}))$. $\overline{\mathbf{V}}$ is not a submatrix of a given $\mathbf{V}$. The way $\mathbf{X}$ contains both simulation data and experimental data is user dependent. In the proposed example, we are using a derivative extended proper orthogonal decomposition (see Schmidt et al. [36]) as explained in Section 5.2.

When the reduced basis contains empirical modes and few FE shape functions located in Ω_R, the method is termed hybrid hyper-reduction [29,35]. The hybrid FE/reduced approximation is obtained by adding few columns of the identity matrix to $\overline{\mathbf{V}}$. In this hybrid approximation, we only add FE degrees of freedom that are not connected to the degrees of freedom in $\mathcal{I}^\star$. The resulting set of degrees of freedom is denoted by $\mathcal{R}^\star$ (see Figure 2). In [29] it has been shown that this permits to have strong coupling in the resulting hybrid approximation. Let us define the subdomain connected to $\mathcal{I}$:

$$\Omega_{\mathcal{I}} = \cup_{i\in\mathcal{I}}\mathrm{supp}(\boldsymbol{\psi}_i) \cap \Omega_R \tag{28}$$

Then, we get:

$$\mathcal{R} = \left\{ i \in \overline{\mathcal{F}} | \int_{\Omega_I} \boldsymbol{\psi}_i^2(\mathbf{x})\mathrm{d}x = 0 \right\} \tag{29}$$

and $\mathcal{R}^\star$ is such that:

$$\mathcal{R} = \overline{\mathcal{F}}[\mathcal{R}^\star] \tag{30}$$

The hybrid reduced basis is denoted by $\overline{\mathbf{V}}^H$. It reads, by using the Kronecker delta (δ_{ji}):

$$\overline{\mathbf{V}}^H[:,1:N_R] = \overline{\mathbf{V}},\quad \overline{\mathbf{V}}^H[i, N_R+k] = \delta_{\mathcal{R}^\star_k i} \quad k = 1,\ldots,\mathrm{card}(\mathcal{R}) \tag{31}$$

The equations of the hybrid hyper-reduced order model (H^{2}ROM) [35] reads: find $\mathbf{b}^H \in \mathbb{R}^{N_R+\text{card}(\mathcal{R})}$ such that,

$$(\overline{\mathbf{V}}^H[\mathcal{F}^\star,:])^T\, \overline{\mathbf{r}}^{FE}(\overline{\mathbf{V}}^H\, \mathbf{b}^H)[\mathcal{F}^\star] = 0 \tag{32}$$

If the reduced equations do not have a full rank, it is suggested to remove the columns of $\overline{\mathbf{V}}$, in $\overline{\mathbf{V}}^H$, that cause the rank deficiency. When using the SVD to obtain $\overline{\mathbf{V}}$ from data, the last columns have the smallest contribution in the data approximation. These columns must be removed first in case of rank deficiency.

Theorem 3. *When $\Omega_R = \Omega$, then the hybrid hyper-reduced equations are the original FE equations on the full mesh.*

Proof. If $\Omega_R = \Omega$, then $\mathcal{I} = \emptyset$, $\mathcal{F}^\star = \mathcal{R}^\star = \{1, \ldots, N_d\}$ and the reduced mesh is the original mesh. In addition, all the empirical modes have to be removed from $\overline{\mathbf{V}}^H$ to get a full rank system of equations. Hence, $\overline{\mathbf{V}}^H$ is the identity matrix. Thus, the hybrid hyper-reduced equation are exactly the original FE equations. There is no complexity reduction. □

Theorem 4. *If ϵ_{tol} is set to zero; if both hybrid hyper-reduced equations and FE equations have unique solutions; if the FE solution $\mathbf{a}^{FE}$ belongs to the subspace spanned by the data $\mathbf{X}$; and if there exists a matrix $\mathbf{G}$ such that $\|\mathbf{a}^{FE} - \mathbf{X}\,\mathbf{G}\,\mathbf{a}^{FE}[\overline{\mathcal{F}}]\| = 0$ (i.e., the FE solution can be reconstructed by using the FE solution restricted to the RED), with $\mathbf{G} = \overline{\mathbf{W}}\,\overline{\mathbf{S}}^{-1}\,\overline{\mathbf{V}}^T$, then $\mathbf{b}^H[1 : N_R] = \overline{\mathbf{V}}^T\, \mathbf{a}^{FE}[\overline{\mathcal{F}}]$ and $\mathbf{b}^H[1 + N_R : card(\mathcal{R}) + N_R] = \mathbf{0}_{\mathcal{R}}^T$, where $\mathbf{0}_{\mathcal{R}}$ is a vector of zero in $\mathbb{R}^{card(\mathcal{R})}$. This means that the hyper-reduced solution is exact and the FE correction in the hybrid approximation is null.*

Proof. Let us introduce the matrix $\widehat{\mathbf{V}} = \mathbf{X}\,\overline{\mathbf{W}}\,\overline{\mathbf{S}}^{-1}$. Then,

$$\widehat{\mathbf{V}}[\overline{\mathcal{F}},:] = \overline{\mathbf{X}}\,\overline{\mathbf{W}}\,\overline{\mathbf{S}}^{-1} = \overline{\mathbf{V}} \tag{33}$$

If $\|\mathbf{a}^{FE} - \mathbf{X}\,\mathbf{G}\,\mathbf{a}^{FE}[\overline{\mathcal{F}}]\| = 0$ with $\mathbf{G} = \overline{\mathbf{W}}\,\overline{\mathbf{S}}^{-1}\,\overline{\mathbf{V}}^T$, so $\|\mathbf{a}^{FE} - \widehat{\mathbf{V}}\,\overline{\mathbf{V}}^T\,\mathbf{a}^{FE}[\overline{\mathcal{F}}]\| = 0$, then $\|\mathbf{a}^{FE}[\overline{\mathcal{F}}] - \overline{\mathbf{V}}\,\widehat{\mathbf{b}}^{FE}\| = 0$ with $\widehat{\mathbf{b}}^{FE} = \overline{\mathbf{V}}^T\,\mathbf{a}^{FE}[\overline{\mathcal{F}}]$ and $[(\widehat{\mathbf{b}}^{FE})^T, \mathbf{0}_{\mathcal{R}}^T]^T$ fulfills the following equation:

$$\overline{\mathbf{r}}^{FE}(\overline{\mathbf{V}}^H\,[(\widehat{\mathbf{b}}^{FE})^T, \mathbf{0}_{\mathcal{R}}^T]^T)[\mathcal{F}^\star,:] = 0 \tag{34}$$

Then, the balance equations of the hybrid hyper-reduced equations are fulfilled by $[(\widehat{\mathbf{b}}^{FE})^T, \mathbf{0}_{\mathcal{R}}^T]^T$. If both hybrid hyper-reduced equations and FE equations have unique solutions, then the solution of the hybrid hyper-reduced equations is $\mathbf{b}^H = [(\widehat{\mathbf{b}}^{FE})^T, \mathbf{0}_{\mathcal{R}}^T]^T$. □

4. Illustrating Example: Polyurethane Bonded Sand Studied with X-ray CT

4.1. Material and Test Description

The material studied here is a polyurethane bonded sand used in casting foundry to mold the internal cavities of foundry parts. The resin makes bonds between grains and improves drastically the mechanical properties of the cores (stiffness, maximum yield stress, traction strength, etc.). The material has been extensively studied with standards laboratory tests, focusing on macroscopic displacement-force curves. This casting sand was experimentally investigated by Jomaa et al. [37], Bargaoui et al. [38]. These macroscopic data are completed with an in-situ uniaxial compression test studied in X-ray CT on an as-received sample. According to Bargaoui et al. [38], the process used to make the cores (cold box process) guarantees the homogeneity of the material. In the sequel, the resin bonded sand is supposed homogeneous.

The sample is a parallelepiped (20.0 × 22.4 × 22.5 mm^3). The load was increased (with a constant displacement rate of 0.5 mm/min) and the displacement was stopped at several levels, noted P_i. During these stopped displacement periods, the sample was scanned with a tension beam of 80 kV and an intensity of 280 μA. P_0 corresponds to the initial state, before the appliance of the load. Then, seven tomography scans were performed at increasing compressed states. At P_7, the sample is broken. The bottom and top extremities were excluded from the images because of the artifacts induced by the plates. A grayscale image of the tested cemented sand is displayed in Figure 3. During the test, the reaction is measured at the top of the sample. It is plotted in Figure 4. The first six steps (non-broken sample) are situated before the peak of the loading curve.

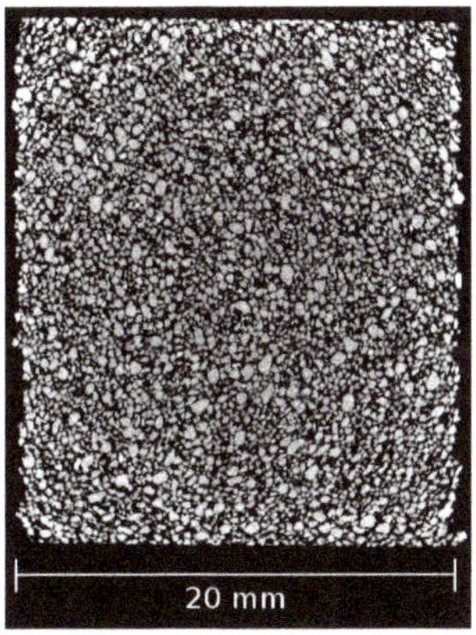

Figure 3. View of the sand.

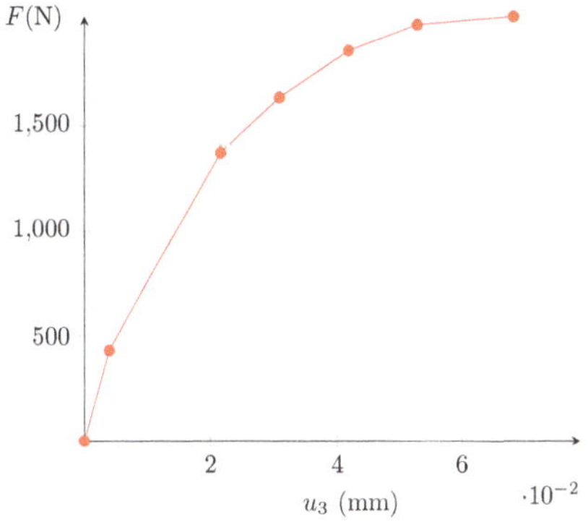

Figure 4. Measured top reaction.

4.2. DVC and Error Estimation

The displacement fields at these different stages were calculated using a 3D-digital image correlation (DVC) software named Ufreckles, developed by LaMCos (see [6]). A finite element continuum method is used to calculate the displacement field with a non-linear least square error minimization method. The chosen element size is near 0.5 mm. The final region of interest is 20.0 × 22.4 × 15.8 mm^3. The top of the sample has been excluded. The DVC is performed on a parallelipedic mesh composed of around 470,000 degrees of freedom.

The DVC showed that the pre-peak displacement field is extremely non-homogeneous, as shown in Figure 5. The test showed a complex and rich behavior of the material tested with a non-homogeneous displacement field and pre-peak strain bifurcations. The experimental data are very suited for testing the ability of a given model to predict such phenomena.

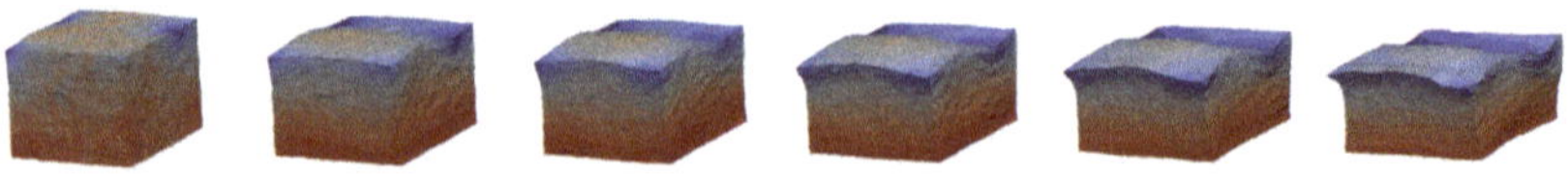

Step 1 Step 2 Step 3 Step 4 Step 5 Step 6

Figure 5. Magnitude of the experimental displacements at the pre-peak steps (deformed × 75).

4.3. Building the Reduced Experimental Basis

For a precise data pruning procedure, the experimental displacement and strain snapshot matrices are computed. The attention is drawn to the fact that the studied test does not have many time steps ($N_t = 7$) and the experimental mesh is not that big. The DVC matrices $\mathbf{Q}_u$ and $\mathbf{Q}_\varepsilon$ are, respectively, 474,405 × 7 and 1,774,080 × 14. If the truncated SVD is applied on these matrices, only six modes are extracted for the displacement and 13 for strain. As the number of time steps is rather small, the use of empirical modes does not reduce the size of the experimental data, as stated before.

In other words, the experimental data are not suited for the dimensionality reduction. This method is efficient on matrices with numerous columns and rather few lines, whereas tomographic data tend to have the exact opposite: few columns (time steps) and a lot of lines (degrees of freedom).

4.4. RED after DVC on the Specimen

During the test, the loading curve was measured at the top of the sample. To compare computed and measured reactions for model assessment, the elements at the top of the mesh are considered as a ZOI. In the remaining, Ω_+ is one layer of elements around $\Omega_u \cup \Omega_\varepsilon \cup \Omega_{user}$.

The RED was determined varying the number k of selected lines in the k-SWIM Algorithm. Its influence is assessed in Figure 6. For $k = 1$, the standard DEIM algorithm selects very few degrees of freedom. Most of the RED is actually the ZOI. This is due to the relatively low number of modes contained in the reduced basis (only 6). This apparent issue can be overcome by selecting more lines during the k-SWIM algorithm. When increasing k, the number of degrees of freedom linearly rises. The attention is drawn on the fact that the resultant RED for k = 25 or k = 50 are discontinuous, as is usually the case when using hyper-reduction methods. The newly selected zones are situated in the sheared regions. For the sake of reproducibility, the binary files related to $\mathbf{V}_u$, $\mathbf{b}_u$ and $\mathcal{P}_u$ are available as supplementary files.

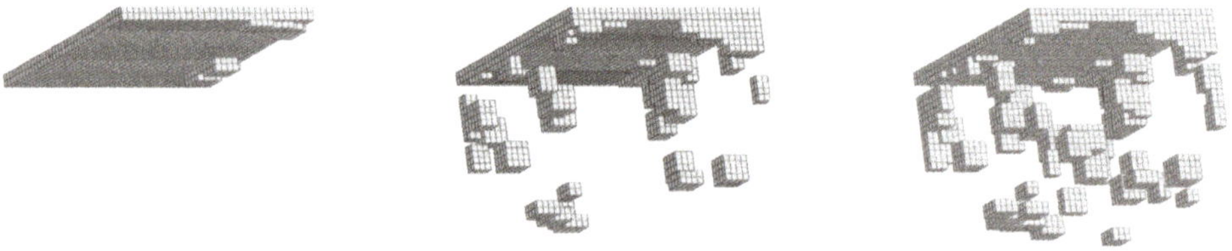

(**a**) $k = 1$ card $(\overline{\mathcal{F}})$= 47,382 (10% of N_d)

(**b**) $k = 25$ card $(\overline{\mathcal{F}})$= 73,911 (15.6% of N_d)

(**c**) $k = 50$ card $(\overline{\mathcal{F}})$= 98,064 (20.5% of N_d)

Figure 6. Influence of k in the k-SWIM algorithm.

The final RED was arbitrarily selected with $k = 25$ (around 15.6% of the total domain Ω). It is displayed in Figure 6b. The reduced domain construction is analyzed in Figure 7 where the subdomains Ω_u, Ω_ε, Ω_{user} and Ω_I are displayed.

(a) Ω_u (b) Ω_ε (c) Ω_{user} (d) Ω_+

Figure 7. Different subdomains for the selected RED for k = 25.

A summary of the different matrix sizes at each step is displayed in Table 1. As stated before, it is clear that for this kind of data, the PCA analysis does not reduce significantly the memory usage. The hyper-reduction scheme used allowed saving up to 85% of the memory space for the illustrating example.

Table 1. Size of the matrix stored at each step of the data pruning.

Experimental Data		Empirical Modes		Pruned Data	
$\mathbf{Q}_u$	474,405 × 7	$\mathbf{V}_u$	474,405 × 6	$\overline{\mathbf{V}}_u$	73,911 × 6
		$\mathbf{b}_u$	6 × 7	$\mathbf{b}_u$	6 × 7
Memory Saved		15%		85%	

5. Assessing the Relevance of the Pruned Data via Finite Element Model Updating-H^{2}ROM

In this section, the relevance of the pruned data for further usage is discussed. The experimental data extracted from computed tomography can have various purposes. This paper focuses on its use for model calibration, and is illustrated with the in-situ compressive test of a resin bonded sand presented in the previous section. The main aim of this part is to prove that the RED computed thanks to a model free procedure is relevant to assess or calibrate an arbitrary constitutive model.

The model used for the illustrating example is a constitutive elastoplastic model with m unknown parameters to calibrate. The procedure employed is a Finite Element Model Updating (FEMU) technique, coupled with an hybrid hyper-reduction method for the solution of approximate balance equations. The use of such method is straightforward as the input data are actually hyper-reduced. This approach is termed FEMU-H^{2}ROM.

The FEMU-H^{2}ROM method is resumed in the flowchart in Figure 8. The FEMU-H^{2}ROM aims to find the best parameter $\boldsymbol{\mu}^*$ that replicate the experimental data available on the RED by using hyper-reduced equations. During the optimization procedure, the parameters are updated via hybrid hyper-reduced simulations. After few adaptation steps, the optimality of the parameter is checked by using a full FE simulation. If required, the reduced basis involved in the hyper-reduced simulation are updated.

5.1. Constitutive Model MC-CASM

5.1.1. Presentation

The resin-bonded sand behavior is modeled with a relatively simple constitutive model based on the Cemented Clay and Sand Model (C-CASM). It consists in the extension of the Clay And Sand Model developed by Yu [39] for unbonded sand and clay to bonded geomaterials within the framework developed by Gens and Nova [40]. The C-CASM has been extensively described in [41]. The Modified Cemented Clay And Sand Model (MC-CASM) presented here has some modifications of the C-CASM:

- Addition of a damage law whose equation is phenomenological (based on cycled compressive tests).

- The hardening law of the bonding parameter b is different: A first hardening precedes the softening. It is supposed here that the polyurethane resin goes through a first hardening before breaking.

It is supposed here that the yield function was previously calibrated with standard laboratory tests. The calibration concerns the parameters involved in the different damage and hardening laws that can be more difficult to assess with macroscopic loading curves. In the continuation of the paper, the equivalent von Mises stress is denoted q and the mean pressure p. The MC-CASM equations are summarized hereafter.

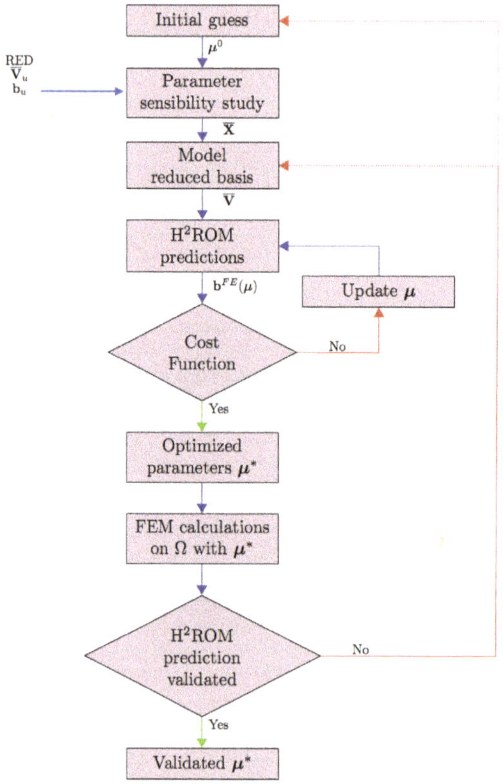

Figure 8. Flowchart of the FEMU-H^{2}ROM.

5.1.2. Yield Function and Plastic Flow

The yield function, f, of the constitutive model is defined by:

$$f(\sigma; p_c, b) = \left(\frac{q}{M(p+p_t)}\right)^n + \frac{1}{\ln r}\ln\left(\frac{p+p_t}{p_c(1+b)+p_t}\right) \tag{35}$$

where M, r, and n are constant parameters that control the shape of the yield function. p_c is the preconsolidation pressure, that is to say the maximum yield pressure during an isotropic compressive test (see Roscoe et al. [42]). b is the bounding parameter modeling the amplification of the yield surface due to intergranular bonding. p_t is the traction resistance of the soil defined by Gens and Nova [40] as:

$$p_t = \alpha b p_c \tag{36}$$

where α is a constant parameter modeling the influence of the binder on the traction resistance. The yield function is supposed to be calibrated. This means that M, r, n, α and the initial values of p_c and b are known. The yield surfaces of the unbonded (blue) and bonded sand (red) are plotted in Figure 9.

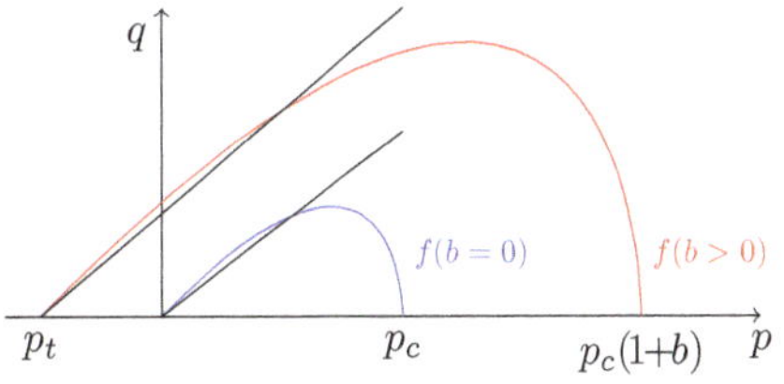

Figure 9. Yield surfaces in the (p, q) plane.

5.1.3. Hardening and Damage Laws

The model has two hardening variables: the preconsolidation pressure p_c and the bonding parameter b. The evolution of p_c is directly controlled by the incremental plastic volumetric strain $\dot{\varepsilon}_v^p$, whereas b relies on a plastic strain damage measure h:

$$\frac{\dot{p}_c}{p_c} = \mu_1 \dot{\varepsilon}_v^p \tag{37}$$

$$\dot{b} = (-be^{-h} + \mu_6 \mu_7 e^{-\mu_7 h})\dot{h} \tag{38}$$

The incremental value of h is defined as a weighting of the effects of the incremental plastic shear strain and the incremental plastic volumetric strain:

$$\dot{h} = \mu_2 |\dot{\varepsilon}_s^p| + \mu_3 |\dot{\varepsilon}_v^p| \tag{39}$$

The model also includes a damage law whose formulation is purely phenomenological:

$$E = E_0(1 - D) \tag{40}$$

$$D = \mu_4 h^{\mu_5} \tag{41}$$

The hardening and damage laws provide $m = 7$ unknown parameters to calibrate.

5.2. Calibration Protocol by Using the Hybrid Hyper-Reduction Method

The FEMU-H^{2}ROM is preceded by an off-line phase similar to an unsupervised machine learning phase. It consists in building the empirical reduced basis $\overline{\mathbf{V}}$ that is mandatory to set up the hybrid hyper-reduced equations. It is similar to the first step of the data pruning method: a snapshot matrix is constructed based on simulations and experimental results (and not on experiments only).

The starting point of the off-line phase is to assess the parameter sensibilities of the model starting from an initial guess $\boldsymbol{\mu}^0 = \{\mu_1^0, \ldots, \mu_m^0\}$. This guess can come from a previous calibration, or a calibration done using macroscopic force–displacement curves of standard tests without predicting strain localization.

The off-line calculations are performed on the full domain Ω and thus can be time consuming. The boundary conditions are the experimental displacements taken from the computed tomography imposed at the top and the bottom of the sample. The displacement field is not imposed inside the sample because one of the aims of the model is to correctly capture the strain localization appearing inside the sample during the test, under the constraint of balance equations. Imposing the displacement field inside the specimen gives less balance equation to fulfill. m calculations are made on Ω. Attention is drawn to the fact that these calculations can be done in parallel. Only the displacement snapshot matrices are needed. A total of $m+1$ independent calculations are performed:

- One initial calculation where $\boldsymbol{\mu} = \boldsymbol{\mu}^0$, which gives $\mathbf{Q}_u^{FE}(\boldsymbol{\mu}^0)$;
- m parameters sensibility calculations where $\boldsymbol{\mu} = \boldsymbol{\mu}^i = \{\mu_1^0, \ldots, \mu_i^0 + \delta\mu_i^0, \ldots, \mu_m^0\}$, which give $\mathbf{Q}_u^{FE}(\boldsymbol{\mu}^i)$ for $i = 1, \ldots, m$

Once done, these calculations are restricted to the reduced experimental domain Ω_R. They are denoted $\overline{\mathbf{Q}}_u^{FE}(\boldsymbol{\mu}^i)$ for $i = 0, \ldots, m$. All these results have to be aggregated in one snapshot matrix $\overline{\mathbf{X}}$ before the computation of the empirical modes $\overline{\mathbf{V}}$. Instead of concatenating the $m+1$ matrices into one, a DEPOD method is used (see Schmidt et al. [36]). This approach has been validated in previous works on model calibration with hyper-reduction (see Ryckelynck and Missoum Benziane [43]). This allows capturing the effects of each parameter variation.

$$\overline{\mathbf{X}} = [\alpha \overline{\mathbf{V}}_u \mathbf{b}_u, \overline{\mathbf{Q}}_u^{FE}(\boldsymbol{\mu}^0), \frac{\|\overline{\mathbf{Q}}_u^{FE}(\boldsymbol{\mu}^0)\|_F}{2\,\|\overline{\mathbf{Q}}_u^{FE}(\boldsymbol{\mu}^1) - \overline{\mathbf{Q}}_u^{FE}(\boldsymbol{\mu}^0)\|_F}(\overline{\mathbf{Q}}_u^{FE}(\boldsymbol{\mu}^1) - \overline{\mathbf{Q}}_u^{FE}(\boldsymbol{\mu}^0)), \ldots,$$
$$\frac{\|\overline{\mathbf{Q}}_u^{FE}(\boldsymbol{\mu}^0)\|_F}{2\,\|\overline{\mathbf{Q}}_u^{FE}(\boldsymbol{\mu}^m) - \overline{\mathbf{Q}}_u^{FE}(\boldsymbol{\mu}^0)\|_F}(\overline{\mathbf{Q}}_u^{FE}(\boldsymbol{\mu}^m) - \overline{\mathbf{Q}}_u^{FE}(\boldsymbol{\mu}^0))] \quad (42)$$

where $\|\cdot\|_F$ is the Frobenius norm. The first term $\alpha \overline{\mathbf{V}}_u \mathbf{b}_u$ corresponds to the pruned experimental data. It is weighted by a custom parameter α that enables giving more impact to the experimental fluctuations in the empirical modes. The finite element methods tends to smooth these fluctuations, thus provoking a certain loss of information.

Empirical modes depending on the factor α are displayed in Figure 10. For $\alpha = 0$, that is to say without experimental data in the bulk, the empirical modes have strong fluctuations only at the top and the bottom of the specimen, where the experimental boundary conditions are imposed. This can be explained by the natural smoothing that ensures the finite element method with rather elliptic equations. Increasing the importance of the experimental data tends to naturally perturb the displacement field inside the sample. Even for strongly perturbed modes ($\alpha = 10$), the last empirical mode is roughly smooth: this is due to the POD algorithm that filters the data. In the sequel, we choose $\alpha = 1$. The experimental data are as important as simulation data related to FE balance equations.

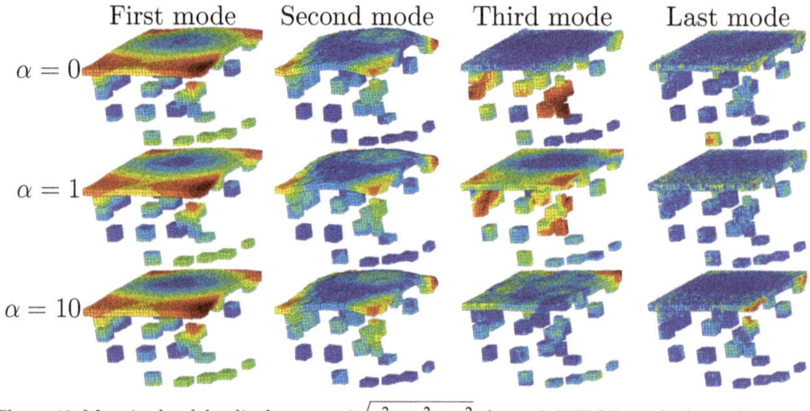

Figure 10. Magnitude of the displacement ($\sqrt{u_1^2 + u_2^2 + u_3^2}$) for each DEPOD mode depending on α.

Once $\overline{\mathbf{V}}$ is available, the hybrid reduced basis $\overline{\mathbf{V}}^H$ can be defined. Then, the experimental reduced coordinates are projected on the empirical reduced basis to be compared during the optimization loop:

$$\mathbf{b}_u^H = (\overline{\mathbf{V}}^H)^T \, \overline{\mathbf{V}}_u \, \mathbf{b}_u \quad (43)$$

For the proposed example, there is a fast decay of the singular value (see Figure 11 where ϵ_{POD} is set to 10^{-4}). When this decay is not sufficient to provide a small number of empirical modes, we refer the reader to [44–46] to cluster the data in order to divide the time interval and construct local reduced basis in time.

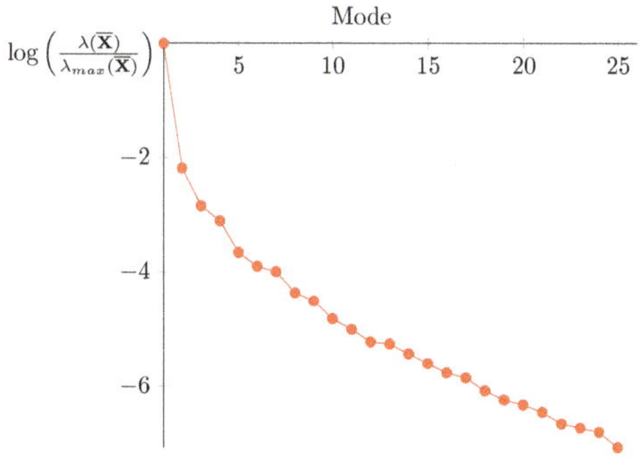

Figure 11. Singular values of $\overline{\mathbf{X}}$ verifying $\lambda(\overline{\mathbf{X}}) > \epsilon_{POD}\lambda_{max}(\overline{\mathbf{X}})$.

5.3. Discussion on Dirichlet Boundary Conditions

After the data pruning, experimental data are available in all Ω_R. When displacements are constrained to follow the experimental data, we loose FE balance equations. The following theorem helps to discuss the Dirichlet boundary conditions.

Theorem 5. *If $\alpha > 0$, $\epsilon_{tol} = 0$; if the experimental data $\overline{\mathbf{Q}}_u = \overline{\mathbf{V}}_u\, \mathbf{b}_u$ fulfill the FE equations on Ω_R with the following additional Dirichlet boundary conditions:*

$$\mathbf{a}^{FE}(t_j, \boldsymbol{\mu})[\mathcal{I}] = \overline{\mathbf{Q}}_u[\mathcal{I}^*, j]; \tag{44}$$

and if both hybrid hyper-reduced equations and FE equations on Ω_R are unique, then the solution of the hybrid hyper-reduced equation is the exact projection of the experimental data on the empirical reduced basis $\mathbf{b}^H(\boldsymbol{\mu}) = [(\overline{\mathbf{V}}^T \overline{\mathbf{Q}}_u)^T, \mathbf{0}_{\mathcal{R}}^T]^T$, with $\|\overline{\mathbf{Q}}_u - \overline{\mathbf{V}}\,\overline{\mathbf{V}}^T \overline{\mathbf{Q}}_u\| = 0$.

Proof. If the solution of the FE equations in Ω_R is unique with Dirichlet boundary conditions on $\mathcal{I}^*$ equal to $\mathbf{a}^{FE}(t_j, \boldsymbol{\mu})[\mathcal{I}]$, then this solution is $\mathbf{a}^{FE}(t_j, \boldsymbol{\mu})[\mathcal{F}]$. If $\overline{\mathbf{Q}}_u$ fulfills the FE equations on Ω_R, with the additional Dirichlet boundary conditions, then:

$$\mathbf{a}^{FE}(t_j, \boldsymbol{\mu})[\mathcal{F}] = \overline{\mathbf{Q}}_u[:, j] \quad j = 1, \ldots, M$$

and

$$\overline{\mathbf{r}}^{FE}(\overline{\mathbf{Q}}_u[:, j])[\mathcal{F}^\star, :] = 0 \quad j = 1, \ldots, M$$

If $\alpha > 0$ and $\epsilon_{tol} = 0$, then

$$\mathbf{a}^{FE}(t_j, \boldsymbol{\mu})[\mathcal{F}] = \overline{\mathbf{V}}\, \mathbf{b}^{FE}(t_j, \boldsymbol{\mu}) \quad j = 1, \ldots, M,$$

with

$$\mathbf{b}^{FE}(t_j, \boldsymbol{\mu}) = \overline{\mathbf{V}}^T \overline{\mathbf{Q}}_u[:, j] = \overline{\mathbf{V}}^T \overline{\mathbf{V}}_u\, \mathbf{b}_u(t_j) \quad j = 1, \ldots, M$$

Then,

$$\overline{\mathbf{r}}^{FE}(\overline{\mathbf{V}}^H \mathbf{b}^H(t_j, \boldsymbol{\mu}))[\mathcal{F}^\star, :] = 0 \quad j = 1, \ldots, M$$

with $\mathbf{b}^H(t_j,\boldsymbol{\mu}) = [(\mathbf{b}^{FE}(t_j,\boldsymbol{\mu}))^T, \mathbf{0}_R^T]^T$. Thus, $\mathbf{b}^H(t_j,\boldsymbol{\mu})$ is the unique solution of the hybrid hyper-reduced equations, and the exact projection of the restrained FE solution. □

The last theorem does not imply that imposing $\mathbf{a}^{FE}(t_j,\boldsymbol{\mu})[\mathcal{I}] = \overline{\mathbf{Q}}_u[\mathcal{I}^*, j]$ as a boundary condition to degrees of freedom in $\mathcal{I}^*$ is the best way to fulfill FE balance equations on the full mesh. In fact, with the additional boundary conditions on $\mathcal{I}^*$, the maximum of available FE equations is $\text{card}(\mathcal{F}^*)$. Theorem 4 means that if the empirical reduced basis is exact, then all the N_d FE balance equations are fulfilled in Ω. In a sense, in the proposed calibration protocol, we better trust in FE balance equations than in experimental data. Accurate FE balance equations can be obtained by a convenient mesh of Ω, although noise is always present in experimental data.

5.4. Parameters Updating

In the optimization loop (Figure 8), a given set of parameters $\boldsymbol{\mu}$ is assessed. The H^{2}ROM calculations provide the reduced coordinates associated with the empirical basis previously determined on the RED denoted $\mathbf{b}^H(\boldsymbol{\mu})$. The top reaction $\mathbf{F}^{FE}(\boldsymbol{\mu})$ is also calculated as the average axial stress in the ZOI.

In the example, the cost function that must be minimized, evaluates two scales of error: the microscale error between experimental and computed reduced coordinates and the macroscale error between the measured and computed top reactions. These error functions are, respectively, denoted $\chi_u^2(\boldsymbol{\mu})$ and $\chi_F^2(\boldsymbol{\mu})$.

The microscale error is defined as:

$$\chi_u^2(\boldsymbol{\mu}) = (\mathbf{b}^H(\boldsymbol{\mu}) - (\overline{\mathbf{V}}^H)^T\,\overline{\mathbf{V}}_u\,\mathbf{b}_u)^T(\mathbf{b}^H(\boldsymbol{\mu}) - (\overline{\mathbf{V}}^H)^T\,\overline{\mathbf{V}}_u\,\mathbf{b}_u) \tag{45}$$

The choice of the norm is user-dependent. The inverse covariance matrix of the displacement is the best norm for a Gaussian noise according to [47,48] for a Bayesian framework. However, in this present study, to keep the treated problem rather simple, a 2-norm has been chosen. The macroscale error is defined as:

$$\chi_F^2(\boldsymbol{\mu}) = \|\mathbf{F}^{FE}(\boldsymbol{\mu}) - \mathbf{F}\|^2_{\partial_u\Omega} \tag{46}$$

Here, $\partial_u\Omega$ is the top surface of the ZOI, where the experimental load was measured and where the experimental displacements are imposed as Dirichlet boundary conditions. The experimental load measurements are supposed uncorrelated and their variance is denoted by σ_F^2. In a Bayesian framework, for a Gaussian noise corrupting the load measurements [23], the previous equation can be written as:

$$\chi_F^2(\boldsymbol{\mu}) = \frac{1}{N_t\sigma_F^2}(\mathbf{F}^{FE}(\boldsymbol{\mu}) - \mathbf{F})^T(\mathbf{F}^{FE}(\boldsymbol{\mu}) - \mathbf{F}) \tag{47}$$

For the the optimization loop, the final objective function is a weighted sum of the two previous sub-objective functions:

$$\chi^2(\boldsymbol{\mu}) = c_u\chi_u^2(\boldsymbol{\mu}) + c_F\chi_F^2(\boldsymbol{\mu}) \tag{48}$$

where c_u and c_F are the weights. They can be chosen to balance the two cost functions or to privilege one scale to another. In the illustrating example, the cost function is balanced. A classical Levenberg–Marquardt algorithm is employed for the minimization of the error function and the update of the parameters vector $\boldsymbol{\mu}$.

5.5. Model Calibration and FEM Validation

The optimization loop took 53 iterations. The speed ratio between FEM calculations and H^{2}ROM predictions is around 70. Moreover, the H^{2}ROM predictions only needed around 3% of the FEM calculation memory cost. The H^{2}ROM predictions converge way more easily than the FEM calculations. The problem simulated in the optimization loop is a displacement imposed problem. The use of the reduced basis to predict the displacement field facilitates drastically the convergence. That explains also the important speed-up time that does not come only from the reduction of the integration domain.

Figure 12 displays the experimental and the computed top reactions (initial and optimized). At the end of the optimization loop, it is mandatory to assess the relevance of the H^{2}ROM prediction. The FEMU-H^{2}ROM is dependent on the initial guess μ^0. This input determines the relevance of the reduced basis of the model after the parameters sensibility study and the DEPOD analysis. When updating the model, the parameter set may be too different from the initial guess. As a consequence, the empirical reduced basis $\overline{\mathbf{V}}^H$ may not be accurate and the H^{2}ROM predictions will not be admissible. That is to say that the discrepancy between hyper reduced and Finite Element calculations may not be negligible. That is why the optimized parameters set μ^* must be validated with FEM calculations on the full domain Ω. It is worth noting that, if the experimental data are included in the DEPOD, the final H^{2}ROM prediction should be close to the experiments.

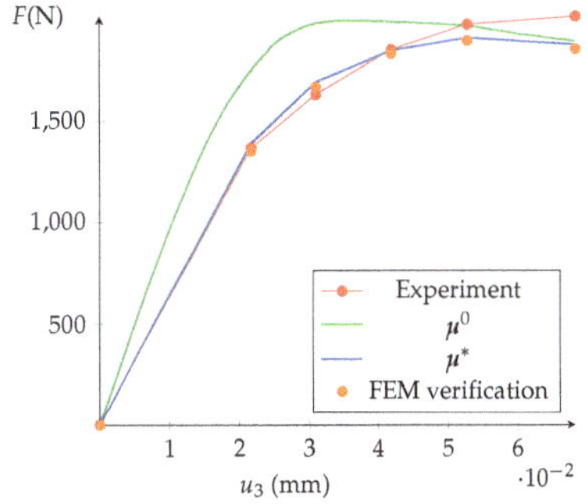

Figure 12. Result of the H^{2}ROM optimization.

In a similar manner to the optimization loop, an error function between both calculations can be defined focusing on the microscale (displacement error) and macroscale (top reactions differences).

Concerning the microscale, the discrepancy is only computed in the RED, as H^{2}ROM predictions are only made on this domain and cannot be reconstructed in the full domain with this particular approach. The microscale discrepancy is estimated by r_u:

$$r_u^2(\mu^*) = \left(\overline{\mathbf{a}}^H(\mu^*) - \mathbf{a}^{FE}(\mu^*)[\overline{\mathcal{F}}]\right)^T \left(\overline{\mathbf{a}}^H(\mu^*) - \mathbf{a}^{FE}(\mu^*)[\overline{\mathcal{F}}]\right), \text{ with } \overline{\mathbf{a}}^H(\mu^*) = \overline{\mathbf{V}}^H \mathbf{b}^H(\mu^*) \tag{49}$$

In the same manner, the macroscale discrepancy measure the norm of the difference between the two prediction of the load applied to the specimen. This indicator is denoted by r_F. The microscale and macroscale errors should not exceed a few percents of the FEM calculations. In Figure 12, the FEM top reaction is plotted in orange. It is clear that its value is extremely close to the one computed thanks to H^{2}ROM. The error is around 1% at each step.

This final verification is purely numerical. If the H^{2}ROM predictions are validated, it is advised to analyze deeper the full field FEM calculation.

In the case of notable differences between H^{2}ROM prediction and FEM calculations, or between FEM calculations and experiment, the FEMU-H^{2}ROM is not validated. Two solutions are possible to overcome this issue:

1. Perform again the whole parameters sensibility study with $\mu^0 = \mu^*$.
2. Concatenate the previously determined matrix $\overline{\mathbf{X}}$ from Equation (42) with $\overline{\mathbf{Q}}_u(\mu^*)$ and perform a new truncated SVD to determine ultimately an enriched reduced basis $\overline{\mathbf{V}}^H$. No new FEM calculations are needed.

The first solution should be performed in the case of strong differences between H^{2}ROM prediction and FEM calculations. The second option "only" costs a FEM calculation. It is also possible to modify the optimization loop to include regularly FEM-H^{2}ROM comparison and enrich $\overline{\mathbf{V}}^H$ incrementally.

6. Discussion

6.1. Limitations of the Pruning Procedure

The present paper focused on DVC sets and not on the images themselves. Since each element covers several voxels, the images are also known to be particularly heavy and perhaps more problematic than the DVC data. The pruning procedure considers that they can be deleted. Actually, it can be problematic. For instance, new DVC algorithm could improve the determination of the displacement field (for example for complex problems involving cracks).

The images could be pruned too, in the sense that the only the pixels of the images inside the determined RED can be conserved. However, we preconize to store only the reduced DVC data when the data storage is an issue.

In the case of non homogeneous materials, the data concerning the inhomogeneity outside the RED must be saved as well.

6.2. About the Reconstruction of Data outside the RED

Because of the proposed data pruning, experimental data outside the RED are no more available. However, the finite element verification gives access to an estimation of these data via the finite element model and the optimal parameters $\boldsymbol{\mu}^*$. For instance, the shear strain distribution can be estimated by the finite element model with the optimal values of the parameter. In the illustrating example, the computed and measured shear strain distributions, over the integration points in Ω, were compared. The analysis is summarized in the histograms displayed in Figure 13 for the last pre-peak step. The discrepancy between computed (via FE verification) and measured distributions was considered here as satisfying.

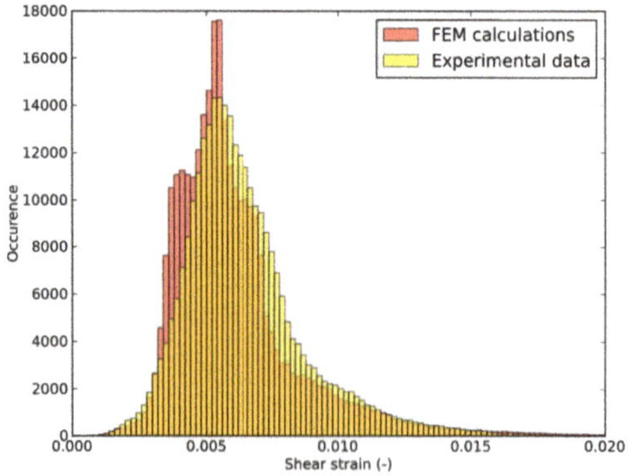

Figure 13. Probability distribution of shear strain at the last pre-peak step in the whole domain Ω, comparing FEM calculation (verification step) and experimental data.

6.3. Shear Strain Distributions in the RED

We can also consider the shear strain distributions is inside the whole domain Ω and the RED Ω_R for the illustrating example. It would be preferable that the pruning procedure stores in the RED the most different configurations. The shear strain distributions in the whole domain and in the RED might be different (not the same mean value for example). Figures 14a and 15a present the shear strain distributions at the first and last pre-peak step. It appears that the statistical distribution of the shear strain inside the RED is not the same than the one inside the full domain. Nevertheless, zooms at both histograms in Figures 14b and 15b reveal that the extremum values of the shear strain are conserved. One can see that the RED contains nearly all the elements where the shear is maximal. Even if the proposed procedure is model-free, it is intimately linked with the mechanics of solids: it will store preferably the data that are mechanically more relevant. For strain localization phenomenon, it is the most sheared zone. The proposed method is not statistical: it actually induces a sampling bias.

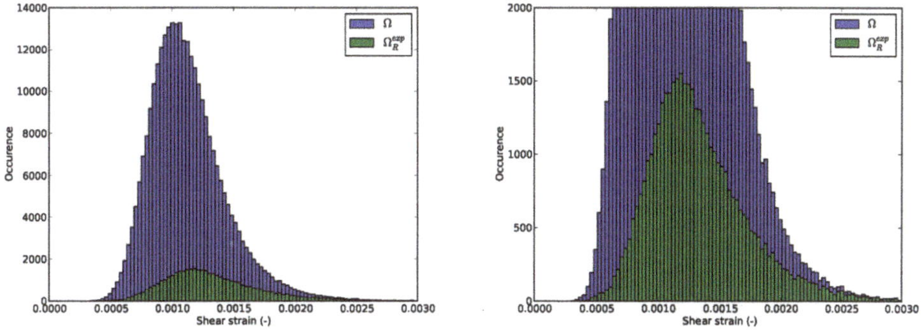

(a) (b)

Figure 14. Shear strain distributions (**a**) in the whole domain and (**b**) in the RED at the first step.

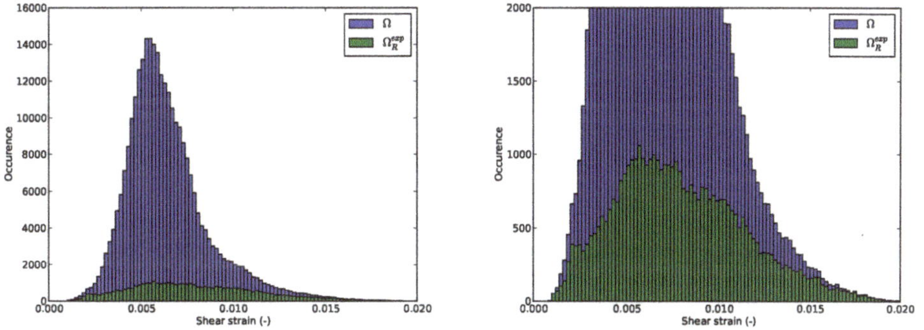

(a) (b)

Figure 15. Shear strain distributions (**a**) in the whole domain and (**b**) in the RED at the last step.

7. Conclusions

The present paper proposes a data pruning procedure for DVC data that is model free and versatile. The k-SWIM algorithm, through its parameter k, enables the user to define the size of the stored data.

The resultant data can still be used afterwards, for instance for calibration. The use of hybrid hyper-reduction is particularly suitable for the pruned data as it enables a non-negligible reduction of memory and time costs in the FEMU optimization loop. The FEMU-H^{2}ROM method is thus a new way to use massive DVC data for deeper mechanical studies.

Supplementary Materials: The following are available online at http://www.mdpi.com/2297-8747/24/1/18/s1 as supplementary files to make the output of Algorithm 1 reproducible. The ASCII file Node-iXYZ.txt contains the node indices and the related coordinates. The files Vu.npy, bu.npy and Pu_reference.npy, are binary files related to $\mathbf{V}_u$, $\mathbf{b}_u$ and $\mathcal{P}_u$, respectively. They have been generated by using the NumPy instruction "save". The ASCII file k_swim.py contains Algorithm 1 written with SciPy instructions. In the ASCII file run_kswim.py, this algorithm is applied to the data $\mathbf{V}_u$.

Author Contributions: Conceptualization, D.R.; methodology, D.R. and W.H.; experimental data, C.M.; data analysis, W.H.; writing, D.R. and W.H.

Funding: This research was funded by Agence Nationale de la Recherche, in France; grant number is ANR-14-CE07-0038-03 FIMALIPO.

Acknowledgments: The authors would like to acknowledge the Agence Nationale de la Recherche for their financial support for the FIMALIPO project.

Conflicts of Interest: The authors declare no conflict of interest.

References

1. Chu, T.; Ranson, W.; Sutton, M.; Peters, W. Application of digital-image-dorrelation techniques to experimental mechanics. *Exp. Mech.* **1985**, *25*, 232–244. [CrossRef]
2. Bay, B.; Smith, T.; Fyhrie, D.; Saad, M. Digital volume correlation: Three-dimensional strain mapping using X-ray tomography. *Exp. Mech.* **1999**, *39*, 217–226. [CrossRef]
3. Van Ooijen, P.M.A.; Broekema, A.; Oudkerk, M. Use of a Thin-Section Archive and Enterprise 3-Dimensional Software for Long-Term Storage of Thin-Slice CT Datasets—A Reviewers' Response. *J. Digit. Imag.* **2008**, *21*, 188–192. [CrossRef] [PubMed]
4. Pan, D. A tutorial on MPEG/audio compression. *IEEE MultiMed.* **1995**, *2*, 60–74. [CrossRef]
5. Cioaca, A.; Sandu, A. Low-rank approximations for computing observation impact in 4D-Var data assimilation. *Comput. Math. Appl.* **2014**, *67*, 2112–2126. [CrossRef]
6. Réthoré, J.; Roux, S.; Hild, F. From pictures to extended finite elements: Extended digital image correlation (X-DIC). *C. R. Méc.* **2007**, *335*, 131–137. [CrossRef]
7. Rojanaarpa, T.; Kataeva, I. Density-Based Data Pruning Method for Deep Reinforcement Learning. In Proceedings of the 2016 15th IEEE International Conference on Machine Learning and Applications (ICMLA), Anaheim, CA, USA, 18–20 December 2016; pp. 266–271.
8. Hu, Y.; Chen, H.; Li, G.; Li, H.; Xu, R.; Li, J. A statistical training data cleaning strategy for the PCA-based chiller sensor fault detection, diagnosis and data reconstruction method. *Energy Build.* **2016**, *112*, 270–278. [CrossRef]
9. Hong, Y.; Kwong, S.; Chang, Y.; Ren, Q. Unsupervised data pruning for clustering of noisy data. *Knowl. Based Syst.* **2008**, *21*, 612–616. [CrossRef]
10. Kavanagh, K.T.; Clough, R.W. Finite element applications in the characterization of elastic solids. *Int. J. Solids Struct.* **1971**, *7*, 11–23. [CrossRef]
11. Kavanagh, K.T. Extension of classical experimental techniques for characterizing composite-material behavior. *Exp. Mech.* **1972**, *12*, 50–56. [CrossRef]
12. Ienny, P.; Caro-Bretelle, A.S.; Pagnacco, E. Identification from measurements of mechanical fields by finite element model updating strategies. *Eur. J. Comput. Mech.* **2009**, *18*, 353–376. [CrossRef]
13. Lecompte, D.; Smits, A.; Sol, H.; Vantomme, J.; Hemelrijck, D.V. Mixed numerical–experimental technique for orthotropic parameter identification using biaxial tensile tests on cruciform specimens. *Int. J. Solids Struct.* **2007**, *44*, 1643–1656. [CrossRef]
14. Molimard, J.; Le Riche, R.; Vautrin, A.; Lee, J.R. Identification of the four orthotropic plate stiffnesses using a single open-hole tensile test. *Exp. Mech.* **2005**, *45*, 404–411. [CrossRef]
15. Meijer, R.; Douven, L.F.A.; Oomens, C.W.J. Characterisation of Anisotropic and Non-linear Behaviour of Human Skin In Vivo. *Comput. Methods Biomech. Biomed. Eng.* **1999**, *2*, 13–27. [CrossRef] [PubMed]
16. Bruno, L. Mechanical characterization of composite materials by optical techniques: A review. *Opt. Lasers Eng.* **2018**, *104*, 192–203. [CrossRef]
17. Mahnken, R.; Stein, E. A unified approach for parameter identification of inelastic material models in the frame of the finite element method. *Comput. Methods Appl. Mech. Eng.* **1996**, *136*, 225–258. [CrossRef]

18. Forestier, R.; Massoni, E.; Chastel, Y. Estimation of constitutive parameters using an inverse method coupled to a 3D finite element software. *J. Mater. Process. Technol.* **2002**, *125–126*, 594–601. [CrossRef]
19. Giton, M.; Caro-Bretelle, A.S.; Ienny, P. Hyperelastic Behaviour Identification by a Forward Problem Resolution: Application to a Tear Test of a Silicone-Rubber. *Strain* **2006**, *42*, 291–297. [CrossRef]
20. Cugnoni, J.; Botsis, J.; Janczak-Rusch, J. Size and Constraining Effects in Lead-Free Solder Joints. *Adv. Eng. Mater.* **2006**, *8*, 184–191. [CrossRef]
21. Latourte, F.; Chrysochoos, A.; Pagano, S.; Wattrisse, B. Elastoplastic behavior identification for heterogeneous loadings and materials. *Exp. Mech.* **2008**, *48*, 435–449. [CrossRef]
22. Padmanabhan, S.; Hubner, J.P.; Kumar, A.V.; Ifju, P.G. Load and Boundary Condition Calibration Using Full-field Strain Measurement. *Exp. Mech.* **2006**, *46*, 569–578. [CrossRef]
23. Neggers, J.; Allix, O.; Hild, F.; Roux, S. Big Data in Experimental Mechanics and Model Order Reduction: Today's Challenges and Tomorrow's Opportunities. *Arch. Comput. Methods Eng.* **2017**. [CrossRef]
24. Cugnoni, J.; Gmür, T.; Schorderet, A. Inverse method based on modal analysis for characterizing the constitutive properties of thick composite plates. *Comput. Struct.* **2007**, *85*, 1310–1320. [CrossRef]
25. Aubry, N.; Holmes, P.; Lumley, J.L.; Stone, E. The dynamics of coherent structures in the wall region of a turbulent boundary layer. *J. Fluid Mech.* **1988**, *192*, 115–173. [CrossRef]
26. Passieux, J.C.; Perie, J.N. High resolution digital image correlation using proper generalized decomposition: PGD-DIC. *Int. J. Numer. Methods Eng.* **2012**, *92*, 531–550. [CrossRef]
27. Ryckelynck, D. A priori hyperreduction method: An adaptive approach. *J. Comput. Phys.* **2005**, *202*, 346–366. [CrossRef]
28. Hernandez, J.A.; Caicedo, M.A.; Ferrer, A. Dimensional hyper-reduction of nonlinear finite element models via empirical cubature. *Comput. Methods Appl. Mech. Eng.* **2017**, *313*, 687–722. [CrossRef]
29. Baiges, J.; Codina, R.; Idelson, S. A domain decomposition strategy for reduced order models. Application to the incompressible Navier-Stokes equations. *Comput. Methods Appl. Mech. Eng.* **2013**, *267*, 23–42. [CrossRef]
30. Ryckelynck, D.; Lampoh, K.; Quilici, S. Hyper-reduced predictions for lifetime assessment of elasto-plastic structures. *Meccanica* **2015**. [CrossRef]
31. Liu, J.; Musialski, P.; Wonka, P.; Ye, J. Tensor Completion for Estimating Missing Values in Visual Data. *IEEE Trans. Pattern Anal. Mach. Intell.* **2013**, *35*, 208–220. [CrossRef] [PubMed]
32. Shang, Q.; Yang, Z.; Gao, S.; Tan, D. An Imputation Method for Missing Traffic Data Based on FCM Optimized by PSO-SVR. *J. Adv. Transp.* **2018**. [CrossRef]
33. Chaturantabut, S.; Sorensen, D.C. Nonlinear Model Reduction via Discrete Empirical Interpolation. *J. Sci. Comput.* **2010**, *32*, 2737–2764. [CrossRef]
34. Chaturantabut, S.; Sorensen, D.C. A state space error estimate for POD-DEIM nonlinear model reduction. *J. Numer. Anal.* **2012**, *50*, 46–63. [CrossRef]
35. Fauque, J.; Ramière, I.; Ryckelynck, D. Hybrid hyper-reduced modeling for contact mechanics problems. *Int. J. Numer. Methods Eng.* **2018**, *115*, 117–139. [CrossRef]
36. Schmidt, A.; Potschka, A.; Koerkel, S.; Bock, H.G. Derivative-Extended POD Reduced-Order Modeling for Parameter Estimation. *J. Sci. Comput.* **2013**, *35*, A2696–A2717. [CrossRef]
37. Jomaa, G.; Goblet, P.; Coquelet, C.; Morlot, V. Kinetic modeling of polyurethane pyrolysis using non-isothermal thermogravimetric analysis. *Thermochim. Acta* **2015**, *612*, 10–18. [CrossRef]
38. Bargaoui, H.; Azzouz, F.; Thibault, D.; Cailletaud, G. Thermomechanical behavior of resin bonded foundry sand cores during casting. *J. Mater. Process. Technol.* **2017**, *246*, 30–41. [CrossRef]
39. Yu, H.S. CASM: A unified state parameter model for clay and sand. *Int. J. Numer. Anal. Methods Geomechan.* **1998**, *22*, 621–653. [CrossRef]
40. Gens, A.; Nova, R. Conceptual bases for a constitutive model for bonded soils and weak rocks. *Geotech. Eng. Hard Soils Soft Rocks* **1993**, *1*, 485–494.
41. Rios, S.; Ciantia, M.; Gonzalez, N.; Arroyo, M.; Viana da Fonseca, A. Simplifying calibration of bonded elasto-plastic models. *Comput. Geotech.* **2016**, *73*, 100–108. [CrossRef]
42. Roscoe, K.; Schofield, A.; Wroth, C. On the yielding of soils. *Geotechnique* **1958**, *8*, 22–52. [CrossRef]
43. Ryckelynck, D.; Missoum Benziane, D. Hyper-reduction framework for model calibration in plasticity-induced fatigue. *Adv. Model. Simul. Eng. Sci.* **2016**, *3*, 15. [CrossRef]
44. Ghavamian, F.; Tiso, P.; Simone, A. POD-DEIM model order reduction for strain-softening viscoplasticity. *Comput. Methods Appl. Mech. Eng.* **2017**, *317*, 458–479. [CrossRef]

45. Peherstorfer, B.; Butnaru, D.; Willcox, K.; Bungartz, H.J. Localized Discrete Empirical Interpolation Method. *J. Sci. Comput.* **2014**, *36*, 168–192. [CrossRef]
46. Haasdonk, B.; Dihlmann, M.; Ohlberger, M. A training set and multiple bases generation approach for parameterized model reduction based on adaptive grids in parameter space. *Math. Comput. Model. Dyn. Syst.* **2011**, *17*, 423–442. [CrossRef]
47. Tarantola, A. *Inverse Problem Theory: Methods For Data Fitting and Model Parameter Estimation*; Elsevier: Amsterdam, The Netherlands, 1987.
48. Kaipio, J.; Somersalo, E. Statistical Inverse Problems: Discretization, Model Reduction and Inverse Crimes. *J. Comput. Appl. Math.* **2007**, *198*, 493–504. [CrossRef]

Mathematical and Computational Applications

MDPI

Article

Reduced-Order Modelling and Homogenisation in Magneto-Mechanics: A Numerical Comparison of Established Hyper-Reduction Methods

Benjamin Brands *, **Denis Davydov**, **Julia Mergheim and Paul Steinmann**

Chair of Applied Mechanics, Friedrich-Alexander Universität Erlangen-Nürnberg, Egerlandstraße 5, 91058 Erlangen, Germany; denis.davydov@fau.de (D.D.); julia.mergheim@fau.de (J.M.); paul.steinmann@fau.de (P.S.)

* Correspondence: benjamin.brands@fau.de; Tel.: +49-(0)9131-85-64406

Received: 10 December 2018; Accepted: 29 January 2019; Published: 1 February 2019

Abstract: The simulation of complex engineering structures built from magneto-rheological elastomers is a computationally challenging task. Using the FE^2 method, which is based on computational homogenisation, leads to the repetitive solution of micro-scale FE problems, causing excessive computational effort. In this paper, the micro-scale FE problems are replaced by POD reduced models of comparable accuracy. As these models do not deliver the required reductions in computational effort, they are combined with hyper-reduction methods like the Discrete Empirical Interpolation Method (DEIM), Gappy POD, Gauss–Newton Approximated Tensors (GNAT), Empirical Cubature (EC) and Reduced Integration Domain (RID). The goal of this work is the comparison of the aforementioned hyper-reduction techniques focusing on accuracy and robustness. For the application in the FE^2 framework, EC and RID are favourable due to their robustness, whereas Gappy POD rendered both the most accurate and efficient reduced models. The well-known DEIM is discarded for this application as it suffers from serious robustness deficiencies.

Keywords: model order reduction; POD; DEIM; gappy POD; GNAT; ECSW; empirical cubature; hyper-reduction; reduced integration domain; computational homogenisation

1. Introduction

The ongoing development of so-called smart materials over the last decades has given rise to the quest for numerical models which enable predictive, fast and accurate simulations of engineering structures. For smart materials, the desired constitutive behaviour is frequently architectured by tailoring the microstructure of said material, e.g., fibre-reinforced composites, auxetic materials, metal foams and many more. An established approach to model such structures is denoted multiscale modelling for which commonly two scales, the micro- and macro-scale, are introduced. The geometric complexities and advanced boundary conditions of engineering structures are modelled on the macro-scale, whereas the microstructure is represented on the micro-scale. One way to consistently couple micro- and macro-scale is the so-called FE^2 method [1]. FE^2 is a multi-level finite element method that derives the constitutive response in every quadrature point of the macro-scale from an FE simulation incorporating the microstructure using the framework of computational homogenisation [2].

Even though the everlasting increase in computational resources following Moore's law has enabled scientists to solve FE problems with 10^{13} DoFs [3], the numerical cost of FE^2 simulations is still prohibitive for most realistic problems. The idea of replacing the micro-scale FE simulation by a less expensive model has brought together the fields of multiscale and reduced-order modelling.

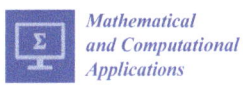

In the last several years, several viable models targeted at reducing the multiscale FE simulation have been developed. In this contribution, we focus on projection-based models using a reduced

basis, but there are also alternatives like the Nonuniform Transformation Field Analysis [4,5] and Proper Generalized Decomposition [6,7]. In projection-based reduced models, the reduced basis is a set of few functions with global support that is constructed to approximate the solution manifold of the problem in question. Projecting the governing equations onto the reduced basis yields a considerable reduction in the number of unknowns compared to using the locally supported FE basis functions. The most commonly used methods to construct the reduced basis are Proper Orthogonal Decomposition (POD) [8–11] and the Reduced Basis Method [12–15]. Both rely on solutions of the parametrised partial differential equation (pPDE), for POD, the pPDE is solved for a set of given parameters, whereas the Reduced Basis Method employs a greedy algorithm equipped with an *a posteriori* error estimator to determine the parameters adaptively. As there are hardly any efficient and reliable error estimators for coupled nonlinear multi-physic problems, POD is the method of choice in this work. In [16], a POD reduced basis was used for the first time in multiscale analysis of nonlinear elasticity at finite strains, namely by reducing the micro-scale model. This was extended in [17] by introducing the computation of a consistent tangent operator based on the reduced model.

For problems with nonlinearities or non-affine parameter dependence, the sole application of a reduced basis does not render the desired computational savings as the nonlinearity or non-affine parameter dependence has to be evaluated for the original model and subsequently projected onto the reduced basis. A widely used method accelerating the computation of the nonlinearity is the (Discrete) Empirical Interpolation Method (D)EIM [18,19]. DEIM approximates the nonlinearity by a linear combination of collateral basis functions. The coefficients are computed using interpolation based on values of the nonlinearity sampled at a relatively small number of points. In order to improve the approximation, interpolation is replaced by linear regression for Gappy POD [20]. In [21], Petrov–Galerkin projection is used to increase the stability of reduced models. Together with Gappy POD and possibly differing approximations of the reduced system matrix, this is referred to as GNAT (Gauss–Newton Approximated Tensors). As DEIM, Gappy POD and GNAT use collateral basis functions to approximate the nonlinearity, they are classified as collateral basis methods. Both POD and DEIM have been applied previously to various mechanical problems: a simplified beam model for multiscale modelling at small strains including damage [22], strain-softening viscoplasticity at small strains [23] and structural mechanics using a variant of DEIM based on the unassembled nonlinearity [24]. A collateral basis for the stresses instead of the nonlinearity was used in [25] for homogenisation of elasto-plastic materials at small strains, together with Gappy POD and a tailored method to determine the locations at which the stresses are evaluated in the reduced model. A detailed survey of DEIM, Gappy POD and GNAT for homogenisation of hyper-elastic materials at finite strains that focus on accuracy and robustness was performed in [26].

Cubature methods are another possibility of reducing the cost of computing the nonlinearity. In this sense, a problem specific quadrature rule replaces the quadrature used to integrate the weak form, e.g., Gaussian quadrature. This empirically determined quadrature uses only a subset of the support points or elements of the original FE model and computes the weights accordingly. This idea was put forward in [27] and later introduced to the field of computational homogenisation as Energy-Conserving Sampling and Weighting (ECSW) [28]. A possibility to reduce the cost of constructing the cubature was introduced in [29] together with the term Empirical Cubature (EC). The accuracy and efficiency of EC was compared to a variant of DEIM/Gappy POD used in [25] for homogenisation of elasto-viscoplastic materials in small strains [30].

The Hyper-Reduction method [31] makes use of a reduced integration domain (RID) to speed up the computation of the nonlinearity. It defines test functions with support confined to the RID, which in combination with trial functions obtained by POD results in a Petrov–Galerkin projection. The expression hyper-reduction was coined in [31] but is now used as a term encompassing all methods aiming at accelerating the computation of nonlinearities in the field of model reduction. Therefore, we will refer to the Hyper-Reduction method [31] as RID to avoid any notational confusion. The RID was used for the simulation of elasto-plasticity [32], the simulation of nonlinear thermal and mechanical problems involving internal variables [33] and the lifetime assessment of elasto-plastic structures [34]. An algorithmic comparison with DEIM for the nonlinear heat equation was carried out in [35]. Similarly, the Missing Point Estimation (MPE) method [36] computes the Galerkin projection in a small subset of the computational domain to accelerate the assembly of the reduced problem. An investigation of the MPE method is beyond the scope of this article and accordingly we refer the interested reader to [37], where a detailed comparison of MPE, DEIM and Gappy POD was performed for a predator–prey model.

In this contribution, we will show the first application of reduced-order modelling for computational homogenisation in magneto-mechanics at finite strains. We will focus on reducing the problem at the micro-scale, using POD to compute the reduced basis and applying following hyper-reduction methods: DEIM, Gappy POD, GNAT, EC and RID. Through various numerical studies, a thorough comparison between the techniques with emphasis on accuracy and robustness will be drawn.

2. Homogenisation in Magneto-Mechanics

The simulation of engineering structures requires evaluations of a material law at the engineering/macro-scale (≈mm–m). For magneto-rheological elastomers (MREs), the constitutive behaviour on the macro-scale is determined by the underlying microstructure (≈nm–μm). Usually, MREs are composite materials consisting of an elastomeric matrix with embedded magneto-active particles [38] which induce changes in stiffness or deformations due to applied magnetic fields. Due to the scale separation, a resolution of the microstructure in the discretisation of the macrostructure is computationally not feasible. The tools of computational homogenisation offer an expedient to the issue as the constitutive behaviour for any point on the macro-scale is computed from the solution of a boundary value problem (BVP) representative for the microstructure. The material composition of the microstructure is described by an RVE (Representative Volume Element) for which the constitutive behaviour of the constituents is prescribed. In the context of magneto-mechanics, the macroscopic deformation gradient $\overline{\boldsymbol{F}}$ and the magnetic field $\overline{\mathbb{H}}$ are the input variables for the microstructural BVP [39,40]. In the remainder, we use an over-bar to denote macro variables $\overline{(\bullet)}$.

As is common in homogenisation, the micro displacement and scalar magnetic potential are additively split into two parts, the macroscopic fields and the fluctuations:

$$\boldsymbol{u} = \overline{\boldsymbol{F}} \cdot \boldsymbol{X} + \tilde{\boldsymbol{u}} \qquad \mathbb{y} = \overline{\mathbb{H}} \cdot \boldsymbol{X} + \tilde{\mathbb{y}}. \tag{1}$$

The macroscopic fields depend linearly on the macroscopic deformation gradient $\overline{\boldsymbol{F}}$, the macroscopic magnetic field $\overline{\mathbb{H}}$ and the position vector $\boldsymbol{X}$.

We use linear boundary conditions to fulfill the Hill–Mandel condition. Using the fluctuations $\tilde{\boldsymbol{u}}$ and $\tilde{\mathbb{y}}$ as primary variables allows us to transform the linear into homogeneous boundary conditions. The RVE occupies the domain $\mathcal{B}_0 \subset \mathbb{R}^{\mathrm{d}}$ with its boundary $\partial\mathcal{B}_0$, where d denotes the space dimension. The energy density $\Psi(\boldsymbol{F}, \mathbb{H})$ is expressed in terms of the deformation gradient $\boldsymbol{F}$ and the magnetic field $\mathbb{H}$ and used to define the constitutive relations for the Piola stress $\boldsymbol{P}$ and the magnetic induction $\mathbb{B}$. The balance of linear momentum and Gauss's law for magnetism (also known as conservation of magnetic flux) complete the strong form of magneto-mechanics on the micro-scale [41]:

$$\begin{aligned}
&\boldsymbol{F} = \nabla_{\boldsymbol{X}} \boldsymbol{u} = \overline{\boldsymbol{F}} + \nabla_{\boldsymbol{X}} \tilde{\boldsymbol{u}} && \mathbb{H} = \nabla_{\boldsymbol{X}} \mathbb{y} = \overline{\mathbb{H}} + \nabla_{\boldsymbol{X}} \tilde{\mathbb{y}} && \text{in } \mathcal{B}_0, \\
&\boldsymbol{P} = \frac{\partial \Psi(\boldsymbol{F}, \mathbb{H})}{\partial \boldsymbol{F}} && \mathbb{B} = -\frac{\partial \Psi(\boldsymbol{F}, \mathbb{H})}{\partial \mathbb{H}} && \text{in } \mathcal{B}_0, \\
&\operatorname{Div} \boldsymbol{P} = \boldsymbol{0} && \operatorname{Div} \mathbb{B} = 0 && \text{in } \mathcal{B}_0, \\
&\tilde{\boldsymbol{u}} = \boldsymbol{0} && \tilde{\mathbb{y}} = 0 && \text{on } \partial\mathcal{B}_0.
\end{aligned} \tag{2}$$

For the sake of an FE solution, the weak form

$$
\begin{aligned}
&\int_{\mathcal{B}_0} \nabla_X \delta\tilde{\boldsymbol{u}} : \boldsymbol{P} \, \mathrm{d}V = 0 \quad \forall \delta\tilde{\boldsymbol{u}} \in \left\{ \delta\tilde{\boldsymbol{u}} \in H^1(\mathcal{B}_0) : \delta\tilde{\boldsymbol{u}} = \boldsymbol{0} \text{ on } \partial\mathcal{B}_0 \right\} \\
\text{and } &\int_{\mathcal{B}_0} \nabla_X \delta\tilde{\mathrm{y}} \cdot \mathbb{B} \, \mathrm{d}V = 0 \quad \forall \delta\tilde{\mathrm{y}} \in \left\{ \delta\tilde{\mathrm{y}} \in H^1(\mathcal{B}_0) : \delta\tilde{\mathrm{y}} = 0 \text{ on } \partial\mathcal{B}_0 \right\}
\end{aligned} \tag{3}
$$

is derived using the test functions $\delta\tilde{\boldsymbol{u}}$ and $\delta\tilde{\mathrm{y}}$.

For the spatial discretisation, the standard Bubnov–Galerkin FEM is used. The continuum body is approximated by a mesh $\mathcal{B}_0 \approx \mathcal{T} = \bigcup_{e=1}^{\mathrm{M}} \Omega_e$ with $\Omega_i \cap \Omega_j = \emptyset$ for $i \neq j$ and $i, j \in [1, \ldots, \mathrm{M}]$, where M denotes the number of elements. The displacement and potential fields in any finite element Ω_e are approximated by the piecewise continuous vector-valued polynomials $\boldsymbol{N}_i^{\boldsymbol{u}}(\boldsymbol{X})$ and scalar polynomials $N_i^{\mathrm{y}}(\boldsymbol{X})$, respectively:

$$
\begin{aligned}
\tilde{\boldsymbol{u}}_{|\Omega_e} &:= \sum_{i=1}^{\mathrm{d}_e^{\boldsymbol{u}}} \tilde{u}_i \boldsymbol{N}_i^{\boldsymbol{u}}(\boldsymbol{X}), & \delta\tilde{\boldsymbol{u}}_{|\Omega_e} &:= \sum_{i=1}^{\mathrm{d}_e^{\boldsymbol{u}}} \delta\tilde{u}_i \boldsymbol{N}_i^{\boldsymbol{u}}(\boldsymbol{X}), \\
\tilde{\mathrm{y}}_{|\Omega_e} &:= \sum_{i=1}^{\mathrm{d}_e^{\mathrm{y}}} \tilde{\mathrm{y}}_i N_i^{\mathrm{y}}(\boldsymbol{X}), & \delta\tilde{\mathrm{y}}_{|\Omega_e} &:= \sum_{i=1}^{\mathrm{d}_e^{\mathrm{y}}} \delta\tilde{\mathrm{y}}_i N_i^{\mathrm{y}}(\boldsymbol{X}).
\end{aligned} \tag{4}
$$

The scalars $\mathrm{d}_e^{\boldsymbol{u}}$ and $\mathrm{d}_e^{\mathrm{y}}$ are the numbers of mechanical and magnetic DoFs in the element Ω_e. Using these approximations results in the discrete weak form

$$
\hat{\boldsymbol{R}} = \begin{bmatrix} \hat{\boldsymbol{R}}^{\boldsymbol{u}} \in \mathbb{R}^{\mathrm{N}_{\boldsymbol{u}}} \\ \hat{\boldsymbol{R}}^{\mathrm{y}} \in \mathbb{R}^{\mathrm{N}_{\mathrm{y}}} \end{bmatrix} = \underset{e=1}{\overset{\mathrm{M}}{\mathcal{A}}} \begin{bmatrix} {}^e\hat{\boldsymbol{R}}^{\boldsymbol{u}} \in \mathbb{R}^{\mathrm{d}_e^{\boldsymbol{u}}} \\ {}^e\hat{\boldsymbol{R}}^{\mathrm{y}} \in \mathbb{R}^{\mathrm{d}_e^{\mathrm{y}}} \end{bmatrix} = \boldsymbol{0} \quad \text{with} \quad \begin{bmatrix} {}^e\hat{\boldsymbol{R}}^{\boldsymbol{u}}[i] = \int_{\Omega_e} \nabla_X \boldsymbol{N}_i^{\boldsymbol{u}} : \boldsymbol{P} \, \mathrm{d}V \\ {}^e\hat{\boldsymbol{R}}^{\mathrm{y}}[i] = \int_{\Omega_e} \nabla_X N_i^{\mathrm{y}} \cdot \mathbb{B} \, \mathrm{d}V \end{bmatrix} \quad \text{for } \Omega_e \in \mathcal{T}. \tag{5}
$$

The sub-/superscripts $(\bullet)^{\boldsymbol{u}}/(\bullet)_{\boldsymbol{u}}$ and $(\bullet)^{\mathrm{y}}/(\bullet)_{\mathrm{y}}$ encode whether a variable is associated with the mechanical or magnetic component and are used throughout the remainder of the paper. The notation $(\hat{\bullet})$ is consistently used to differentiate between a continuous field and its discrete FE counterpart, e.g., $\tilde{\boldsymbol{u}}$ is the displacement fluctuation and $\hat{\boldsymbol{u}}$ is the vector containing the nodal coefficients for the FE discretisation. To refer to single elements of any vector/first-order tensor $\boldsymbol{X}$ and of any matrix/second-order tensor $\boldsymbol{Y}$, the notation $\boldsymbol{X}[i]$ and $\boldsymbol{Y}[i,j]$ are used. The operator $\underset{e=1}{\overset{\mathrm{M}}{\mathcal{A}}}$ represents the assembly of the element contributions and the scalars $\mathrm{N}_{\boldsymbol{u}}$, N_{y} and $\mathrm{N} = \mathrm{N}_{\boldsymbol{u}} + \mathrm{N}_{\mathrm{y}}$ are the numbers of DoFs employed in the FE discretisation.

The numerical solution of the system of nonlinear Equations (5) using the iterative Newton–Raphson scheme requires its linearisation

$$
\boldsymbol{K}_k \Delta\hat{\boldsymbol{y}}_k = -\hat{\boldsymbol{R}}_k \text{ with } \hat{\boldsymbol{y}}_{k+1} = \hat{\boldsymbol{y}}_k + \Delta\hat{\boldsymbol{y}}_k \text{ and } \hat{\boldsymbol{y}}_k = \begin{bmatrix} \hat{\tilde{\boldsymbol{u}}}_k \\ \hat{\tilde{\mathrm{y}}}_k \end{bmatrix}, \tag{6}
$$

introducing the iteration count k.

For the sake of notational clarity, the dependences of $\hat{\boldsymbol{R}}_k\left(\hat{\boldsymbol{y}}_k; \overline{\boldsymbol{F}}, \overline{\mathbb{H}}\right)$ and $\boldsymbol{K}_k\left(\hat{\boldsymbol{y}}_k; \overline{\boldsymbol{F}}, \overline{\mathbb{H}}\right)$ are dropped. The tangent stiffness matrix $\boldsymbol{K}$ is given as

$$
\begin{aligned}
&\boldsymbol{K} = \begin{bmatrix} \boldsymbol{K}^{uu} \in \mathbb{R}^{\mathrm{N}_u \times \mathrm{N}_u} & \boldsymbol{K}^{uy} \in \mathbb{R}^{\mathrm{N}_u \times \mathrm{N}_y} \\ \boldsymbol{K}^{yu} \in \mathbb{R}^{\mathrm{N}_y \times \mathrm{N}_u} & \boldsymbol{K}^{yy} \in \mathbb{R}^{\mathrm{N}_y \times \mathrm{N}_y} \end{bmatrix} = \overset{\mathrm{M}}{\underset{e=1}{\mathcal{A}}} \begin{bmatrix} {}^e\boldsymbol{K}^{uu} \in \mathbb{R}^{\mathrm{d}_e^u \times \mathrm{d}_e^u} & {}^e\boldsymbol{K}^{uy} \in \mathbb{R}^{\mathrm{d}_e^u \times \mathrm{d}_e^y} \\ {}^e\boldsymbol{K}^{yu} \in \mathbb{R}^{\mathrm{d}_e^y \times \mathrm{d}_e^u} & {}^e\boldsymbol{K}^{yy} \in \mathbb{R}^{\mathrm{d}_e^y \times \mathrm{d}_e^y} \end{bmatrix} \\
&\text{with} \quad
\begin{aligned}
{}^e\boldsymbol{K}^{uu}[i,j] &= \int_{\Omega_e} \nabla_{\boldsymbol{X}} \boldsymbol{N}_i^u : \frac{\partial \boldsymbol{P}}{\partial \boldsymbol{F}} : \nabla_{\boldsymbol{X}} \boldsymbol{N}_j^u \,\mathrm{d}V \\
{}^e\boldsymbol{K}^{uy}[i,j] &= \int_{\Omega_e} \nabla_{\boldsymbol{X}} \boldsymbol{N}_i^u : \frac{\partial \boldsymbol{P}}{\partial \mathbb{H}} \cdot \nabla_{\boldsymbol{X}} N_j^y \,\mathrm{d}V \\
{}^e\boldsymbol{K}^{yu}[i,j] &= \int_{\Omega_e} \nabla_{\boldsymbol{X}} N_i^y \cdot \frac{\partial \mathbb{B}}{\partial \boldsymbol{F}} : \nabla_{\boldsymbol{X}} \boldsymbol{N}_j^u \,\mathrm{d}V \\
{}^e\boldsymbol{K}^{yy}[i,j] &= \int_{\Omega_e} \nabla_{\boldsymbol{X}} N_i^y \cdot \frac{\partial \mathbb{B}}{\partial \mathbb{H}} \cdot \nabla_{\boldsymbol{X}} N_j^y \,\mathrm{d}V
\end{aligned}
\quad \text{for } \Omega_e \in \mathcal{T}.
\end{aligned} \tag{7}
$$

Once the solution of (5) is obtained, the output quantities $\overline{\boldsymbol{P}}$ and $\overline{\mathbb{B}}$ are computed using

$$
\overline{\boldsymbol{P}} = \frac{1}{V} \int_{\mathcal{B}_0} \boldsymbol{P} \,\mathrm{d}V \quad \text{and} \quad \overline{\mathbb{B}} = \frac{1}{V} \int_{\mathcal{B}_0} \mathbb{B} \,\mathrm{d}V, \tag{8}
$$

where the volume of the RVE is denoted by V.

3. Reduced-Order Modelling

3.1. Reduced Basis

Instead of using a large number N of compact trial functions, projection-based ROMs are built upon a small number n of global functions spanning the space in which the solution manifold of the pPDE resides. Consequently, the unknown fluctuation fields $\tilde{\boldsymbol{u}}$ and $\tilde{y}$ are expressed as linear combinations of the global trial functions with the reduced coefficients $\tilde{u}_i^r$ and $\tilde{y}_i^r$:

$$
\tilde{\boldsymbol{u}} = \sum_{i=1}^{\mathrm{n}_u} \boldsymbol{\varphi}_i^u \tilde{u}_i^r \quad \text{and} \quad \tilde{y} = \sum_{i=1}^{\mathrm{n}_y} \varphi_i^y \tilde{y}_i^r. \tag{9}
$$

The reducibility of the problem, namely the conditions $\mathrm{n}_u \ll \mathrm{N}_u$ and $\mathrm{n}_y \ll \mathrm{N}_y$, is accepted implicitly but has to be confirmed by numerical studies. To avoid scaling issues due to differently chosen units, separate reduced bases are used for the mechanical and the magnetic fluctuation fields. The numbers n_u, n_y and $\mathrm{n} = \mathrm{n}_u + \mathrm{n}_y$ are the numbers of reduced basis functions to be taken into account.

A common method to compute the reduced basis for a pPDE is POD [9,42]. In order to do so, we define the parameter domain of the microscopic problem

$$
\begin{aligned}
\mathcal{P} = &\left(\overline{\boldsymbol{F}}^{\min}[1,1], \overline{\boldsymbol{F}}^{\max}[1,1]\right) \times \cdots \times \left(\overline{\boldsymbol{F}}^{\min}[\mathrm{d},\mathrm{d}], \overline{\boldsymbol{F}}^{\max}[\mathrm{d},\mathrm{d}]\right) \\
&\times \left(\overline{\mathbb{H}}^{\min}[1], \overline{\mathbb{H}}^{\max}[1]\right) \times \cdots \times \left(\overline{\mathbb{H}}^{\min}[\mathrm{d}], \overline{\mathbb{H}}^{\max}[\mathrm{d}]\right) \subset \mathbb{R}^{\mathrm{d}^2+\mathrm{d}}
\end{aligned} \tag{10}
$$

with reasonably chosen limits for the components of the macroscopic loading parameters. Each element $\mathbf{p}_i = \left(\overline{\boldsymbol{F}}_i, \overline{\mathbb{H}}_i\right) \in \mathcal{P}$ comprises an instance of the macroscopic deformation gradient and magnetic field. The parameter domain $\mathcal{P}$ is sampled using n_s parameters gathered in the set

$$
\mathcal{S} = \left\{\mathbf{p}_1, \ldots, \mathbf{p}_{\mathrm{n}_s}\right\} \subset \mathcal{P} \tag{11}
$$

and the full-order model (FOM) (5) is solved for all elements in $\mathcal{S}$. The solutions are collected in the snapshot matrices

$$
\boldsymbol{S}_u = \left[\hat{\boldsymbol{u}}(\mathbf{p}_1), \ldots, \hat{\boldsymbol{u}}(\mathbf{p}_{\mathrm{n}_s})\right] \in \mathbb{R}^{\mathrm{N}_u \times \mathrm{n}_s} \quad \text{and} \quad \boldsymbol{S}_y = \left[\hat{\tilde{y}}(\mathbf{p}_1), \ldots, \hat{\tilde{y}}(\mathbf{p}_{\mathrm{n}_s})\right] \in \mathbb{R}^{\mathrm{N}_y \times \mathrm{n}_s} \tag{12}
$$

and the subsequent application of

$$\mathrm{POD}\left(S_u\right) \rightarrow B_u = \left[\hat{\varphi}^u_1, \ldots, \hat{\varphi}^u_{\mathrm{n}_u}\right] \in \mathbb{R}^{\mathrm{N}_u \times \mathrm{n}_u} \quad \text{and} \quad \mathrm{POD}\left(S_{\mathrm{y}}\right) \rightarrow B_{\mathrm{y}} = \left[\hat{\varphi}^{\mathrm{y}}_1, \ldots, \hat{\varphi}^{\mathrm{y}}_{\mathrm{n}_{\mathrm{y}}}\right] \in \mathbb{R}^{\mathrm{N}_{\mathrm{y}} \times \mathrm{n}_{\mathrm{y}}} \tag{13}$$

gives the discrete reduced bases contained in the matrices B_u and B_{y}. For details on POD, we refer to [9,43].

3.2. Galerkin ROM

In the Galerkin reduced model, the same ansatz (9) as for the solution is used for the test functions

$$\delta\tilde{u} = \sum_{i=1}^{\mathrm{n}_u} \delta\tilde{u}^r_i \varphi^u_i \quad \text{and} \quad \delta\tilde{\mathrm{y}} = \sum_{i=1}^{\mathrm{n}_{\mathrm{y}}} \delta\tilde{\mathrm{y}}^r_i \varphi^{\mathrm{y}}_i . \tag{14}$$

Inserting (9) and (14) into (5) results in the weak form of the Galerkin reduced model

$$\int_{\mathcal{B}_0} \nabla_X \varphi^u_i : P \,\mathrm{d}V = 0 \quad \forall i = 1, \ldots, \mathrm{n}_u \quad \text{and} \quad \int_{\mathcal{B}_0} \nabla_X \varphi^{\mathrm{y}}_i \cdot \mathbb{B} \,\mathrm{d}V = 0 \quad \forall i = 1, \ldots, \mathrm{n}_{\mathrm{y}}, \tag{15}$$

where the dependences $P\left(\tilde{u}, \tilde{\mathrm{y}}; \bar{F}, \overline{\mathbb{H}}\right)$ and $\mathbb{B}\left(\tilde{u}, \tilde{\mathrm{y}}; \bar{F}, \overline{\mathbb{H}}\right)$ are dropped for notational brevity.

Analogously, the discrete weak (16) form and its linearisation (17) are derived:

$$B^\top \hat{R} = 0 \quad \text{with } B := \begin{bmatrix} B_u & 0 \\ 0 & B_{\mathrm{y}} \end{bmatrix} \in \mathbb{R}^{\mathrm{N} \times \mathrm{n}}, \tag{16}$$

$$B^\top K_k B \Delta\tilde{y}^r_k = -B^\top \hat{R}_k \quad \text{with } \tilde{y}^r_{k+1} = \tilde{y}^r_k + \Delta\tilde{y}^r_k \text{ and } \tilde{y}^r_k = \begin{bmatrix} \tilde{u}^r_k \in \mathbb{R}^{\mathrm{n}_u} \\ \tilde{\mathrm{y}}^r_k \in \mathbb{R}^{\mathrm{n}_{\mathrm{y}}} \end{bmatrix}. \tag{17}$$

Even though the size of the system of linear Equations (17) is significantly smaller than in Equation (6) and hence the cost of the linear solver reduces from $\mathcal{O}(\mathrm{N}^2)$ to $\mathcal{O}(\mathrm{n}^3)$, the speed-up is only marginal as the assembly of (17) depends on the original problem size. The cost for evaluating the constitutive law for every quadrature point is roughly $\mathcal{O}(\mathrm{Nn} + \mathrm{n}^{\mathrm{el}}_{\mathrm{qp}}\mathrm{M})$, where $\mathrm{n}^{\mathrm{el}}_{\mathrm{qp}}$ is the number of quadrature points per element. The computational complexities of assembling and projecting $B^\top \hat{R}_k$ and $B^\top K_k B$ are proportional to $\mathcal{O}(\mathrm{nN})$ and $\mathcal{O}(\mathrm{n}^2\mathrm{N} + \mathrm{nN})$, respectively. Therefore, the application of hyper-reduction methods is imperative.

4. Hyper-Reduction

4.1. Discrete Empirical Interpolation Method

The Discrete Empirical Interpolation Method [19] is the standard hyper-reduction method for non-affine or nonlinear pPDEs, for some problems even equipped with *a posteriori* and *a priori* error estimators [44,45]. The first step is to approximate the discrete residuum

$$\begin{aligned} &\hat{R} = \begin{bmatrix} \hat{R}^u \\ \hat{R}^{\mathrm{y}} \end{bmatrix} \approx \begin{bmatrix} H_{R^u} & 0 \\ 0 & H_{R^{\mathrm{y}}} \end{bmatrix} \begin{bmatrix} r_u \\ r_{\mathrm{y}} \end{bmatrix} =: H_R r, \\ &\text{with } H_{R^u} = \left[\hat{\phi}^{R^u}_1, \ldots, \hat{\phi}^{R^u}_{\mathrm{r}_u}\right] \in \mathbb{R}^{\mathrm{N}_u \times \mathrm{r}_u}, \; H_{R^{\mathrm{y}}} = \left[\hat{\phi}^{R^{\mathrm{y}}}_1, \ldots, \hat{\phi}^{R^{\mathrm{y}}}_{\mathrm{r}_{\mathrm{y}}}\right] \in \mathbb{R}^{\mathrm{N}_{\mathrm{y}} \times \mathrm{r}_{\mathrm{y}}}, \\ &r_u \in \mathbb{R}^{\mathrm{r}_u}, \; r_{\mathrm{y}} \in \mathbb{R}^{\mathrm{r}_{\mathrm{y}}}, \; H_R \in \mathbb{R}^{\mathrm{N} \times \mathrm{r}} \text{ and } r \in \mathbb{R}^{\mathrm{r}} \end{aligned} \tag{18}$$

by a linear combination of collateral basis vectors contained in H_R with r being the vector of coefficients. Due to a different number range, it is advisable to approximate the mechanical and magnetic residua by two separate collateral bases H_{R^u} and H_{R^y}. The collateral basis is computed based on snapshots of the residuum. For that purpose, (16) is solved for every parameter $\mathbf{p}_i \in \mathcal{S}$ and the residua

$$T_i^u = \left[\hat{R}_1^u(\mathbf{p}_i), \ldots, \hat{R}_{k_u}^u(\mathbf{p}_i)\right] \quad \text{and} \quad T_i^y = \left[\hat{R}_1^y(\mathbf{p}_i), \ldots, \hat{R}_{k_y}^y(\mathbf{p}_i)\right] \tag{19}$$

are collected in the course of the Newton–Raphson process to build the matrices

$$S_{R^u} = \left[T_1^u, \ldots, T_{\mathrm{n_s}}^u\right] \quad \text{and} \quad S_{R^y} = \left[T_1^y, \ldots, T_{\mathrm{n_s}}^y\right]. \tag{20}$$

As $\hat{R}_j^u(\mathbf{p}_i)$ and $\hat{R}_j^y(\mathbf{p}_i)$ converge to the null vector during the iterative solution of (16), only residua fulfilling $\left\|\hat{R}_j^u(\mathbf{p}_i)\right\| / \left\|\hat{R}_1^u(\mathbf{p}_i)\right\| > \text{tol}$ and $\left\|\hat{R}_j^y(\mathbf{p}_i)\right\| / \left\|\hat{R}_1^y(\mathbf{p}_i)\right\| > \text{tol}$ are taken into account. The subsequent application of POD $(S_{R^u}) \rightarrow H_{R^u}$ and POD $(S_{R^y}) \rightarrow H_{R^y}$ gives the collateral bases.

The coefficients in (18) are determined using interpolation

$$\begin{aligned} \mathbf{P}_{R^u}^\top \hat{R}^u &= \mathbf{P}_{R^u}^\top H_{R^u} r_u \qquad \text{with } \mathbf{P}_{R^u} = \left[e_{\rho_1^u}, \ldots, e_{\rho_{\mathrm{r}_u}^u}\right] \in \mathbb{N}^{\mathrm{N}_u \times \mathrm{r}_u}, \\ \mathbf{P}_{R^y}^\top \hat{R}^y &= \mathbf{P}_{R^y}^\top H_{R^y} r_y \qquad \text{with } \mathbf{P}_{R^y} = \left[e_{\rho_1^y}, \ldots, e_{\rho_{\mathrm{r}_y}^y}\right] \in \mathbb{N}^{\mathrm{N}_y \times \mathrm{r}_y}, \end{aligned} \tag{21}$$

meaning the approximation has to be equal to the residuum at the interpolation indices. The matrices $\mathbf{P}_{R^u}$ and $\mathbf{P}_{R^y}$ are sampling matrices, where $e_{\rho_i^u}$ for $i = 1, \ldots, \mathrm{r}_u$ and $e_{\rho_j^y}$ for $j = 1, \ldots, \mathrm{r}_y$ are unit vectors with only one non-zero component in the ρ_i^u-th and ρ_j^y-th entry.

Application of the DEIM Algorithm A1, provided in Appendix A, to the collateral bases $\left\{\hat{\phi}_i^{R^u}\right\}_{i=1}^{\mathrm{r}_u}$ and $\left\{\hat{\phi}_i^{R^y}\right\}_{i=1}^{\mathrm{r}_y}$ returns the interpolation indices and guarantees the matrix products $\left[\mathbf{P}_{R^u}^\top H_{R^u}\right]$ and $\left[\mathbf{P}_{R^u}^\top H_{R^u}\right]$ to be non-singular. Consequently, the interpolation coefficients are calculated by

$$r_u = \left(\mathbf{P}_{R^u}^\top H_{R^u}\right)^{-1} \mathbf{P}_{R^u}^\top \hat{R}^u \quad \text{and} \quad r_y = \left(\mathbf{P}_{R^y}^\top H_{R^y}\right)^{-1} \mathbf{P}_{R^y}^\top \hat{R}^y. \tag{22}$$

Introducing (18) and (21) into (16) gives the hyper-reduced weak form

$$R_k^r := B^\top H_R \left(\mathbf{P}_R^\top H_R\right)^{-1} \mathbf{P}_R^\top \hat{R}_k = 0 \quad \text{with } \mathbf{P}_R := \begin{bmatrix} \mathbf{P}_{R^u} & 0 \\ 0 & \mathbf{P}_{R^y} \end{bmatrix} \in \mathbb{N}^{\mathrm{N} \times \mathrm{r}} \tag{23}$$

and its linearisation becomes

$$\underbrace{B^\top H_R \left(\mathbf{P}_R^\top H_R\right)^{-1}}_{\text{precomputed: } \mathbb{R}^{\mathrm{n} \times \mathrm{r}}} \mathbf{P}_R^\top K_k B \Delta \tilde{y}_k^r = -\underbrace{B^\top H_R \left(\mathbf{P}_R^\top H_R\right)^{-1}}_{\text{precomputed: } \mathbb{R}^{\mathrm{n} \times \mathrm{r}}} \mathbf{P}_R^\top \hat{R}_k. \tag{24}$$

The cost for evaluating the constitutive law for every quadrature point in elements containing interpolation indices is reduced to approximately $\mathcal{O}(\mathrm{N_{eval}}\mathrm{n} + \mathrm{n_{qp}^{el}}\mathrm{m})$, where m is the number of elements containing DEIM indices and $\mathrm{N_{eval}}$ the number of DoFs associated with this elements. The computational complexities of computing the residuum and tangent stiffness matrix are proportional to $\mathcal{O}(\mathrm{rN_{eval}})$ and $\mathcal{O}(\mathrm{nrN_{eval}} + \mathrm{rN_{eval}})$, respectively. It is to be noted that an efficient computation of the stiffness matrix utilises the sparsity of the FE matrix. Consequently, the assembly and solution of (24) do not depend on the size of the FOM and should therefore result in the desired speed-ups.

4.2. Gappy POD

Instead of interpolation, Gappy POD uses linear regression to determine the collateral basis coefficients, meaning the residual is evaluated at more indices than coefficients. This is particularly beneficial for hyper-reduced models originating from FE models, as for the calculation of $\hat{\boldsymbol{R}}^u$ or $\hat{\boldsymbol{R}}^{\mathrm{y}}$ at each evaluation index ρ_i^u or ρ_i^{y} the solution $(\tilde{\boldsymbol{u}}, \tilde{\boldsymbol{y}})$ and the respective constitutive components have to be computed for every finite element containing the index. Hence, it is more economical to use all DoFs attached to a node instead of possibly only one as for DEIM. The collateral basis coefficients are computed solving

$$\begin{aligned} \boldsymbol{r}_u &= \underset{\boldsymbol{a}\in\mathbb{R}^{\mathrm{r}_u}}{\arg\min} \left\| \mathbf{P}_{R^u}^\top \hat{\boldsymbol{R}}^u - \mathbf{P}_{R^u}^\top \boldsymbol{H}_{R^u} \boldsymbol{a} \right\|_2^2 \quad \text{with } \mathbf{P}_{R^u} = \left[\boldsymbol{e}_{\rho_1^u}, \ldots, \boldsymbol{e}_{\rho_{\mathrm{p}_u\mathrm{d}}^u} \right] \in \mathbb{N}^{\mathrm{N}_u \times \mathrm{p}_u\mathrm{d}} \\ \text{and } \boldsymbol{r}_{\mathrm{y}} &= \underset{\boldsymbol{a}\in\mathbb{R}^{\mathrm{r}_{\mathrm{y}}}}{\arg\min} \left\| \mathbf{P}_{R^{\mathrm{y}}}^\top \hat{\boldsymbol{R}}^{\mathrm{y}} - \mathbf{P}_{R^{\mathrm{y}}}^\top \boldsymbol{H}_{R^{\mathrm{y}}} \boldsymbol{a} \right\|_2^2 \quad \text{with } \mathbf{P}_{R^{\mathrm{y}}} = \left[\boldsymbol{e}_{\rho_1^{\mathrm{y}}}, \ldots, \boldsymbol{e}_{\rho_{\mathrm{p}_{\mathrm{y}}}^{\mathrm{y}}} \right] \in \mathbb{N}^{\mathrm{N}_{\mathrm{y}} \times \mathrm{p}_{\mathrm{y}}}. \end{aligned} \tag{25}$$

The integers p_u and p_{y} are the numbers of FE nodes at which the residua $\hat{\boldsymbol{R}}^u$ and $\hat{\boldsymbol{R}}^{\mathrm{y}}$ are computed and N_u and N_{y} are the dimensions of the underlying FE model. For (25) to have unique solutions, $\mathrm{p}_u\mathrm{d} \geq \mathrm{r}_u$ and $\mathrm{p}_{\mathrm{y}} \geq \mathrm{r}_{\mathrm{y}}$ have to hold. The solutions of (25) are obtained by computing the pseudo-inverses $\left(\mathbf{P}_{R^u}^\top \boldsymbol{H}_{R^u}\right)^+$ and $\left(\mathbf{P}_{R^{\mathrm{y}}}^\top \boldsymbol{H}_{R^{\mathrm{y}}}\right)^+$ rendering the explicit expressions

$$\boldsymbol{r}_u = \left(\mathbf{P}_{R^u}^\top \boldsymbol{H}_{R^u}\right)^+ \mathbf{P}_{R^u}^\top \hat{\boldsymbol{R}}^u \quad \text{and} \quad \boldsymbol{r}_{\mathrm{y}} = \left(\mathbf{P}_{R^{\mathrm{y}}}^\top \boldsymbol{H}_{R^{\mathrm{y}}}\right)^+ \mathbf{P}_{R^{\mathrm{y}}}^\top \hat{\boldsymbol{R}}^{\mathrm{y}} \tag{26}$$

for the collateral basis coefficients.

Inserting (26) into (16) gives the Gappy POD hyper-reduced weak form

$$\boldsymbol{R}_k^r := \boldsymbol{B}^\top \boldsymbol{H}_R \left(\mathbf{P}_R^\top \boldsymbol{H}_R\right)^+ \mathbf{P}_R^\top \hat{\boldsymbol{R}}_k = \boldsymbol{0} \tag{27}$$

and the linearisation becomes

$$\underbrace{\boldsymbol{B}^\top \boldsymbol{H}_R \left(\mathbf{P}_R^\top \boldsymbol{H}_R\right)^+}_{\text{precomputed: } \mathbb{R}^{\mathrm{n}\times\mathrm{p}}} \mathbf{P}_R^\top \boldsymbol{K}_k \boldsymbol{B} \Delta\tilde{\boldsymbol{y}}_k^r = -\underbrace{\boldsymbol{B}^\top \boldsymbol{H}_R \left(\mathbf{P}_R^\top \boldsymbol{H}_R\right)^+}_{\text{precomputed: } \mathbb{R}^{\mathrm{n}\times\mathrm{p}}} \mathbf{P}_R^\top \hat{\boldsymbol{R}}_k, \tag{28}$$

introducing $\mathrm{p} = \mathrm{p}_u\mathrm{d} + \mathrm{p}_{\mathrm{y}}$. In (28), the only difference to the DEIM hyper-reduced system (24) is the appearance of the pseudo-inverse $(\bullet)^+$ instead of the inverse $(\bullet)^{-1}$. The cost for evaluating the constitutive law is the same as for DEIM $\mathcal{O}(\mathrm{N}_{\mathrm{eval}}\mathrm{n} + \mathrm{n}_{\mathrm{qp}}^{\mathrm{el}}\mathrm{m})$. The computational complexities of computing the residuum and tangent stiffness matrix are proportional to $\mathcal{O}(\mathrm{pN}_{\mathrm{eval}})$ and $\mathcal{O}(\mathrm{npN}_{\mathrm{eval}} + \mathrm{pN}_{\mathrm{eval}})$.

To determine the FE nodes/indices, Algorithm A2, given in Appendix B, which is an advancement of the algorithm proposed in [21] for multi-physic problems, is applied. Algorithm A2 uses normalised maxima to cope with distinct domains and different units in multi-physic problems. Algorithm A2 is applied to $\left\{\hat{\boldsymbol{\phi}}_1^{R^u}, \ldots, \hat{\boldsymbol{\phi}}_{\mathrm{r}_u}^{R^u}\right\}$ and $\left\{\hat{\boldsymbol{\phi}}_1^{R^{\mathrm{y}}}, \ldots, \hat{\boldsymbol{\phi}}_{\mathrm{r}_{\mathrm{y}}}^{R^{\mathrm{y}}}\right\}$ either separately or combined. In the latter case, the same FE nodes are used for the gappy reconstruction of the residua $\hat{\boldsymbol{R}}^u$ and $\hat{\boldsymbol{R}}^{\mathrm{y}}$, resulting in more efficient reduced models.

4.3. GNAT

To improve the accuracy and stability [21] of reduced models, the GNAT hyper-reduced model is not based on Galerkin but on Petrov–Galerkin projection and therefore adopts different spaces for the test and trial functions.

For accuracy reasons, the state-dependent test functions $\boldsymbol{K}(\boldsymbol{B}\boldsymbol{y}^r)\,\boldsymbol{B}$ are chosen and the discrete weak form

$$\boldsymbol{B}^\top \boldsymbol{K}^\top \hat{\boldsymbol{R}} = \boldsymbol{0} \tag{29}$$

and its linearisation

$$\boldsymbol{B}^{\top}\boldsymbol{K}_k^{\top}\boldsymbol{K}_k\boldsymbol{B}\Delta\tilde{\boldsymbol{y}}^r = -\boldsymbol{B}^{\top}\hat{\boldsymbol{R}}_k \quad \text{with} \quad \sum_i \hat{\boldsymbol{R}}[i]\frac{\partial^2\hat{\boldsymbol{R}}[i]}{\partial\hat{\boldsymbol{y}}^2} \approx \boldsymbol{0} \tag{30}$$

are obtained. Equation (30) is the normal equation for the associated least-squares problem

$$\Delta\tilde{\boldsymbol{y}}_k^r = \underset{\boldsymbol{a}\in\mathbb{R}^{\mathrm{n}}}{\arg\min}\left\|\boldsymbol{K}_k\boldsymbol{B}\boldsymbol{a} + \hat{\boldsymbol{R}}_k\right\|_2^2 \tag{31}$$

and therefore the solution of (29) is equivalent to solving the minimisation problem

$$\underset{\boldsymbol{a}\in\mathbb{R}^{\mathrm{n}}}{\text{minimise}}\left\|\hat{\boldsymbol{R}}\left(\boldsymbol{B}\boldsymbol{a}\right)\right\|_2 \tag{32}$$

with the Gauss–Newton method.

As the computation of (31) still depends on the original problem size, GNAT similarly to Gappy POD uses collateral bases to approximate the nonlinearities and linear regression to determine the coefficients:

$$\begin{aligned}&\hat{\boldsymbol{R}} = \boldsymbol{H}_R\boldsymbol{r} \quad \text{and} \quad \boldsymbol{K}\boldsymbol{B} = \boldsymbol{H}_K\boldsymbol{k},\\ &\text{with } \boldsymbol{r} = \left(\mathbf{P}_R^{\top}\boldsymbol{H}_R\right)^{+}\mathbf{P}_R^{\top}\hat{\boldsymbol{R}} \quad \text{and} \quad \boldsymbol{k} = \left(\mathbf{P}_K^{\top}\boldsymbol{H}_K\right)^{+}\mathbf{P}_K^{\top}\left[\boldsymbol{K}\boldsymbol{B}\right].\end{aligned} \tag{33}$$

Numerical experiments have shown that the choice $\boldsymbol{H}_R = \boldsymbol{H}_K$ and consequently $\mathbf{P}_R = \mathbf{P}_K$ renders reduced models of superior accuracy compared to models employing a separate basis for $\boldsymbol{H}_K$. Putting (33) into (31) and multiplying from the left with $\boldsymbol{H}_R^{\top}$ renders the least-squares problem to be solved in every Gauss–Newton iteration

$$\Delta\tilde{\boldsymbol{y}}_k^r = \underset{\boldsymbol{a}\in\mathbb{R}^{\mathrm{n}}}{\arg\min}\left\|\underbrace{\left(\mathbf{P}_R^{\top}\boldsymbol{H}_R\right)^{+}\mathbf{P}_R^{\top}\left[\boldsymbol{K}_k\boldsymbol{B}\right]}_{\text{precomputed: }\mathbb{R}^{\mathrm{r}\times\mathrm{p}}}\boldsymbol{a} + \overbrace{\underbrace{\left(\mathbf{P}_R^{\top}\boldsymbol{H}_R\right)^{+}}_{\text{precomputed: }\mathbb{R}^{\mathrm{r}\times\mathrm{p}}}\mathbf{P}_R^{\top}\hat{\boldsymbol{R}}_k}^{\boldsymbol{R}_k^r}\right\|_2^2 \tag{34}$$

and recalling $\mathrm{p} = \mathrm{p}_{\boldsymbol{u}}\mathrm{d} + \mathrm{p}_{\mathrm{y}}$.

The complexity of assembling and solving (34) is similar to (28). As the Gauss–Newton method does not converge quadratically like the classical Newton–Raphson scheme, more iterations are necessary to minimise (32).

The computation of the collateral basis is similar to DEIM except that the residua are gathered during the solution of (31). The matrix $\mathbf{P}_R$ is determined using Algorithm A2 with the collateral bases as input.

4.4. Empirical Cubature

Cubature methods aim at reducing the cost of computing the nonlinearity in (15) by defining an empirical quadrature, which evaluates the integrand only in a limited number of quadrature points or elements. Instead of summing up all element contributions, the nonlinearities are computed only in the elements of the so-called reduced meshes $\mathcal{E}_u$ and $\mathcal{E}_{\mathrm{y}}$ and multiplied by positive weights:

$$\begin{aligned}&\left[\boldsymbol{B}_{\boldsymbol{u}}^{\top}\hat{\boldsymbol{R}}^{\boldsymbol{u}}\right][i] = \sum_{e=1}^{\mathrm{M}}\int_{\Omega_e}\nabla_{\boldsymbol{X}}\boldsymbol{\varphi}_i^{\boldsymbol{u}} : \boldsymbol{P}\,\mathrm{d}V \approx \sum_{e\in\mathcal{E}_{\boldsymbol{u}}}\omega_e^{\boldsymbol{u}}\int_{\Omega_e}\nabla_{\boldsymbol{X}}\boldsymbol{\varphi}_i^{\boldsymbol{u}} : \boldsymbol{P}\,\mathrm{d}V \quad \text{for } i = 1,\ldots,\mathrm{n}_{\boldsymbol{u}}\\ \text{and } &\left[\boldsymbol{B}_{\mathrm{y}}^{\top}\hat{\boldsymbol{R}}^{\mathrm{y}}\right][i] = \sum_{e=1}^{\mathrm{M}}\int_{\Omega_e}\nabla_{\boldsymbol{X}}\varphi_i^{\mathrm{y}}\cdot\mathbb{B}\,\mathrm{d}V \approx \sum_{e\in\mathcal{E}_{\mathrm{y}}}\omega_e^{\mathrm{y}}\int_{\Omega_e}\nabla_{\boldsymbol{X}}\varphi_i^{\mathrm{y}}\cdot\mathbb{B}\,\mathrm{d}V \quad \text{for } i = 1,\ldots,\mathrm{n}_{\mathrm{y}},\\ &\text{with } \mathcal{E}_{\boldsymbol{u}} = \left\{e\in\{1,\ldots,\mathrm{M}\} : \omega_e^{\boldsymbol{u}} > 0\right\} \quad \text{and} \quad \mathcal{E}_{\mathrm{y}} = \left\{e\in\{1,\ldots,\mathrm{M}\} : \omega_e^{\mathrm{y}} > 0\right\}.\end{aligned} \tag{35}$$

In (35), each element e in $\mathcal{E}_u$ or $\mathcal{E}_y$ is equipped with a positive weight ω_e^u or ω_e^y, whereas all the other elements are assigned weights $\omega_e^u = \omega_e^y = 0$.

The approximation in (35) induces the errors

$$
\begin{aligned}
{}^u e_{ij} &= \sum_{e\in\mathcal{E}_u} \omega_e^u \int_{\Omega_e} \nabla_X \boldsymbol{\varphi}_i^u : \boldsymbol{P}_j \,\mathrm{d}V - \sum_{e=1}^{M} \int_{\Omega_e} \nabla_X \boldsymbol{\varphi}_i^u : \boldsymbol{P}_j \,\mathrm{d}V \\
\text{and} \quad {}^y e_{ij} &= \sum_{e\in\mathcal{E}_y} \omega_e^y \int_{\Omega_e} \nabla_X \varphi_i^y \cdot \mathbb{B}_j \,\mathrm{d}V - \sum_{e=1}^{M} \int_{\Omega_e} \nabla_X \varphi_i^y \cdot \mathbb{B}_j \,\mathrm{d}V
\end{aligned} \tag{36}
$$

for snapshots of the stress field $\left\{\boldsymbol{P}_j\right\}_{j=1}^{n_s}$ and the magnetic induction $\left\{\mathbb{B}_j\right\}_{j=1}^{n_s}$. The reduced meshes $\mathcal{E}_u$ and $\mathcal{E}_y$ equipped with the weights $\{\omega_e^u\}_{e=1}^{m_u}$ and $\{\omega_e^y\}_{e=1}^{m_y}$ are constructed by minimising the errors (36), with m_u and m_y being the number of elements in the reduced meshes. Different algorithms for the minimisation of (36) are discussed in [46].

Since this minimisation is numerically expensive, collateral bases [29]

$$
\boldsymbol{P} = \sum_{j=1}^{n_P} c_j^P \boldsymbol{\phi}_j^P \quad \text{and} \quad \mathbb{B} = \sum_{j=1}^{n_{\mathbb{B}}} c_j^{\mathbb{B}} \boldsymbol{\phi}_j^{\mathbb{B}} \tag{37}
$$

for the stress and induction fields are introduced, where $n_P \ll n_s$ and $n_{\mathbb{B}} \ll n_s$ should hold. For that reason, (15) is solved for the parameters in $\mathcal{S}$ (11) and the snapshots of the stress and induction fields are gathered in the matrices $\boldsymbol{S}_P$ and $\boldsymbol{S}_{\mathbb{B}}$. A successive application of POD gives the collateral bases $\boldsymbol{H}_P$ and $\boldsymbol{H}_{\mathbb{B}}$:

$$
\begin{aligned}
\mathrm{POD}(\boldsymbol{S}_P) \rightarrow \boldsymbol{H}_P &= \left[\hat{\boldsymbol{\phi}}_1^P, \ldots, \hat{\boldsymbol{\phi}}_{n_P}^P\right] \in \mathbb{R}^{n_{qp} d^2 \times n_P} \quad \text{with } \boldsymbol{S}_P = \left[\hat{\boldsymbol{P}}(\mathbf{p}_1), \ldots, \hat{\boldsymbol{P}}(\mathbf{p}_{n_s})\right] \in \mathbb{R}^{n_{qp} d^2 \times n_s}, \\
\mathrm{POD}(\boldsymbol{S}_{\mathbb{B}}) \rightarrow \boldsymbol{H}_{\mathbb{B}} &= \left[\hat{\boldsymbol{\phi}}_1^{\mathbb{B}}, \ldots, \hat{\boldsymbol{\phi}}_{n_{\mathbb{B}}}^{\mathbb{B}}\right] \in \mathbb{R}^{n_{qp} d \times n_{\mathbb{B}}} \quad \text{with } \boldsymbol{S}_{\mathbb{B}} = \left[\hat{\mathbb{B}}(\mathbf{p}_1), \ldots, \hat{\mathbb{B}}(\mathbf{p}_{n_s})\right] \in \mathbb{R}^{n_{qp} d \times n_s}.
\end{aligned} \tag{38}
$$

The column vectors $\left\{\hat{\boldsymbol{P}}_i\right\}_{i=1}^{n_s}$ and $\left\{\hat{\mathbb{B}}_i\right\}_{i=1}^{n_s}$ contain the components of $\boldsymbol{P}$ and $\mathbb{B}$ at the n_{qp} quadrature points of the FE model.

By introducing (37) to (36) and recalling that the coefficients in (37) do not depend on the position of the elements in the reduced meshes, we obtain alternative errors

$$
\begin{aligned}
{}^u \breve{e}_{ij} &= \sum_{e\in\mathcal{E}_u} \omega_e^u \int_{\Omega_e} \nabla_X \boldsymbol{\varphi}_i^u : \boldsymbol{\phi}_j^P \,\mathrm{d}V - \sum_{e=1}^{M} \int_{\Omega_e} \nabla_X \boldsymbol{\varphi}_i^u : \boldsymbol{\phi}_j^P \,\mathrm{d}V \\
\text{and} \quad {}^y \breve{e}_{ij} &= \sum_{e\in\mathcal{E}_y} \omega_e^y \int_{\Omega_e} \nabla_X \varphi_i^y \cdot \boldsymbol{\phi}_j^{\mathbb{B}} \,\mathrm{d}V - \sum_{e=1}^{M} \int_{\Omega_e} \nabla_X \varphi_i^y \cdot \boldsymbol{\phi}_j^{\mathbb{B}} \,\mathrm{d}V.
\end{aligned} \tag{39}
$$

For the details on minimisation of (39) in order to obtain the reduced meshes and the weights, the interested reader is referred to Appendix C or [29]. In contrast to the method put forward in [29], the EC introduced here uses elements instead of single Gauss points, resembling the ECSW method [28]. By doing so, the effective number of quadrature points employed in the reduced model increases, but the implementation is less code invasive.

The linearisation of the weak form of the EC hyper-reduced model (35) becomes

$$
\begin{bmatrix}
\sum_{e\in\mathcal{E}^u} \omega_e^u \, {}^e\boldsymbol{B}_u^\top \, {}^e\boldsymbol{K}_k^{uu} \, {}^e\boldsymbol{B}_u & \sum_{e\in\mathcal{E}^u} \omega_e^u \, {}^e\boldsymbol{B}_u^\top \, {}^e\boldsymbol{K}_k^{uy} \, {}^e\boldsymbol{B}_y \\
\sum_{e\in\mathcal{E}^y} \omega_e^y \, {}^e\boldsymbol{B}_y^\top \, {}^e\boldsymbol{K}_k^{yu} \, {}^e\boldsymbol{B}_u & \sum_{e\in\mathcal{E}^y} \omega_e^y \, {}^e\boldsymbol{B}_y^\top \, {}^e\boldsymbol{K}_k^{yy} \, {}^e\boldsymbol{B}_y
\end{bmatrix}
\begin{bmatrix}
\Delta\tilde{\boldsymbol{u}}_k^r \\
\Delta\tilde{\boldsymbol{y}}_k^r
\end{bmatrix}
= -
\begin{bmatrix}
\sum_{e\in\mathcal{E}^u} \omega_e^u \, {}^e\boldsymbol{B}_u^\top \, {}^e\hat{\boldsymbol{R}}_k^u \\
\sum_{e\in\mathcal{E}^y} \omega_e^y \, {}^e\boldsymbol{B}_y^\top \, {}^e\hat{\boldsymbol{R}}_k^y
\end{bmatrix}, \tag{40}
$$

where ${}^{e}\boldsymbol{B}_u$ and ${}^{e}\boldsymbol{B}_y$ are the restrictions of $\boldsymbol{B}_u$ and $\boldsymbol{B}_u$ to the finite element Ω_e. For EC, the reduced system matrix has the same properties, e.g., symmetry and positive definiteness, as the system matrix of the FE model, as the weights are strictly positive. This property is not shared by the collateral basis methods.

The cost for evaluating the constitutive law in the elements of $\mathcal{E}_u \cup \mathcal{E}_y$ is roughly $\mathcal{O}(\mathrm{N_{eval}n} + \mathrm{n_{qp}^{el}m})$, where m is the number of elements in the union of the reduced meshes and $\mathrm{N_{eval}}$ the number of associated DoFs. The assembly of the residuum and the tangent matrix is proportional to $\mathcal{O}(\mathrm{nN_{eval}})$ and $\mathcal{O}(\mathrm{n^2N_{eval}} + \mathrm{nN_{eval}})$, respectively.

4.5. Reduced Integration Domain

For RID, two reduced integration domains $\Omega^u_{\mathrm{RID}} \subset \mathcal{T}$ and $\Omega^y_{\mathrm{RID}} \subset \mathcal{T}$ are introduced, which are used to define test functions $\delta\tilde{\boldsymbol{u}}$ and $\delta\tilde{\boldsymbol{y}}$ with support only in Ω^u_{RID} and Ω^y_{RID}. Hence, the nonlinearities will be computed solely in Ω^u_{RID} and Ω^y_{RID}, which provides the desired reduction of computational cost.

In the discrete setting, the test functions in $\boldsymbol{B}_{\mathrm{RID}}$ with confined support are expressed in terms of the reduced bases $\boldsymbol{B}$ as

$$
\begin{aligned}
\boldsymbol{B}_{\mathrm{RID}} &:= \begin{bmatrix} \mathbf{P}^u_{\mathrm{RID}}(\mathbf{P}^u_{\mathrm{RID}})^\top & \mathbf{0} \\ \mathbf{0} & \mathbf{P}^y_{\mathrm{RID}}\left(\mathbf{P}^y_{\mathrm{RID}}\right)^\top \end{bmatrix} \begin{bmatrix} \boldsymbol{B}_u & \mathbf{0} \\ \mathbf{0} & \boldsymbol{B}_y \end{bmatrix} = \mathbf{P}_{\mathrm{RID}}\mathbf{P}^\top_{\mathrm{RID}}\boldsymbol{B} \\
&\text{with } \mathbf{P}^u_{\mathrm{RID}} = \left[\boldsymbol{e}_{\rho^u_1}, \ldots, \boldsymbol{e}_{\rho^u_{\mathrm{l}_u}}\right] \in \mathbb{N}^{\mathrm{N}_u \times \mathrm{l}_u},\ \mathbf{P}^y_{\mathrm{RID}} = \left[\boldsymbol{e}_{\rho^y_1}, \ldots, \boldsymbol{e}_{\rho^y_{\mathrm{l}_y}}\right] \in \mathbb{N}^{\mathrm{N}_y \times \mathrm{l}_y} \\
&\text{and } \mathbf{P}_{\mathrm{RID}} := \begin{bmatrix} \mathbf{P}^u_{\mathrm{RID}} & \mathbf{0} \\ \mathbf{0} & \mathbf{P}^y_{\mathrm{RID}} \end{bmatrix} \in \mathbb{N}^{\mathrm{N}\times \mathrm{l}},
\end{aligned}
\tag{41}
$$

where $\boldsymbol{e}_{\rho^u_i} \in \mathbb{R}^{\mathrm{N}_u}$ for $i = 1, \ldots, \mathrm{l}_u$ and $\boldsymbol{e}_{\rho^y_j} \in \mathbb{R}^{\mathrm{N}_u}$ for $j = 1, \ldots, \mathrm{l}_y$ are unit vectors with only one non-zero component in the ρ^u_i-th and ρ^y_j-th entry. The indices $\left\{\rho^u_i\right\}_{i=1}^{\mathrm{l}_u}$ and $\left\{\rho^y_i\right\}_{i=1}^{\mathrm{l}_y}$ are the interior DoFs of Ω^u_{RID} and Ω^y_{RID}.

The reduced domains are generated from the gradients of the reduced bases

$$
\boldsymbol{H}_{\nabla x u} := \left[\nabla_{\boldsymbol{X}}\hat{\boldsymbol{\varphi}}^u_1, \ldots, \nabla_{\boldsymbol{X}}\hat{\boldsymbol{\varphi}}^u_{\mathrm{n}_u}\right] \in \mathbb{R}^{\mathrm{n_{qp}d^2} \times \mathrm{n}_u} \quad \text{and} \quad \boldsymbol{H}_{\nabla x y} := \left[\nabla_{\boldsymbol{X}}\hat{\boldsymbol{\varphi}}^y_1, \ldots, \nabla_{\boldsymbol{X}}\hat{\boldsymbol{\varphi}}^y_{\mathrm{n}_y}\right] \in \mathbb{R}^{\mathrm{n_{qp}d} \times \mathrm{n}_y}. \tag{42}
$$

The columns $\left\{\nabla_{\boldsymbol{X}}\hat{\boldsymbol{\varphi}}^u_i\right\}_{i=1}^{\mathrm{n}_u}$ and $\left\{\nabla_{\boldsymbol{X}}\hat{\boldsymbol{\varphi}}^y_i\right\}_{i=1}^{\mathrm{n}_y}$ contain the components of the gradients of $\left\{\boldsymbol{\varphi}^u_i\right\}_{i=1}^{\mathrm{n}_u}$ and $\left\{\varphi^y_i\right\}_{i=1}^{\mathrm{n}_y}$ in the $\mathrm{n_{qp}}$ quadrature points of the FE model. Applying the DEIM Algorithm A1 to $\boldsymbol{H}_{\nabla x u}$ and $\boldsymbol{H}_{\nabla x y}$ returns the indices used to construct the reduced domains Ω^u_{RID} and Ω^y_{RID}. Strictly speaking, the reduced domains are the unions of elements containing these indices.

Using (41) in (16) renders the discrete weak from of the RID hyper-reduced model

$$
\boldsymbol{R}^r_k := \boldsymbol{B}^\top \mathbf{P}_{\mathrm{RID}}\mathbf{P}^\top_{\mathrm{RID}}\hat{\boldsymbol{R}}_k = \mathbf{0} \tag{43}
$$

and its linearisation becomes

$$
\underbrace{\boldsymbol{B}^\top \mathbf{P}_{\mathrm{RID}}}_{\text{precomputed: } \mathbb{R}^{\mathrm{n}\times \mathrm{l}}} \mathbf{P}^\top_{\mathrm{RID}}\boldsymbol{K}_k\boldsymbol{B}\Delta\tilde{\boldsymbol{y}}^r_k = -\underbrace{\boldsymbol{B}^\top \mathbf{P}_{\mathrm{RID}}}_{\text{precomputed: } \mathbb{R}^{\mathrm{n}\times \mathrm{l}}} \mathbf{P}^\top_{\mathrm{RID}}\hat{\boldsymbol{R}}_k. \tag{44}
$$

The cost for evaluating the constitutive law is roughly $\mathcal{O}(\mathrm{N_{eval}n} + \mathrm{n_{qp}^{el}m})$, where m is the number of elements of the union $\Omega^u_{\mathrm{RID}} \cup \Omega^y_{\mathrm{RID}}$ and $\mathrm{N_{eval}}$ the number of DoFs associated with those elements. The approximate costs of assembling the residuum and tangent stiffness matrix are $\mathcal{O}(\mathrm{lN_{eval}})$ and $\mathcal{O}(\mathrm{nlN_{eval}} + \mathrm{lN_{eval}})$, respectively.

For (43) to be a well-posed problem, $\boldsymbol{B}_{\mathrm{RID}}$ is required to have full column rank. If that is not fulfilled or the accuracy of the model is poor, the l element layers surrounding $\Omega^{u}_{\mathrm{RID}}$ and $\Omega^{y}_{\mathrm{RID}}$ are included (cf. Algorithm A4 in Appendix D).

5. Numerical Results

5.1. Test Problem

The magneto-mechanical material model chosen for the numerical studies is of Neo-Hookean type

$$\begin{aligned} \Psi(\boldsymbol{F},\mathbb{H}) =& \frac{1}{2}\lambda_2\left[\boldsymbol{F}:\boldsymbol{F}-\mathrm{d}-2\ln J\right]+\frac{1}{2}\lambda_1\ln^2 J-\frac{1}{2}\mu J\mathbb{H}\cdot\boldsymbol{C}^{-1}\cdot\mathbb{H} \\ &\text{using } J=\det\boldsymbol{F} \text{ and } \boldsymbol{C}=\boldsymbol{F}^{\top}\cdot\boldsymbol{F}, \end{aligned} \tag{45}$$

combining isotropic elastic with linear isotropic magnetic material properties. For further details including expressions of the Piola stress $\boldsymbol{P}$ and the magnetic induction $\mathbb{B}$, see [41].

The mesh used for all numerical tests is displayed in Figure 1, where 1840 quadratic finite elements are used to discretise the continuum body $\mathcal{B}_0$, resulting in $\mathrm{N}=\mathrm{N}_u+\mathrm{N}_y=14{,}882+7441=22{,}323$ DoFs. For numerical integration, a 4×4 Gaussian Quadrature is employed. The implementation of the tests is based on the open–source FE library deal.II [47].

Figure 1. FE mesh of a Unit Cell discretised using $\mathrm{M}=1840$ elements with quadratic Ansatz functions resulting in $\mathrm{N}_u=14,882$ and $\mathrm{N}_y=7441$ DoFs.

In Table 1, the dimensionless Lamé parameters λ_1 and λ_2 and magnetic permeability μ are given. The inclusion/particle has ten times stronger material parameters than the matrix.

Table 1. Material parameters.

	Matrix	**Inclusion**
λ_1	12	120
λ_2	8	80
μ	0.001	0.01

Additionally, the parameter domain for the two-dimensional problem to be investigated is prescribed by

$$\mathcal{P}=\underbrace{(0.9,1.2)}_{\overline{F}[1,1]}\times\underbrace{(-0.2,0.2)}_{\overline{F}[1,2]}\times\underbrace{(-0.2,0.2)}_{\overline{F}[2,1]}\times\underbrace{(0.9,1.2)}_{\overline{F}[2,2]}\times\underbrace{(-10,10)}_{\overline{\mathbb{H}}[1]}\times\underbrace{(-10,10)}_{\overline{\mathbb{H}}[2]}\subset\mathbb{R}^6. \tag{46}$$

The output of interest from the reduced-models are the homogenised Piola stress $\overline{\boldsymbol{P}}$ and magnetic induction $\overline{\mathbb{B}}$, for which the relative error measures

$$\mathrm{Err}_{\overline{\boldsymbol{P}}}(\mathcal{V}) = \mathrm{median}\left\{\frac{\left\|\overline{\boldsymbol{P}}_i^{\mathrm{FOM}} - \overline{\boldsymbol{P}}_i^{\mathrm{ROM}}\right\|_F}{\left\|\overline{\boldsymbol{P}}_i^{\mathrm{FOM}}\right\|_F}\right\}_{i=1}^{|\mathcal{V}|} \quad \text{and} \quad \mathrm{Err}_{\overline{\mathbb{B}}}(\mathcal{V}) = \mathrm{median}\left\{\frac{\left\|\overline{\mathbb{B}}_i^{\mathrm{FOM}} - \overline{\mathbb{B}}_i^{\mathrm{ROM}}\right\|_2}{\left\|\overline{\mathbb{B}}_i^{\mathrm{FOM}}\right\|_2}\right\}_{i=1}^{|\mathcal{V}|} \tag{47}$$

are defined for any set of validation parameters $\mathcal{V}$.

To validate the accuracy and robustness of the reduced models, two sets of randomly chosen parameters

$$\mathcal{V}_{\mathrm{I}} = \{\mathbf{p}_1, \ldots, \mathbf{p}_{200}\} \subset \mathcal{P} \quad \text{and} \quad \mathcal{V}_{\mathrm{II}} = \{\mathbf{p}_1, \ldots, \mathbf{p}_{2000}\} \subset \mathcal{P} \tag{48}$$

are defined and used in combination with (47).

The computation of $\overline{\boldsymbol{P}}^{\mathrm{ROM}}$ and $\overline{\mathbb{B}}^{\mathrm{ROM}}$ (8) requires $\tilde{\boldsymbol{u}}$ and $\tilde{\mathrm{y}}$ to be computed in all cells of the FE mesh using (9). This can be done more efficiently using an auxiliary basis for $\boldsymbol{P}$ and $\mathbb{B}$ together with gappy reconstruction, but this renders an additional error. As our focus is on the numerical study of the performance of the hyper-reduction methods, $\tilde{\boldsymbol{u}}$ and $\tilde{\mathrm{y}}$ are computed for the whole mesh and the constitutive law is subsequently employed to obtain $\overline{\boldsymbol{P}}^{\mathrm{ROM}}$ and $\overline{\mathbb{B}}^{\mathrm{ROM}}$.

It is established in the field of computational homogenisation that the application of linear boundary conditions overestimates the energy compared to e.g., periodic boundary conditions, in particular for small RVEs like the Unit Cell. This has been investigated extensively in [40] for magneto-mechanics. As the choice of boundary conditions does not affect the hyper-reduction methods, the findings from the numerical studies are expected to be valid for different types of boundary conditions. Therefore, due to their simplicity, linear boundary conditions have been chosen to carry out the numerical studies.

5.2. Validation of Galerkin ROM

In order to construct the reduced basis, the parameter space $\mathcal{P}$ has to be sampled. As the number of sampling points increases exponentially with the dimension of the parameter domain for full tensor grids, sparse grids [48,49] are employed. Sparse grids are based on one-dimensional quadrature rules and a sparse tensor product, which alleviates the curse of dimensionality. For that reason, sparse grids are frequently used in sampling, interpolation and integration of high dimensional functions.

In Figure 2, the sampling of the unit square using a full tensor grid and sparse grids is displayed. Sparse grids built from the one-dimensional Gauss–Legendre quadrature are used to sample the six-dimensional parameter domain $\mathcal{P}$ (46).

As the hyper-reduced models are built on top of an existing reduced basis, the accuracy of the Galerkin reduced model (16) for varying cardinalities n_u and n_{y} of the reduced bases for the fluctuation fields is displayed in Figure 3. For increasing n_u and n_{y}, the errors $\mathrm{Err}_{\overline{\boldsymbol{P}}}(\mathcal{V}_{\mathrm{I}})$ and $\mathrm{Err}_{\overline{\mathbb{B}}}(\mathcal{V}_{\mathrm{I}})$ decrease monotonously, though the impact of n_{y} on $\mathrm{Err}_{\overline{\boldsymbol{P}}}(\mathcal{V}_{\mathrm{I}})$ becomes negligible for $\mathrm{n}_{\mathrm{y}} \geq 10$. The reduced basis was constructed from $\mathrm{n}_s = 4541$ training simulations. As a compromise between accuracy and efficiency, a reduced basis with $\mathrm{n}_u = 20$ and $\mathrm{n}_{\mathrm{y}} = 10$ is chosen for all following numerical studies.

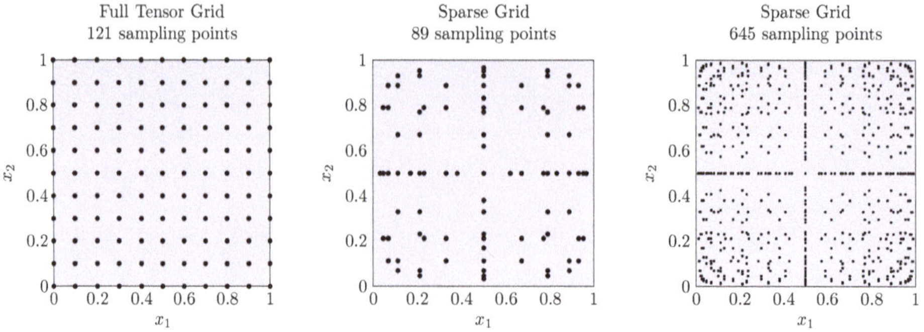

Figure 2. Sampling of the two-dimensional parameter domain $[0,1] \times [0,1]$ using a full tensor grid and two sparse grids based on a one-dimensional Gauss–Legendre quadrature with different sampling densities.

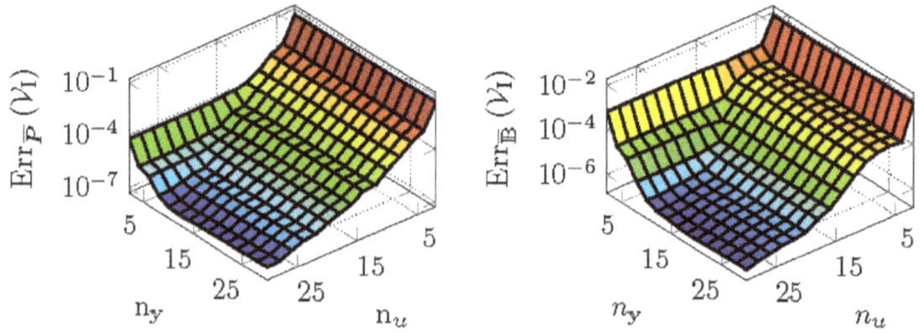

Figure 3. Errors in homogenised quantities of interest for varying sizes of reduced bases n_u and n_y computed from $n_s = 4541$ training simulations.

The error of the Galerkin ROM is not only caused by the truncation of the POD basis but also by an insufficient selection of training parameters $\mathcal{S}$. In this case, insufficient refers to a too sparse sampling of the parameter domain $\mathcal{P}$. In Table 2, the errors $\mathrm{Err}_{\overline{P}}(\mathcal{V}_I)$ and $\mathrm{Err}_{\overline{\mathbb{B}}}(\mathcal{V}_I)$ for $n_u = 20$ and $n_y = 10$ for three different training sets are given. The training set $\mathcal{S}$ with $n_s = 4541$ is considered sufficiently large as the errors for the two sets with greater cardinality are not significantly smaller.

Table 2. Output error for different numbers of training parameters n_s for a reduced basis of fixed size $n_u = 20$ and $n_y = 10$.

$\lvert\mathcal{S}\rvert$	4541	12,841	33,193
$\mathrm{Err}_{\overline{P}}(\mathcal{V}_I)$	2.75×10^{-6}	2.59×10^{-6}	2.53×10^{-6}
$\mathrm{Err}_{\overline{\mathbb{B}}}(\mathcal{V}_I)$	1.55×10^{-6}	1.51×10^{-6}	1.62×10^{-6}

The results of a ROM for one element in $\mathcal{V}_I$ are displayed in Figure 4. The errors in the homogenised quantities are small $\mathcal{O}(10^{-6})$ and differences between the ROM and FOM in the primary fields $\tilde{u}$ and $\tilde{y}$ can not be seen with the unaided eye. Due to the inclusions ten times larger mechanical and magnetic material parameters, the Piola stress and magnetic induction inside the inclusion are significantly larger.

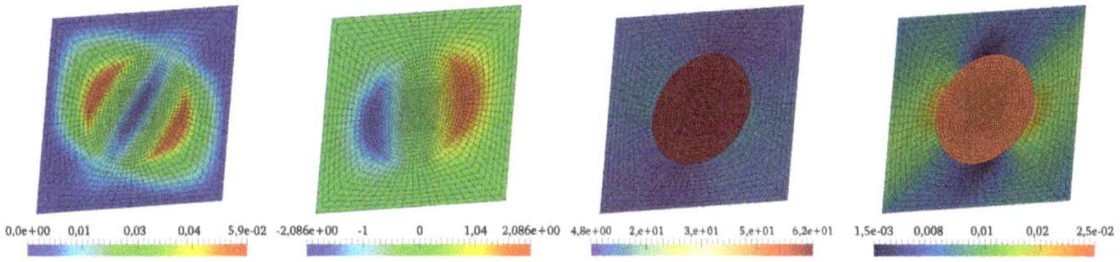

a) $\|\tilde{\boldsymbol{u}}\|_2$ b) $\tilde{y}$ c) $\|\boldsymbol{P}\|_F$ d) $\|\mathbb{B}\|_2$

Figure 4. Results of a reduced model for $\overline{\boldsymbol{F}} = \begin{bmatrix} 1.132 & 0.182 \\ 0.144 & 1.145 \end{bmatrix}$ and $\overline{\mathbb{H}} = \begin{bmatrix} -8.231 \\ 0.129 \end{bmatrix}$ rendering $\overline{\boldsymbol{P}}^{\text{ROM}} = \begin{bmatrix} 5.875 & 2.943 \\ 2.756 & 6.026 \end{bmatrix}$ and $\overline{\mathbb{H}}^{\text{ROM}} = \begin{bmatrix} -1.123 \times 10^{-2} \\ 3.520 \times 10^{-3} \end{bmatrix}$ using a reduced basis with $\mathrm{n}_u = 20$ and $\mathrm{n}_y = 10$. The norm of the displacement fluctuations (**a**) the potential fluctuations; (**b**) the Frobenius norm of the element averaged Piola stress; (**c**) and the norm of the element averaged magnetic induction; (**d**) are depicted in the deformed Unit Cell. The errors for the homogenised quantities $\frac{\|\overline{\boldsymbol{P}}^{\text{FOM}} - \overline{\boldsymbol{P}}^{\text{ROM}}\|_F}{\|\overline{\boldsymbol{P}}^{\text{FOM}}\|_F} = 3.4 \times 10^{-6}$ and $\frac{\|\overline{\mathbb{B}}^{\text{FOM}} - \overline{\mathbb{B}}^{\text{ROM}}\|_2}{\|\overline{\mathbb{B}}^{\text{FOM}}\|_2} = 3.1 \times 10^{-6}$ are small. For better quality, we point to the online version of the article.

5.3. DEIM

Based on a reduced basis with $\mathrm{n}_u = 20$ and $\mathrm{n}_y = 10$, the results of the DEIM hyper-reduced model (23) are summarised in Figure 5, where the collateral basis $\boldsymbol{H_R}$ is computed based on residua collected during the solution of (16) for $\mathrm{n}_s = 4541$ training parameters. For the POD computations, only residua fulfilling $\left\|\hat{\boldsymbol{R}}^u_j(\mathrm{p}_i)\right\| / \left\|\hat{\boldsymbol{R}}^u_1(\mathrm{p}_i)\right\|, \left\|\hat{\boldsymbol{R}}^y_j(\mathrm{p}_i)\right\| / \left\|\hat{\boldsymbol{R}}^y_1(\mathrm{p}_i)\right\| > 10^{-4}$ with $i \in [1, 4541]$ and the iteration index j were taken into account.

It is well-established that DEIM models lack robustness for nonlinear pPDEs [26], which is exposed in Figure 5a. There is just a small region of combinations of numbers of collateral reduced basis functions $(\mathrm{r}_u, \mathrm{r}_y)$, for which the reduced model converged for all parameters in the validation set $\mathcal{V}_{\text{I}}$. There are two possible causes that prevent the convergence of a model. The first is the occurrence of unphysical deformations expressed by $\det \boldsymbol{F} \leq 0$ during the solution process and the second is an insufficient reduction of the residuum $\boldsymbol{R}^r_{\text{max}} / \boldsymbol{R}^r_0 > 10^{-6}$, where max $= 10$ is the highest admissible number of nonlinear solver iterations. From the 135,200 solutions of (23) for 676 different combinations of $(\mathrm{r}_u, \mathrm{r}_y)$ computed in this study, 11,992 failed to converge. It is remarkable that larger r_u and r_y do not result in more robust models. Figure 5b,c show the output errors, where we have to note that, for the calculation of $\text{Err}_{\overline{\boldsymbol{P}}}(\mathcal{V}_{\text{I}})$ and $\text{Err}_{\overline{\boldsymbol{P}}}(\mathcal{V}_{\text{II}})$, only the converged runs are taken into account. The errors decrease with increasing r_u and r_y but certainly not monotonously. For a ROM with $\mathrm{r}_u = 37$ and $\mathrm{r}_y = 25$, small errors $\text{Err}_{\overline{\boldsymbol{P}}}(\mathcal{V}_{\text{I}}) = 1.2 \times 10^{-3}$ and $\text{Err}_{\overline{\mathbb{B}}}(\mathcal{V}_{\text{I}}) = 5.5 \times 10^{-4}$ are obtained with one simulation failing to converge. It is not reasonable to use greater r_u and r_y as the achievable reduction of the errors is disproportionate to the increasing numerical cost of the ROM.

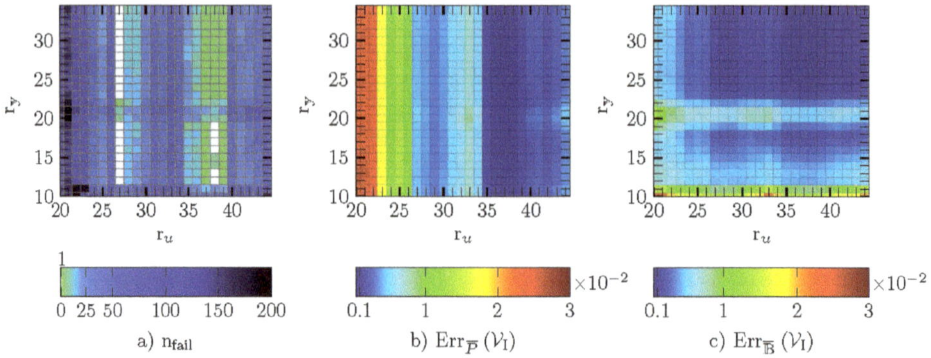

Figure 5. (**a**) robustness and (**b**,**c**) output error analysis for varying numbers of DEIM indices r_u and r_y for a reduced basis of size $n_u = 20$ and $n_y = 10$.

5.4. Gappy POD

For the Gappy POD study, the same collateral basis $\boldsymbol{H}_R$ as in the DEIM study is used. In order to have a fair comparison with DEIM, two sets of gappy points, one for the approximation of $\hat{\boldsymbol{R}}^u$ and the other for $\hat{\boldsymbol{R}}^y$ in (28), are determined by applying Algorithm A2 separately to the collateral bases.

In Figure 6, the robustness and accuracy of Gappy POD is studied for different combinations of (r_u, r_y). To facilitate an adequately accurate approximation of the nonlinearities, large enough numbers of gappy points (p_u, p_y) are employed. Figure 6a shows that linear regression improves the robustness as more combinations of (r_u, r_y) exhibit no convergence failures compared to DEIM (c.f. Figure 5). Nonetheless, 11,339 out of 135,200 simulation runs failed, mostly for smaller values of r_u and r_y. For the failed runs, either the condition $\boldsymbol{R}^r_{10}/\boldsymbol{R}^r_0 < 10^{-6}$ is not fulfilled after ten iterations or unphysical deformations $\det \boldsymbol{F} \leq 0$ are predicted.

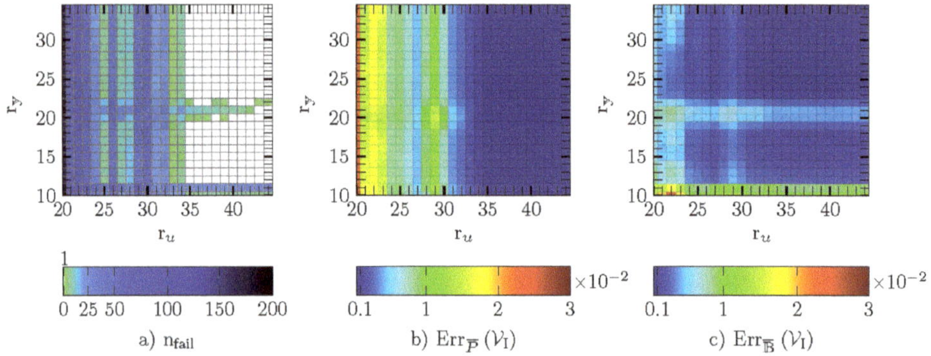

Figure 6. (**a**) robustness and (**b**,**c**) output error analysis for varying numbers of residuum modes r_u and r_y for sufficiently large numbers of gappy points $p_u = \lceil r_u/2 + 10 \rceil$ and $p_y = r_y + 20$ for a reduced basis of size $n_u = 20$ and $n_y = 10$.

For $r_u = 36$ and $r_y = 24$, a combination yielding supposedly robust and accurate models is chosen to investigate the errors for varying numbers of gappy points (p_u, p_y) (see Figure 7). Except for $p_u \leq 20$, no robustness deficiencies occur. The error $Err_{\overline{P}}(\mathcal{V}_I)$ decreases with increasing p_u and is hardly affected by changes of p_y. Similarly, $Err_{\overline{\mathbb{B}}}(\mathcal{V}_I)$ reduces for larger p_y and is only minorly affected by p_u. For a reduced model using $p_u = 50$ and $p_y = 58$ gappy points, the errors $Err_{\overline{P}}(\mathcal{V}_I) = 3.6 \times 10^{-4}$ and $Err_{\overline{\mathbb{B}}}(\mathcal{V}_I) = 3.0 \times 10^{-4}$ were achieved.

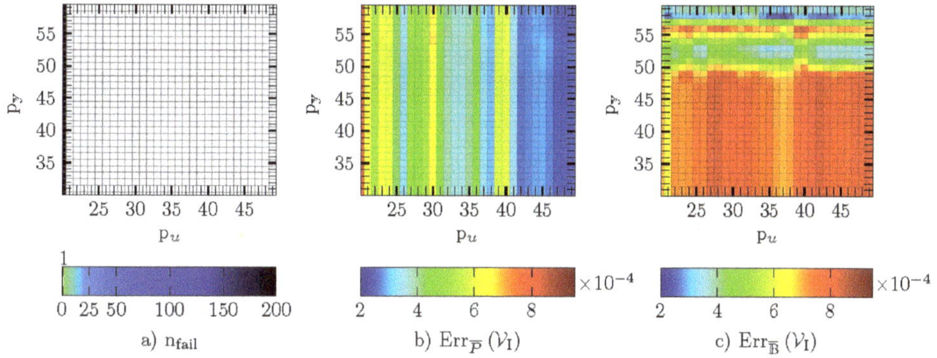

Figure 7. Influence of gappy point numbers p_u and p_y on (**a**) robustness and (**b**,**c**) accuracy for a reduced basis of size $n_u = 20$ and $n_y = 10$ using $r_u = 36$ and $r_y = 24$ residuum modes.

5.5. GNAT

For the study of GNAT, the collateral bases contained in $\boldsymbol{H}_{R^u}$ and $\boldsymbol{H}_{R^y}$ are constructed by gathering the residua from the solution of (32) and a subsequent application of POD. Only residua fulfilling $\left\|\hat{R}^u_j(\mathbf{p}_i)\right\|/\left\|\hat{R}^u_1(\mathbf{p}_i)\right\|, \left\|\hat{R}^y_j(\mathbf{p}_i)\right\|/\left\|\hat{R}^y_1(\mathbf{p}_i)\right\| > 5 \times 10^{-3}$ with $i \in [1, 4541]$ and the iteration index j were taken into account. The gappy points are obtained by applying Algorithm A2 separately to the collateral mechanical and magnetic basis.

Similarly to the previous studies, the robustness and accuracy of GNAT is tested for different combinations of (r_u, r_y) using a large enough number of gappy points. The results are depicted in Figure 8. We never observed unphysical deformations $\det \boldsymbol{F} \leq 0$ for GNAT hyper-reduced models, the unsuccessful runs are due to reduced models failing to sufficiently reduce the residuum $\boldsymbol{R}^r_{15}/\boldsymbol{R}^r_0 < 10^{-3}$. As the Gauss–Newton scheme does not exhibit quadratic convergence, we allow for up to 15 iterations to minimise the residuum. Furthermore, the solution of the nonlinear least squares problem (32) does not render $\boldsymbol{R}^r \equiv \boldsymbol{0}$ in general and consequently the convergence criterion is set to 10^{-3}. While the error $\mathrm{Err}_{\overline{\mathbb{P}}}(\mathcal{V}_\mathrm{I})$ in Figure 8b reduces with increasing r_u and values in the order of 10^{-3} can be achieved, the error $\mathrm{Err}_{\overline{\mathbb{B}}}(\mathcal{V}_\mathrm{I})$ in Figure 8c increases for greater r_u up to a certain point and $\mathrm{Err}_{\overline{\mathbb{B}}}(\mathcal{V}_\mathrm{I}) \approx 5 \times 10^{-2}$ seems feasible.

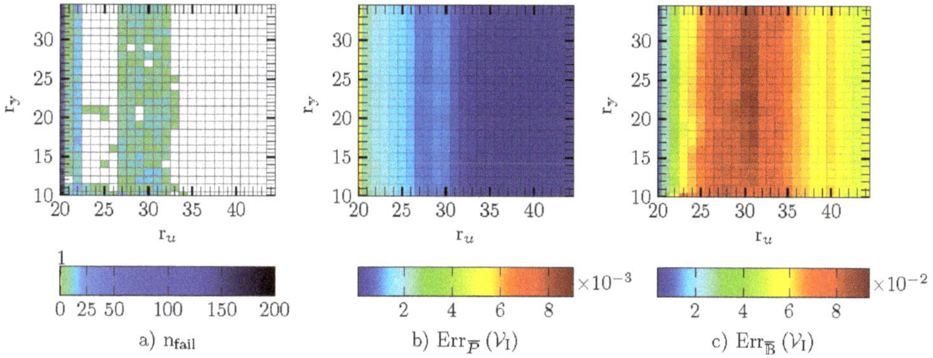

Figure 8. (**a**) robustness and (**b**,**c**) output error analysis for varying numbers of residuum modes r_u and r_y for sufficiently large numbers of gappy points $p_u = \lceil r_u/2 + 10 \rceil$ and $p_y = r_y + 20$ for a reduced basis of size $n_u = 20$ and $n_y = 10$.

5.6. Empirical Cubature

In Figure 9, the accuracy of EC hyper-reduced models for different numbers of elements in the reduced mesh is depicted. To construct the EC model 4541 snapshots of $\boldsymbol{P}$ and $\mathbb{B}$ are taken from the solution of (15) and processed into $n_P = 120$ and $n_{\mathbb{B}} = 100$ POD modes. The singular value distribution of the snapshot matrices is depicted in Figure 10, indicating that $n_P = 120$ and $n_{\mathbb{B}} = 100$ POD modes are sufficient to represent the stress and induction state.

Thereafter, the weights and elements of the reduced meshes are determined by approximately solving (A3) using Algorithm A3, given in Appendix C. Figure 9a shows that $\mathrm{Err}_{\overline{P}}(\mathcal{V}_\mathrm{I})$ decreases for greater m_u and is barely affected by m_y, whereas $\mathrm{Err}_{\overline{\mathbb{B}}}(\mathcal{V}_\mathrm{I})$ decreases for greater values of both m_u and m_y, but the dependence on m_y is more pronounced. As EC does not employ collateral bases combined with linear regression or interpolation, no convergence issues occur for the EC hyper-reduced models.

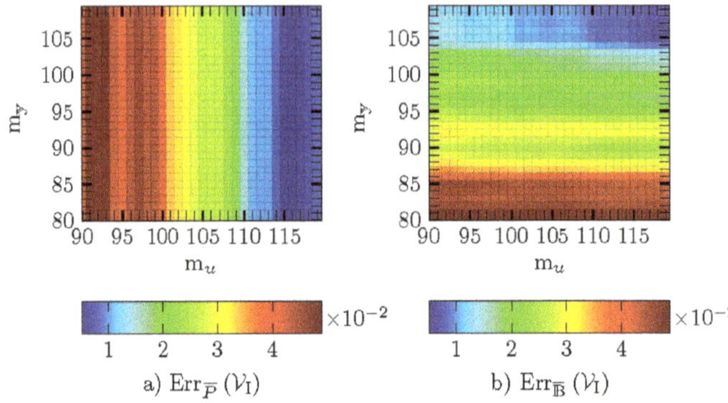

Figure 9. Output error analysis for varying numbers of elements m_u and m_y constituting the reduced meshes $\mathcal{E}_u$ and $\mathcal{E}_y$ built using $n_P = 120$ stress and $n_{\mathbb{B}} = 100$ induction modes for a reduced basis of size $n_u = 20$ and $n_y = 10$.

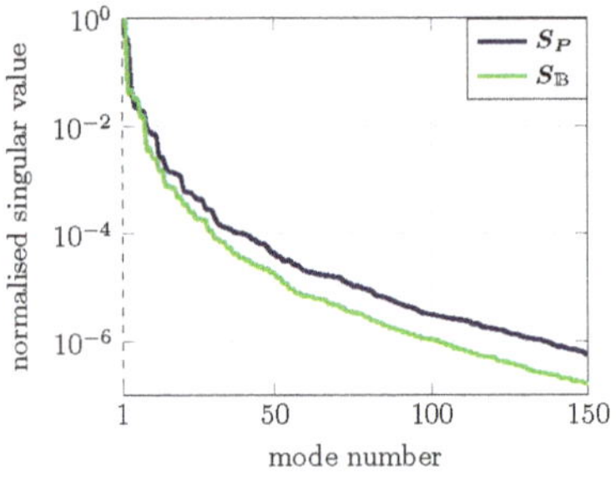

Figure 10. Normalised singular values of the stress S_P and induction snapshot matrix $S_{\mathbb{B}}$.

In Figure 11a, the weights used in a reduced mesh are plotted. Only a small number of elements accumulate more than half of the total weight sum and that is the reason why the errors depicted in Figure 9 decrease slowly with increasing m_u and m_y. The distribution of elements in the reduced meshes with a focus on elements equipped with relatively large weights is shown in Figure 11b,c. It is remarkable that the elements with large weights are all located inside the inclusion, whereas all the other hyper-reduced models (DEIM, GappyPOD and RID) evaluate the nonlinearities in elements at the boundary or in the vicinity of the interface between matrix and inclusion (c.f. Figure 12).

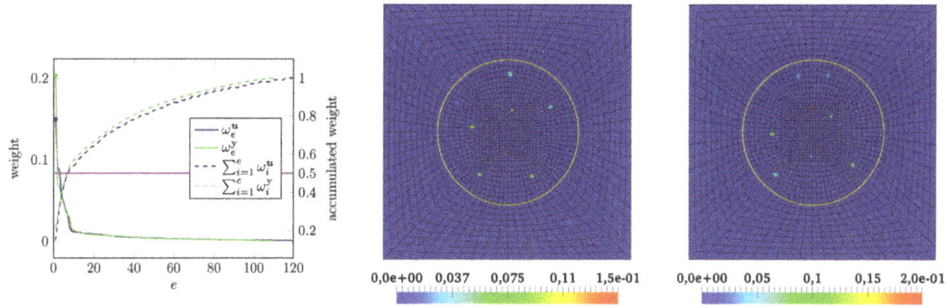

a) Values of $\boldsymbol{\omega}^{\boldsymbol{u}}$ and $\boldsymbol{\omega}^{\mathcal{y}}$ b) Distribution of $\mathcal{E}_{\boldsymbol{u}}$ in $\mathcal{T}$ coloured by values of $\boldsymbol{\omega}^{\boldsymbol{u}}$ c) Distribution of $\mathcal{E}_{\mathrm{y}}$ in $\mathcal{T}$ coloured by values of $\boldsymbol{\omega}^{\mathcal{y}}$

Figure 11. (**a**) values and distribution of weights (**b**) ω^u and (**c**) ω^y in $\mathcal{T}$ for reduced meshes $\mathcal{E}_u$ and $\mathcal{E}_y$ with $\mathrm{m}_u = 120$ and $\mathrm{m}_y = 110$ elements. The yellow circles mark the boundary between matrix and inclusion. For better quality, we point to the online version of the article.

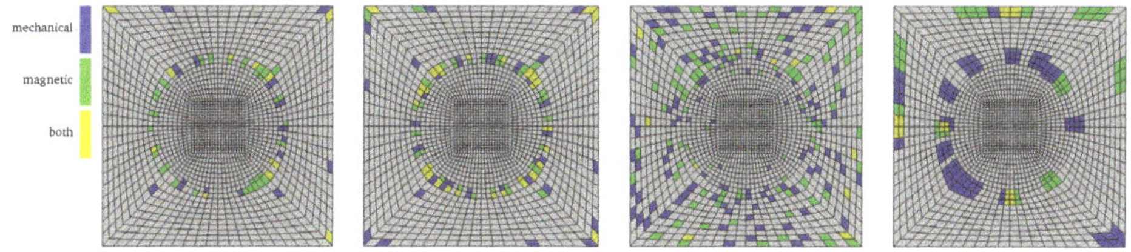

a) DEIM b) Gappy POD c) EC d) RID

Figure 12. Elements in $\mathcal{T}$ relevant for the hyper-reduced models in Table 4.

5.7. Reduced Integration Domain

To construct the reduced integration domain, the gradients of $\{\boldsymbol{\varphi}^u\}_{i=1}^{\mathrm{n}_u=20}$ and $\{\varphi^{\mathrm{y}}\}_{i=1}^{\mathrm{n}_{\mathrm{y}}=10}$ are computed and the application of Algorithm A4 yields the reduced domains Ω^u_{RID} and $\Omega^{\mathrm{y}}_{\mathrm{RID}}$. The errors for three different choices of the number of surrounding layers l are listed in Table 3. As expected, the errors decrease with increasing l but solving (43) becomes computationally more expensive. If no surrounding layers are included, the reduced bases $\{\boldsymbol{\varphi}^u\}_{i=1}^{\mathrm{n}_u=20}$ and $\{\varphi^{\mathrm{y}}\}_{i=1}^{\mathrm{n}_{\mathrm{y}}=10}$ in Ω^u_{RID} and $\Omega^{\mathrm{y}}_{\mathrm{RID}}$ are linearly dependent and consequently (43) is not well-posed.

Table 3. Accuracy of RID depending on the number of neighbouring layers l for $\mathrm{n}_u = 20$ and $\mathrm{n}_{\mathrm{y}} = 10$.

l	1	2	3
$\mathrm{Err}_{\overline{\mathbb{P}}}(\mathcal{V}_{\mathrm{I}})$	9.5×10^{-4}	3.3×10^{-4}	1.5×10^{-4}
$\mathrm{Err}_{\overline{\mathbb{B}}}(\mathcal{V}_{\mathrm{I}})$	8.0×10^{-4}	3.2×10^{-4}	1.8×10^{-4}

As for EC, no convergence problems have been observed for RID hyper-reduced models.

5.8. Comparison of the Hyper-Reduction Methods

In Table 4, the performance statistics of reduced models for each hyper-reduction method except GNAT are provided. The parameters for the models, which are listed in Table 5, were chosen based on the results from the previous sections to achieve high accuracy at moderate computational cost. We except GNAT from the comparison as these reduced models are inferior to Gappy POD models with respect to robustness and accuracy. Most importantly, the error $\mathrm{Err}_{\overline{\mathbb{B}}}(\mathcal{V})$ produced by GNAT models is too large.

The number of elements in which either the mechanical or magnetic nonlinearity have to be computed are denoted by m_u and m_y, respectively. The aforementioned elements are highlighted in Figure 12. It is possible to use the same elements for the evaluation of the nonlinearities for all hyper-reduction methods except DEIM, rendering supposedly slightly less accurate but more efficient reduced models. However, for the sake of comparison of all introduced Hyper-Reduction methods, the nonlinearities are treated separately, resulting in two distinct sets of elements. In all these elements, the solution $(\tilde{\boldsymbol{u}}, \tilde{y})$ has to be computed based on the reduced solution $(\tilde{\boldsymbol{u}}^r, \tilde{y}^r)$ with $N^{\boldsymbol{u}}_{\text{eval}}$ and N^{y}_{eval} denoting the number of DoFs involved in these operation performed in every iteration of the nonlinear solver.

The errors obtained by DEIM, Gappy POD and RID are in the same range, whilst the errors for EC are at least of one order of magnitude larger. The accuracy of Gappy POD is superior to DEIM as linear regression performs better than interpolation and additionally increases the robustness. For DEIM, 18 out of 2000 runs failed to converge, whereas no such deficiencies are observed for the other methods.

The computation times were measured on a single core (AMD™ Ryzen™ 1950X CPU @4 GHz) without using any kind of parallelisation. The speed-up is defined as the ratio of time needed to solve the FOM and the ROM for the 2000 parameters in $\mathcal{V}_{\text{II}}$, which does not include the time to calculate the homogenised quantities. The speed-up for DEIM is the greatest as the least number of solution and nonlinearity evaluations had to be performed, with Gappy POD being second due to more evaluations. The EC and RID reduced models are considerably slower as both methods need to evaluate the nonlinearities for a larger number of elements to gain comparable accuracy.

Table 4. Comparison of selected hyper-reduced models using a reduced basis with $n_u = 20$ and $n_y = 10$.

	DEIM	Gappy POD	EC	RID
m_u & $m_y \to m$	50 & 33 → 71	69 & 75 → 112	120 & 110 → 219	129 & 72 → 183
N^{u}_{eval} & N^{y}_{eval}	1000 & 500	1468 & 734	3400 & 1700	1938 & 969
$\text{Err}_{\overline{\mathbb{P}}}(\mathcal{V}_{\text{II}})$	1.14×10^{-3}	3.86×10^{-4}	1.85×10^{-2}	5.99×10^{-4}
$\text{Err}_{\overline{\mathbb{B}}}(\mathcal{V}_{\text{II}})$	4.71×10^{-4}	2.78×10^{-4}	4.63×10^{-3}	5.18×10^{-4}
n_{fail}	18	0	0	0
speed-up	208	131	23	32

Table 5. Parameters of the hyper-reduced models used in Table 4.

DEIM	$r_u = 37$ and $r_y = 25$
Gappy POD	$r_u = 36$ and $r_y = 24$, $p_u = 50$ and $p_y = 58$
EC	$m_u = 120$ and $m_y = 110$
RID	$l = 1$

6. Conclusions

In this work, we applied the tools of reduced-order modelling to the problems arising in computational homogenisation in magneto-mechanics. The main focus was the investigation and comparison of different hyper-reduction techniques with respect to accuracy and robustness. Collateral basis methods like DEIM and Gappy POD are the fastest and most accurate, but are susceptible to robustness deficiencies. This is particularly true for DEIM, for which we could not build sufficiently robust models in the course of this work. Gappy POD diminishes that issue by using linear regression, providing adequately robust models. Additionally, unlike Gappy POD, DEIM does not offer the option to evaluate the mechanical and magnetic nonlinearity in the same elements, resulting in more expensive reduced models. For those reasons, DEIM will not be considered in future studies. A disadvantage shared among the collateral basis methods is that instances of the FE residuum have to be collected from non-equilibrated states, which results in more expensive POD computations.

EC and RID solve the weak form in a subdomain, to which the first refers to reduced mesh and the latter to reduced integration domain. To obtain similar accuracy as collateral basis methods, EC and RID have to perform more function evaluations and are therefore more expensive. However, their robustness is superior to the collateral basis methods and hence they are particularly suitable for multi-query frameworks like the FE^2 method. For problems with stronger nonlinearities, e.g., due to complex material models, rate-dependence, plasticity and many more, the robustness superiority will be even more pronounced.

The next step is to equip the reduced models with an auxiliary basis to efficiently compute the homogenised Piola stress and magnetic induction, which can be utilised in a perturbation-type method [50] to obtain the macroscopic tangent moduli. Similarly, the macroscopic tangent moduli could be computed adapting the method described in [51] for the reduced model.

Author Contributions: Conceptualization, B.B.; Investigation, B.B.; Software, B.B., D.D.; Supervision, P.S.; Writing—Original Draft, B.B.; Writing—Review and Editing, D.D., J.M. and P.S.

Funding: Benjamin Brands was partially supported by the Indo-German exchange programm "Multiscale Modeling, Simulation and Optimization for Energy, Advanced Materials and Manufacturing" (MMSO) funded by DAAD (Germany) and UGC (India). Denis Davydov was supported by the German Research Foundation (DFG), grant DA 1664/2-1.

Conflicts of Interest: The authors declare no conflict of interest.

Abbreviations

The following abbreviations are used in this manuscript:

MRE	Magneto-Rheological Elastomer
BVP	Boundary Value Problem
RVE	Representative Volume Element
ROM	Reduced-Order Model
FEM	Finite Element Method
DoF	Degree of Freedom
FOM	Full-Order Model
pPDE	parametrised Partial Differential Equation
POD	Proper Orthogonal Decomposition
SVD	Singular Value Decomposition
EIM	Empirical Interpolation Method
DEIM	Discrete Empirical Interpolation Method
GNAT	Gauss–Newton with Approximated Tensors
EC	Empirical Cubature
ECSW	Energy-Conserving Sampling and Weighting
RID	Reduced Integration Domain

Appendix A. DEIM

The classical DEIM Algorithm A1 [19] determines the interpolation indices iteratively. In iteration i, the index ρ_i to be added is the entry where the approximation of the basis vector $\boldsymbol{\phi}_i$ by the preceding vectors $\left\{\boldsymbol{\phi}_j\right\}_{j=1}^{i}$ exhibits the largest error. The basis vectors $\{\boldsymbol{\phi}_i\}_{i=1}^{r}$ have to be linearly independent, which is guaranteed by using a basis computed by POD.

Algorithm A1: DEIM Algorithm

Input: $\{\boldsymbol{\phi}_i\}_{i=1}^{r} \subset \mathbb{R}^{\mathrm{N}}$
Output: $\rho = [\rho_1, \ldots, \rho_r] \in \mathbb{N}^r$, $\mathbf{P}$
$\rho_1 = \underset{a=1,\ldots,\mathrm{N}}{\arg\max} \left|\boldsymbol{\phi}_1[a]\right|$
$S = [\boldsymbol{\phi}_1]$, $\mathbf{P} = \left[e_{\rho_1}\right]$, $\rho = [\rho_1]$
for i=2 **to** r **do**
  Solve for **c** in $\left(\mathbf{P}^\top S\right) \mathbf{c} = \mathbf{P}^\top \boldsymbol{\phi}_i$
  $r = \boldsymbol{\phi}_i - S\mathbf{c}$
  $\rho_i = \underset{a=1,\ldots,\mathrm{N}}{\arg\max} \left|\boldsymbol{\phi}_i[a]\right|$
  $\rho \leftarrow [\rho, \rho_i]$, $\mathbf{P} \leftarrow \left[\mathbf{P}, e_{\rho_i}\right]$, $S \leftarrow [S, \boldsymbol{\phi}_i]$
end

Appendix B. Gappy POD and GNAT

Like the DEIM Algorithm A1, the adapted point search Algorithm A2 [21] seeks to minimise the error in approximating the bases $\left\{\left\{\boldsymbol{\phi}_i^c\right\}_{i=1}^{r_c}\right\}_{c=1}^{n_{\text{components}}}$ and chooses the gappy points accordingly. The algorithm processes nodes instead of indices and is therefore well-suited for vector-valued or multi-physic problems. Differences in scale are taken care of by using normalised maxima for the different fields.

Algorithm A2: Greedy Point Search

Input: reduced bases $\left\{ \left\{ \boldsymbol{\phi}_i^c \right\}_{i=1}^{r_c} \right\}_{c=1}^{n_{\text{components}}}$, number of sampling points p

Output: $\{\mathbf{P}_c\}_{c=1}^{n_{\text{components}}}$

number of greedy iterations: $g = \min\left(\left\{r_1, \ldots, r_{n_{\text{components}}}, \mathrm{p}\right\}\right)$

number of sampling points at iteration i: $s(i)$

$$s(i) = \left\lfloor \frac{w \cdot \mathrm{p}}{g} \right\rfloor \text{ with } w = \left\lceil \frac{g}{\mathrm{p}} \right\rceil; \text{ if } w = 1 \text{ and } i \leq (\mathrm{p} \bmod g) : s(i) \leftarrow s(i) + 1$$

number of basis vectors to be added/processed at iteration i:

$$q_c(i) = \left\lfloor \frac{r_c}{g} \right\rfloor; \text{ if } i \leq (r_c \bmod g) : q_c(i) \leftarrow q_c(i) + 1 \text{ for } c = 1, \ldots, n_{\text{components}}$$

for c=1 **to** $n_{\text{components}}$ **do**

$\quad \left[\mathbf{S}_1^c, \ldots, \mathbf{S}_{q_c(1)}^c\right] \leftarrow \left[\boldsymbol{\phi}_1^c, \ldots, \boldsymbol{\phi}_{q_c(1)}^c\right]$ `// vectors to be processed in first iteration`

end

for i=1 **to** g **do**

$\quad$ **for** j=1 **to** $s(i)$ **do**

$\quad\quad$ **for** c=1 **to** $n_{\text{components}}$ **do**

$$n_{\max}^c \leftarrow \underset{l \in \{1,\ldots,n_{\text{points}}\}}{\arg\max} \sum_{q=1}^{q_c(i)} \left\| \mathbf{S}_q^c[l] \right\|^2 \qquad \text{// location of component maximum}$$

$\quad\quad$ **end**

$$n \leftarrow \underset{l \in \{1,\ldots,n_{\text{points}}\} \setminus \mathcal{N}}{\arg\max} \sum_{c=1}^{n_{\text{components}}} \sum_{q=1}^{q_c(i)} \frac{\left\| \mathbf{S}_q^c[l] \right\|^2}{\left\| \mathbf{S}_q^c[n_{\max}^c] \right\|^2} \qquad \text{// location of combined maximum}$$

$\quad\quad$ **for** c=1 **to** $n_{\text{components}}$ **do**

$\quad\quad\quad I_c = \mathrm{DoFs}_c(n)$ `// get set of component DoFs attached to point`

$\quad\quad\quad \mathbf{P}_c \leftarrow \left[\mathbf{P}_c, \left[\mathbf{e}_{I_c[1]}, \ldots, \mathbf{e}_{I_c[|I_c|]}\right]\right]$

$\quad\quad$ **end**

$\quad\quad \mathcal{N} \leftarrow \mathcal{N} + n$ `// update set of selected points`

$\quad$ **end**

$\quad$ **for** c=1 **to** $n_{\text{components}}$ **do**

$\quad\quad$ **for** j=1 **to** $q_c(i)$ **do**

$$\tilde{\boldsymbol{\phi}}_{Q_c+j}^c = \underset{\mathbf{a}}{\arg\min} \left\| \mathbf{P}_c^\top \left[\boldsymbol{\phi}_1^c, \ldots, \boldsymbol{\phi}_{Q_c}^c\right] \mathbf{a} - \mathbf{P}_c^\top \boldsymbol{\phi}_{Q_c+j}^c \right\|_2^2$$

$$\mathbf{S}_j^c \leftarrow \boldsymbol{\phi}_{Q_c+j}^c - \left[\boldsymbol{\phi}_1^c, \ldots, \boldsymbol{\phi}_{Q_c}^c\right] \tilde{\boldsymbol{\phi}}_{Q_c+j}^c \qquad \text{// vectors processed in next iteration}$$

$\quad\quad$ **end**

$\quad\quad Q_c \leftarrow Q_c + q_c(i)$

$\quad$ **end**

end

Appendix C. Empirical Cubature

To determine the reduced meshes and the weights, the minimisation problems based on (39)

$$(\omega^u, \mathcal{E}_u) = \underset{\substack{w \in \mathbb{R}_+^{m_u} \\ \mathcal{E} \subset \mathcal{T}}}{\arg\min} \sqrt{\sum_{j=1}^{n_P} \sum_{i=1}^{n_u} \left({}^{u}\check{e}_{ij}(w, \mathcal{E})\right)^2} \quad \text{and} \quad (\omega^y, \mathcal{E}_y) = \underset{\substack{w \in \mathbb{R}_+^{m_y} \\ \mathcal{E} \subset \mathcal{T}}}{\arg\min} \sqrt{\sum_{j=1}^{n_\mathbb{B}} \sum_{i=1}^{n_y} \left({}^{y}\check{e}_{ij}(w, \mathcal{E})\right)^2} \tag{A1}$$

have to be solved and read in matrix format as

$$(\omega^u, \mathcal{E}_u) = \underset{\substack{w \in \mathbb{R}_+^{m_u} \\ \mathcal{E} \in \mathcal{T}}}{\arg\min} \left\| \check{J}_\mathcal{E}^u w - \check{b}^u \right\|_2 \quad \text{and} \quad (\omega^y, \mathcal{E}_y) = \underset{\substack{\omega \in \mathbb{R}_+^{m_y} \\ \mathcal{E} \in \mathcal{T}}}{\arg\min} \left\| \check{J}_\mathcal{E}^y w - \check{b}^y \right\|_2, \tag{A2}$$

with $\check{J}^u \in \mathbb{R}^{n_u n_P \times M}, \check{J}^y \in \mathbb{R}^{n_y n_\mathbb{B} \times M}, \check{b}^u \in \mathbb{R}^{n_u n_P}$ and $\check{b}^y \in \mathbb{R}^{n_y n_\mathbb{B}}$,

$$\check{J}^u = \begin{bmatrix} \int\limits_{\Omega_1} \nabla_X \boldsymbol{\varphi}_1^u : \boldsymbol{\phi}_1^P \, dV & \cdots & \int\limits_{\Omega_M} \nabla_X \boldsymbol{\varphi}_1^u : \boldsymbol{\phi}_1^P \, dV \\ \vdots & \ddots & \vdots \\ \int\limits_{\Omega_1} \nabla_X \boldsymbol{\varphi}_{n_u}^u : \boldsymbol{\phi}_1^P \, dV & \cdots & \int\limits_{\Omega_M} \nabla_X \boldsymbol{\varphi}_{n_u}^u : \boldsymbol{\phi}_1^P \, dV \\ \vdots & \ddots & \vdots \\ \int\limits_{\Omega_1} \nabla_X \boldsymbol{\varphi}_{n_u}^u : \boldsymbol{\phi}_{n_P}^P \, dV & \cdots & \int\limits_{\Omega_M} \nabla_X \boldsymbol{\varphi}_{n_u}^u : \boldsymbol{\phi}_{n_P}^P \, dV \end{bmatrix} \qquad \check{b}^u = \sum_{e=1}^{M} \begin{bmatrix} \int\limits_{\Omega_e} \nabla_X \boldsymbol{\varphi}_1^u : \boldsymbol{\phi}_1^P \, dV \\ \vdots \\ \int\limits_{\Omega_e} \nabla_X \boldsymbol{\varphi}_{n_u}^u : \boldsymbol{\phi}_1^P \, dV \\ \vdots \\ \int\limits_{\Omega_e} \nabla_X \boldsymbol{\varphi}_{n_u}^u : \boldsymbol{\phi}_{n_P}^P \, dV \end{bmatrix}$$

$$\check{J}^y = \begin{bmatrix} \int\limits_{\Omega_1} \nabla_X \varphi_1^y : \boldsymbol{\phi}_1^\mathbb{B} \, dV & \cdots & \int\limits_{\Omega_M} \nabla_X \varphi_1^y \cdot \boldsymbol{\phi}_1^\mathbb{B} \, dV \\ \vdots & \ddots & \vdots \\ \int\limits_{\Omega_1} \nabla_X \varphi_{n_y}^y \cdot \boldsymbol{\phi}_1^\mathbb{B} \, dV & \cdots & \int\limits_{\Omega_M} \nabla_X \varphi_{n_y}^y \cdot \boldsymbol{\phi}_1^\mathbb{B} \, dV \\ \vdots & \ddots & \vdots \\ \int\limits_{\Omega_1} \nabla_X \varphi_{n_y}^y \cdot \boldsymbol{\phi}_{n_\mathbb{B}}^\mathbb{B} \, dV & \cdots & \int\limits_{\Omega_M} \nabla_X \varphi_{n_y}^y \cdot \boldsymbol{\phi}_{n_\mathbb{B}}^\mathbb{B} \, dV \end{bmatrix} \qquad \check{b}^y = \sum_{e=1}^{M} \begin{bmatrix} \int\limits_{\Omega_e} \nabla_X \varphi_1^y \cdot \boldsymbol{\phi}_1^\mathbb{B} \, dV \\ \vdots \\ \int\limits_{\Omega_e} \nabla_X \varphi_{n_y}^y \cdot \boldsymbol{\phi}_1^\mathbb{B} \, dV \\ \vdots \\ \int\limits_{\Omega_e} \nabla_X \varphi_{n_y}^y \cdot \boldsymbol{\phi}_{n_\mathbb{B}}^\mathbb{B} \, dV \end{bmatrix}.$$

Note that the problems in (A2) allow for the trivial solutions $\omega^u = \mathbf{0}$ and $\omega^u = \mathbf{0}$. Any POD mode $\hat{\boldsymbol{\phi}}_j^P$ or $\hat{\boldsymbol{\phi}}_j^\mathbb{B}$ is a linear combination of the snapshots $\left\{\hat{\boldsymbol{P}}_1, \ldots, \hat{\boldsymbol{P}}_{n_s}\right\}$ or $\left\{\hat{\mathbb{B}}_1, \ldots, \hat{\mathbb{B}}_{n_s}\right\}$ and, as the snapshots are taken from states of equilibrium, the right-hand sides become $\check{b}^u = \mathbf{0}$ and $\check{b}^y = \mathbf{0}$. Therefore, problems (A2) are regularised by adding the constraints $\sum_{e \in \mathcal{E}_u} \omega_e^u = V$ and $\sum_{e \in \mathcal{E}_y} \omega_e^y = V$. By subtracting the volume averaged row-sums, the regularised minimisation problems (A3) are obtained and approximately solved using Algorithm A3:

$$(\omega^u, \mathcal{E}_u) = \underset{\substack{w \in \mathbb{R}_+^{m_u} \\ \mathcal{E} \in \mathcal{T}}}{\arg\min} \left\| \boldsymbol{J}_{\mathcal{E}}^u w - \boldsymbol{b}^u \right\|_2 \quad \text{and} \quad (\omega^y, \mathcal{E}_y) = \underset{\substack{\omega \in \mathbb{R}_+^{m_y} \\ \mathcal{E} \in \mathcal{T}}}{\arg\min} \left\| \boldsymbol{J}_{\mathcal{E}}^y w - \boldsymbol{b}^y \right\|_2, \tag{A3}$$

with $\boldsymbol{J}^u = \begin{bmatrix} \bar{\boldsymbol{J}}^u \\ \boldsymbol{1}^\top \end{bmatrix} \in \mathbb{R}^{(\mathrm{n}_u \mathrm{n}_P + 1) \times \mathrm{M}}$, $\boldsymbol{J}^y = \begin{bmatrix} \bar{\boldsymbol{J}}^y \\ \boldsymbol{1}^\top \end{bmatrix} \in \mathbb{R}^{(\mathrm{n}_y \mathrm{n}_{\mathbb{B}} + 1) \times \mathrm{M}}$, $\boldsymbol{b}^u = \begin{bmatrix} \boldsymbol{0} \\ V \end{bmatrix} \in \mathbb{R}^{\mathrm{n}_u \mathrm{n}_P + 1}$

and $\boldsymbol{b}^y = \begin{bmatrix} \boldsymbol{0} \\ V \end{bmatrix} \in \mathbb{R}^{\mathrm{n}_y \mathrm{n}_{\mathbb{B}} + 1}$

$$\bar{\boldsymbol{J}}^u = \begin{bmatrix} \int\limits_{\Omega_1} \nabla_X \boldsymbol{\varphi}_1^u : \boldsymbol{\phi}_1^P \,\mathrm{d}V - \frac{1}{V} \int\limits_{\Omega} \nabla_X \boldsymbol{\varphi}_1^u : \boldsymbol{\phi}_1^P \,\mathrm{d}V & \cdots & \int\limits_{\Omega_\mathrm{M}} \nabla_X \boldsymbol{\varphi}_1^u : \boldsymbol{\phi}_1^P \,\mathrm{d}V - \frac{1}{V} \int\limits_{\Omega} \nabla_X \boldsymbol{\varphi}_1^u : \boldsymbol{\phi}_1^P \,\mathrm{d}V \\ \vdots & \ddots & \vdots \\ \int\limits_{\Omega_1} \nabla_X \boldsymbol{\varphi}_{\mathrm{n}_u}^u : \boldsymbol{\phi}_1^P \,\mathrm{d}V - \frac{1}{V} \int\limits_{\Omega} \nabla_X \boldsymbol{\varphi}_{\mathrm{n}_u}^u : \boldsymbol{\phi}_1^P \,\mathrm{d}V & \cdots & \int\limits_{\Omega_\mathrm{M}} \nabla_X \boldsymbol{\varphi}_{\mathrm{n}_u}^u : \boldsymbol{\phi}_1^P \,\mathrm{d}V - \frac{1}{V} \int\limits_{\Omega} \nabla_X \boldsymbol{\varphi}_{\mathrm{n}_u}^u : \boldsymbol{\phi}_1^P \,\mathrm{d}V \\ \vdots & \ddots & \vdots \\ \int\limits_{\Omega_1} \nabla_X \boldsymbol{\varphi}_{\mathrm{n}_u}^u : \boldsymbol{\phi}_{\mathrm{n}_P}^P \,\mathrm{d}V - \frac{1}{V} \int\limits_{\Omega} \nabla_X \boldsymbol{\varphi}_{\mathrm{n}_u}^u : \boldsymbol{\phi}_{\mathrm{n}_P}^P \,\mathrm{d}V & \cdots & \int\limits_{\Omega_\mathrm{M}} \nabla_X \boldsymbol{\varphi}_{\mathrm{n}_u}^u : \boldsymbol{\phi}_{\mathrm{n}_P}^P \,\mathrm{d}V - \frac{1}{V} \int\limits_{\Omega} \nabla_X \boldsymbol{\varphi}_{\mathrm{n}_u}^u : \boldsymbol{\phi}_{\mathrm{n}_P}^P \,\mathrm{d}V \end{bmatrix},$$

$$\bar{\boldsymbol{J}}^y = \begin{bmatrix} \int\limits_{\Omega_1} \nabla_X \varphi_1^y : \boldsymbol{\phi}_1^{\mathbb{B}} \,\mathrm{d}V - \frac{1}{V} \int\limits_{\Omega} \nabla_X \varphi_1^y : \boldsymbol{\phi}_1^{\mathbb{B}} \,\mathrm{d}V & \cdots & \int\limits_{\Omega_\mathrm{M}} \nabla_X \varphi_1^y \cdot \boldsymbol{\phi}_1^{\mathbb{B}} \,\mathrm{d}V - \frac{1}{V} \int\limits_{\Omega} \nabla_X \varphi_1^y : \boldsymbol{\phi}_1^{\mathbb{B}} \,\mathrm{d}V \\ \vdots & \ddots & \vdots \\ \int\limits_{\Omega_1} \nabla_X \varphi_{\mathrm{n}_y}^y \cdot \boldsymbol{\phi}_1^{\mathbb{B}} \,\mathrm{d}V - \frac{1}{V} \int\limits_{\Omega} \nabla_X \varphi_{\mathrm{n}_y}^y : \boldsymbol{\phi}_1^{\mathbb{B}} \,\mathrm{d}V & \cdots & \int\limits_{\Omega_\mathrm{M}} \nabla_X \varphi_{\mathrm{n}_y}^y \cdot \boldsymbol{\phi}_1^{\mathbb{B}} \,\mathrm{d}V - \frac{1}{V} \int\limits_{\Omega} \nabla_X \varphi_{\mathrm{n}_y}^y : \boldsymbol{\phi}_1^{\mathbb{B}} \,\mathrm{d}V \\ \vdots & \ddots & \vdots \\ \int\limits_{\Omega_1} \nabla_X \varphi_{\mathrm{n}_y}^y \cdot \boldsymbol{\phi}_{\mathrm{n}_{\mathbb{B}}}^{\mathbb{B}} \,\mathrm{d}V - \frac{1}{V} \int\limits_{\Omega} \nabla_X \varphi_{\mathrm{n}_y}^y : \boldsymbol{\phi}_{\mathrm{n}_{\mathbb{B}}}^{\mathbb{B}} \,\mathrm{d}V & \cdots & \int\limits_{\Omega_\mathrm{M}} \nabla_X \varphi_{\mathrm{n}_y}^y \cdot \boldsymbol{\phi}_{\mathrm{n}_{\mathbb{B}}}^{\mathbb{B}} \,\mathrm{d}V - \frac{1}{V} \int\limits_{\Omega} \nabla_X \varphi_{\mathrm{n}_y}^y : \boldsymbol{\phi}_{\mathrm{n}_{\mathbb{B}}}^{\mathbb{B}} \,\mathrm{d}V \end{bmatrix}.$$

Algorithm A3: Greedy Mesh Sampling

Input: $\boldsymbol{J}$, tol, m

Output: ω, E

initialisation: $E \leftarrow \emptyset$, set of candidates $C \leftarrow \{1, \ldots, \mathrm{M}\}$, $\boldsymbol{r} \leftarrow \boldsymbol{b}$

do

- $e = \underset{\tilde{e} \in C}{\arg\max} \left\langle \boldsymbol{J}[\tilde{e}] / \|\boldsymbol{J}[\tilde{e}]\|_2, \boldsymbol{r} / \|\boldsymbol{r}\|_2 \right\rangle$ `// determine new element`
- $E \leftarrow E \cup e$, $C \leftarrow C \setminus e$
- build $\boldsymbol{J}_E$ from columns of $\boldsymbol{J}$ based on E
- $\omega = \underset{w \in \mathbb{R}^{|E|}}{\arg\min} \|\boldsymbol{J}_E w - \boldsymbol{b}\|_2^2$ `// solve least-squares`
- **if** $\omega[i] < 0$ *for* $i = 1, \ldots, |E|$ **then**
  - $\omega = \underset{w \in \mathbb{R}_+^{|E|}}{\arg\min} \|\boldsymbol{J}_E w - \boldsymbol{b}\|_2^2$ `// solve non-negative least-squares`
  - $E \leftarrow E \setminus E_0$ with $E_0 = \{e \in E : \omega[e] = 0\}$
  - $C \leftarrow C \cup E_0$
  - $\omega \leftarrow \omega(E)$
- **end**
- $\boldsymbol{r} \leftarrow \boldsymbol{b} - \boldsymbol{J}_E \omega$ `// update residual`

while $\dfrac{\|\boldsymbol{r}\|}{\|\boldsymbol{b}\|} > tol \wedge |E| < m$

Appendix D. Reduced Integration Domain

Algorithm A4 constructs the reduced integration domain for a basis $\{\boldsymbol{\phi}_i\}_{i=1}^r$. The application of the DEIM Algorithm A1 helps to identify areas of interest, e.g., where the basis exhibits significant gradients. At first, all elements containing DEIM indices form the reduced integration domain. Hereafter, the elements in the surrounding layers can be included for accuracy reasons. It is noteworthy that Algorithm A2 or alternatives can be used to determine the initial reduced domain.

Algorithm A4: Determining Reduced Integration Domain

Input: POD basis $\{\boldsymbol{\phi}_i\}_{i=1}^r \subset \mathbb{R}^N$, number of neighbouring element layers l
Output: $\mathbf{P}_{\text{RID}}$, Ω_{RID}

get DEIM indices $\boldsymbol{\rho} \in \mathbb{N}^r$ by applying Algorithm A1 to $\{\boldsymbol{\phi}_i\}_{i=1}^r$
$\Omega_{\text{RID}} \leftarrow$ ContainingElements $(\boldsymbol{\rho})$ `// collect elements containing the DEIM indices`
for *i=1* **to** *l* **do**
  $\Omega_{\text{RID}} \leftarrow \Omega_{\text{RID}} \cup$ NeighbouringElements (Ω_{RID}) `// add neighbouring elements`
end
setup $\mathbf{P}_{\text{RID}}$ based on interior DoFs (primary) in Ω_{RID}

References

1. Feyel, F.; Chaboche, J.L. FE2 multiscale approach for modelling the elastoviscoplastic behaviour of long fibre SiC/Ti composite materials. *Comput. Methods Appl. Mech. Eng.* **2000**, *183*, 309–330. [CrossRef]
2. Saeb, S.; Steinmann, P.; Javili, A. Aspects of computational homogenization at finite deformations: A unifying review from Reuss' to Voigt's bound. *Appl. Mech. Rev.* **2016**, *68*, 050801. [CrossRef]
3. Bartuschat, D.; Gmeiner, B.; Thoennes, D.; Kohl, N.; Rüde, U.; Drzisga, D.; Huber, M.; John, L.; Waluga, C.; Wohlmuth, B.I.; et al. A Finite Element Multigrid Framework for Extreme-Scale Earth Mantle Convection Simulations. In Proceedings of the SIAM Conference on Parallel Processing for Scientific Computing (SIAM PP 18), Tokyo, Japan, 7–10 March 2018.
4. Michel, J.C.; Suquet, P. A model-reduction approach in micromechanics of materials preserving the variational structure of constitutive relations. *J. Mech. Phys. Solids* **2016**, *90*, 254–285. [CrossRef]
5. Fritzen, F.; Leuschner, M. Reduced basis hybrid computational homogenization based on a mixed incremental formulation. *Comput. Methods Appl. Mech. Eng.* **2013**, *260*, 143–154. [CrossRef]
6. Chinesta, F.; Ammar, A.; Leygue, A.; Keunings, R. An overview of the proper generalized decomposition with applications in computational rheology. *J. Non-Newton. Fluid Mech.* **2011**, *166*, 578–592. [CrossRef]
7. Cremonesi, M.; Néron, D.; Guidault, P.A.; Ladevèze, P. A PGD-based homogenization technique for the resolution of nonlinear multiscale problems. *Comput. Methods Appl. Mech. Eng.* **2013**, *267*, 275–292. [CrossRef]
8. Sirovich, L. Turbulence and the Dynamics of Coherent Structures Part I: Coherent Structures. *Q. Appl. Math.* **1987**, *45*, 561–571. [CrossRef]
9. Himpe, C.; Leibner, T.; Rave, S. Hierarchical Approximate Proper Orthogonal Decomposition. *SIAM J. Sci. Comput.* **2018**, *40*, A3267–A3292. [CrossRef]
10. Paul-Dubois-Taine, A.; Amsallem, D. An adaptive and efficient greedy procedure for the optimal training of parametric reduced-order models. *Int. J. Numer. Methods Eng.* **2014**, *102*, 1262–1292. [CrossRef]
11. Brands, B.; Mergheim, J.; Steinmann, P. Reduced-order modelling for linear heat conduction with parametrised moving heat sources. *GAMM-Mitteilungen* **2016**, *39*, 170–188. [CrossRef]
12. Hesthaven, J.S.; Rozza, G.; Stamm, B. *Certified Reduced Basis Methods for Parametrized Partial Differential Equations*; Springer: Cham, Switzerland, 2017.
13. Quarteroni, A.; Manzoni, A.; Negri, F. *Reduced Basis Methods for Partial Differential Equations: An Introduction*; Springer: Cham, Switzerland, 2015; Volume 92.

14. Rozza, G.; Huynh, D.B.P.; Patera, A.T. Reduced basis approximation and a posteriori error estimation for affinely parametrized elliptic coercive partial differential equations. *Arch. Comput. Methods Eng.* **2008**, *15*, 229–275. [CrossRef]
15. Grepl, M.A.; Patera, A.T. A posteriori error bounds for reduced-basis approximations of parametrized parabolic partial differential equations. *ESAIM Math. Model. Numer. Anal.* **2005**, *39*, 157–181. [CrossRef]
16. Yvonnet, J.; He, Q.C. The reduced model multiscale method (R3M) for the nonlinear homogenization of hyperelastic media at finite strains. *J. Comput. Phys.* **2007**, *223*, 341–368. [CrossRef]
17. Radermacher, A.; Bednarcyk, B.A.; Stier, B.; Simon, J.; Zhou, L.; Reese, S. Displacement-based multiscale modeling of fiber-reinforced composites by means of proper orthogonal decomposition. *Adv. Model. Simul. Eng. Sci.* **2016**, *3*, 29. [CrossRef]
18. Grepl, M.A.; Maday, Y.; Nguyen, N.C.; Patera, A.T. Efficient reduced-basis treatment of nonaffine and nonlinear partial differential equations. *ESAIM M2AN* **2007**, *41*, 575–605. [CrossRef]
19. Chaturantabut, S.; Sorensen, D.C. Nonlinear Model Reduction via Discrete Empirical Interpolation. *SIAM J. Sci. Comput.* **2010**, *32*, 2737–2764. [CrossRef]
20. Everson, R.; Sirovich, L. Karhunen-Loeve procedure for gappy data. *J. Opt. Soc. Am. A* **1995**, *12*, 1657–1664. [CrossRef]
21. Carlberg, K.; Farhat, C.; Cortial, J.; Amsallem, D. The GNAT method for nonlinear model reduction: Effective implementation and application to computational fluid dynamics and turbulent flows. *J. Comput. Phys.* **2013**, *242*, 623–647. [CrossRef]
22. Goury, O.; Amsallem, D.; Bordas, S.P.A.; Liu, W.K.; Kerfriden, P. Automatised selection of load paths to construct reduced-order models in computational damage micromechanics: From dissipation-driven random selection to Bayesian optimization. *Comput. Mech.* **2016**, *58*, 213–234. [CrossRef]
23. Ghavamian, F.; Tiso, P.; Simone, A. POD–DEIM model order reduction for strain-softening viscoplasticity. *Comput. Methods Appl. Mech. Eng.* **2017**, *317*, 458–479. [CrossRef]
24. Tiso, P.; Rixen, D.J. Discrete Empirical Interpolation Method for Finite Element Structural Dynamics. In *Topics in Nonlinear Dynamics, Volume 1: Proceedings of the 31st IMAC, A Conference on Structural Dynamics, Garden Grove, CA, USA, 11–14 February 2013*; Kerschen, G., Adams, D., Carrella, A., Eds.; Springer: New York, NY, USA, 2013; pp. 203–212.
25. Hernández, J.; Oliver, J.; Huespe, A.; Caicedo, M.; Cante, J. High-performance model reduction techniques in computational multiscale homogenization. *Comput. Methods Appl. Mech. Eng.* **2014**, *276*, 149–189. [CrossRef]
26. Soldner, D.; Brands, B.; Zabihyan, R.; Steinmann, P.; Mergheim, J. A numerical study of different projection-based model reduction techniques applied to computational homogenisation. *Comput. Mech.* **2017**, *60*, 613–625. [CrossRef]
27. An, S.S.; Kim, T.; James, D.L. Optimizing Cubature for Efficient Integration of Subspace Deformations. *ACM Trans. Graph.* **2008**, *27*, 165. [CrossRef] [PubMed]
28. Farhat, C.; Avery, P.; Chapman, T.; Cortial, J. Dimensional reduction of nonlinear finite element dynamic models with finite rotations and energy-based mesh sampling and weighting for computational efficiency. *Int. J. Numer. Methods Eng.* **2014**, *98*, 625–662. [CrossRef]
29. Hernández, J.; Caicedo, M.; Ferrer, A. Dimensional hyper-reduction of nonlinear finite element models via empirical cubature. *Comput. Methods Appl. Mech. Eng.* **2017**, *313*, 687–722. [CrossRef]
30. van Tuijl, R.A.; Remmers, J.J.C.; Geers, M.G.D. Integration efficiency for model reduction in micro-mechanical analyses. *Comput. Mech.* **2017**, *62*, 151–169. [CrossRef]
31. Ryckelynck, D. A priori hyperreduction method: An adaptive approach. *J. Comput. Phys.* **2005**, *202*, 346–366. [CrossRef]
32. Ryckelynck, D. Hyper-reduction of mechanical models involving internal variables. *Int. J. Numer. Methods Eng.* **2009**, *77*, 75–89. [CrossRef]
33. Ryckelynck, D.; Benziane, D.M. Multi-level A Priori Hyper-Reduction of mechanical models involving internal variables. *Comput. Methods Appl. Mech. Eng.* **2010**, *199*, 1134–1142. [CrossRef]
34. Ryckelynck, D.; Lampoh, K.; Quilicy, S. Hyper-reduced predictions for lifetime assessment of elasto-plastic structures. *Meccanica* **2016**, *51*, 309–317. [CrossRef]
35. Fritzen, F.; Haasdonk, B.; Ryckelynck, D.; Schöps, S. An Algorithmic Comparison of the Hyper-Reduction and the Discrete Empirical Interpolation Method for a Nonlinear Thermal Problem. *Math. Comput. Appl.* **2018**, *23*, 8. [CrossRef]

36. Astrid, P.; Weiland, S.; Willcox, K.; Backx, T. Missing point estimation in models described by proper orthogonal decomposition. *IEEE Trans. Autom. Control* **2008**, *53*, 2237–2251. [CrossRef]
37. Dimitriu, G.; Ştefănescu, R.; Navon, I.M. Comparative numerical analysis using reduced-order modeling strategies for nonlinear large-scale systems. *J. Comput. Appl. Math.* **2017**, *310*, 32–43. [CrossRef]
38. Walter, B.L.; Pelteret, J.P.; Kaschta, J.; Schubert, D.W.; Steinmann, P. Preparation of magnetorheological elastomers and their slip-free characterization by means of parallel-plate rotational rheometry. *Smart Mater. Struct.* **2017**, *26*, 085004. [CrossRef]
39. Chatzigeorgiou, G.; Javili, A.; Steinmann, P. Unified magnetomechanical homogenization framework with application to magnetorheological elastomers. *Math. Mech. Solids* **2014**, *19*, 193–211. [CrossRef]
40. Zabihyan, R.; Mergheim, J.; Javili, A.; Steinmann, P. Aspects of computational homogenization in magneto-mechanics: Boundary conditions, RVE size and microstructure composition. *Int. J. Solids Struct.* **2018**, *130*, 105–121. [CrossRef]
41. Javili, A.; Chatzigeorgiou, G.; Steinmann, P. Computational homogenization in magneto-mechanics. *Int. J. Solids Struct.* **2013**, *50*, 4197–4216. [CrossRef]
42. Holmes, P.; Lumley, J.L.; Berkooz, G.; Rowley, C.W. *Turbulence, Coherent Structures, Dynamical Systems and Symmetry*; Cambridge University Press: Cambridge, UK, 2012.
43. Volkwein, S. *Proper Orthogonal Decomposition: Theory and Reduced-Order Modelling*; Lecture Notes; University of Konstanz: Konstanz, Germany, 2013; Volume 4.
44. Chaturantabut, S.; Sorensen, D. A State Space Error Estimate for POD-DEIM Nonlinear Model Reduction. *SIAM J. Numer. Anal.* **2012**, *50*, 46–63. [CrossRef]
45. Wirtz, D.; Sorensen, D.; Haasdonk, B. A Posteriori Error Estimation for DEIM Reduced Nonlinear Dynamical Systems. *SIAM J. Sci. Comput.* **2014**, *36*, A311–A338. [CrossRef]
46. Chapman, T.; Avery, P.; Collins, P.; Farhat, C. Accelerated mesh sampling for the hyper reduction of nonlinear computational models. *Int. J. Numer. Methods Eng.* **2017**, *109*, 1623–1654. [CrossRef]
47. Alzetta, G.; Arndt, D.; Bangerth, W.; Boddu, V.; Brands, B.; Davydov, D.; Gassmöller, R.; Heister, T.; Heltai, L.; Kormann, K.; et al. The deal.II Library, Version 9.0. *J. Numer. Math.* **2018**, *26*, 173–183. [CrossRef]
48. Stoyanov, M. Adaptive Sparse Grid Construction in a Context of Local Anisotropy and Multiple Hierarchical Parents. In *Sparse Grids and Applications—Miami 2016*; Garcke, J., Pflüger, D., Webster, C.G., Zhang, G., Eds.; Springer International Publishing: Cham, Switzerland, 2018; pp. 175–199.
49. Garcke, J.; Griebel, M. *Sparse Grids and Applications*, 1st ed.; Lecture Notes in Computational Science and Engineering; Springer Science & Business Media: Cham, Switzerland, 2012; Volume 88.
50. Miehe, C. Numerical computation of algorithmic (consistent) tangent moduli in large-strain computational inelasticity. *Comput. Methods Appl. Mech. Eng.* **1996**, *134*, 223–240. [CrossRef]
51. Pelteret, J.P.; Steinmann, P. *Magneto-Active Polymers: Fabrication, Characterisation, Modelling and Simulation at the Micro- and Macro-Scale*; In Preparation; de Gruyter Mouton: Berlin, Germany, 2019.

Mathematical and Computational Applications

Article

Time Stable Reduced Order Modeling by an Enhanced Reduced Order Basis of the Turbulent and Incompressible 3D Navier–Stokes Equations

Nissrine Akkari [1,*], Fabien Casenave [1] and Vincent Moureau [2]

1 Safran Tech, Modelling and Simulation, Rue des Jeunes Bois, Châteaufort, 78114 Magny-Les-Hameaux, France; fabien.casenave@safrangroup.com

2 CORIA, CNRS UMR 6614, Université et INSA de Rouen, Campus Universitaire du Madrillet, Saint Etienne du Rouvray, 76800 Rouen, France; vincent.moureau@coria.fr

* Correspondence: nissrine.akkari@safrangroup.com; Tel.: +33-1-61-31-83-80

Received: 6 March 2019; Accepted: 23 April 2019; Published: 24 April 2019

Abstract: In the following paper, we consider the problem of constructing a time stable reduced order model of the 3D turbulent and incompressible Navier–Stokes equations. The lack of stability associated with the order reduction methods of the Navier–Stokes equations is a well-known problem and, in general, it is very difficult to account for different scales of a turbulent flow in the same reduced space. To remedy this problem, we propose a new stabilization technique based on an a priori enrichment of the classical proper orthogonal decomposition (POD) modes with dissipative modes associated with the gradient of the velocity fields. The main idea is to be able to do an a priori analysis of different modes in order to arrange a POD basis in a different way, which is defined by the enforcement of the energetic dissipative modes within the first orders of the reduced order basis. This enables us to model the production and the dissipation of the turbulent kinetic energy (TKE) in a separate fashion within the high ranked new velocity modes, hence to ensure good stability of the reduced order model. We show the importance of this a priori enrichment of the reduced basis, on a typical aeronautical injector with Reynolds number of 45,000. We demonstrate the capacity of this order reduction technique to recover large scale features for very long integration times (25 ms in our case). Moreover, the reduced order modeling (ROM) exhibits periodic fluctuations with a period of 2.2 ms corresponding to the time scale of the precessing vortex core (PVC) associated with this test case. We will end this paper by giving some prospects on the use of this stable reduced model in order to perform time extrapolation, that could be a strategy to study the limit cycle of the PVC.

Keywords: reduced order modeling (ROM); proper orthogonal decomposition (POD); enhanced POD; a priori enrichment; modal analysis; stabilization; dynamic extrapolation

1. Introduction

Reduced order modeling (ROM) of the complete Navier–Stokes equations by the projection of these equations upon a reduced order space, that is generated by a minimal number of spatial modes, is still an attractive research area especially when we consider turbulent flows such as the ones encountered in aeronautical engines and that feature a large range of scales. The most important issue to be addressed when performing an order reduction of the turbulent and unsteady Navier–Stokes equations is the construction of a minimal reduced order space that could cover properly the large range of scales of a turbulent fluid flow. If we find a minimal subspace that could verify these properties, then we can do further studies concerning the efficient adaptivity of the associated reduced order equations in terms of design and optimization of the components of an aeronautical engine, such as the combustion chamber and the fuel injection system.

The large vortices carry the major amount of the turbulent kinetic energy (TKE), while small scales are responsible for the dissipation of TKE. The real time prediction of this physics, under strong unsteadiness and variable constraints, is a big challenge for the industrial design in aeronautical engineering. Reduced order modeling by proper orthogonal decomposition (POD) is a very good candidate for solving such problems. It enables approximation of the high-fidelity (HF) partial differential equations (PDE)s in a subspace of small dimension, which reproduces accurately the energy of the coherent structures of a fluid flow. Nevertheless, the ROMs by POD for the turbulent and incompressible Navier–Stokes equations suffer a time instability due to the misrepresentation of the energy of the small vortices of these convection dominated equations, by the coherent energetic POD modes. Many authors propose techniques to remedy time instability within ROMs by POD of the Navier–Stokes equations. By time stable reduced order model, we mean the capacity of the reduced equations to verify the energy balance and the mass conservation properties of the complete Navier–Stokes equations.

Improvement of the Galerkin reduced order modeling using mathematical approaches such as the stabilization based on the role of the neglected POD modes to enhance the dissipation of the TKE [1], or reduced order models based on deconvolution methods for large-eddy simulation (LES) have been proposed in the literature: Rowley et al. [2] proposed a Galerkin ROM-POD for the compressible Navier–Stokes equations, constructed by projection via an energy-based inner product. Codina et al. [3] proposed the enrichment of the ROM by increasing the dimension of the projection POD subspace, and computing the new temporal weights amplitudes by a least square minimization with respect to the initial temporal snapshots. The later sub-scale approach was applied for an incompressible and turbulent flow around a cylinder of which the Reynolds number varies from 32,000 to 74,000. Balajewicz et al. [4] proposed an enhanced Galerkin approximation by POD of the compressible Navier–Stokes equations based on an a priori implementation of a traditional eddy-viscosity based closure model in order to modify the overall eigenvalue distribution of the dissipative linear operator within the Galerkin reduced order modeling. This technique was applied for a 2D laminar airfoil at Reynolds 500. Xie et al. [5] proposed a deconvolution method for LES-based reduced order models in order to model the subfilter stress-scale tensor within the reduced order modeling. This approach was performed for the 1D Burgers equation and a 3D flow past a cylinder at Reynolds 1000. In [6], the authors propose to study theoretically and numerically the influence of different types of finite element on the ROM mass conservation. They tested the Taylor–Hood (TH) element and the Scott–Vogelius (SV) element. They showed that the SV-ROM yields to more accurate results when applied to a 2D flow past a circular cylinder at a Reynolds number $Re = 100$, especially for coarser meshes and longer time intervals. We can find in literature also stabilization of reduced order models based on operator splitting, specifically the streamline-upwind Petrov–Galerkin (SUPG) stabilization method [7].

In the domain of model order reduction for finite volume numerics for computational fluid dynamics, we cite the work of Carlberg et al. [8], where it has been proposed a method for model reduction of finite-volume models that guarantees the resulting reduced-order model to be conservative. The proposed reduced order model is associated with optimization problems that explicitly enforce conservation over subdomains. In [9], the authors proposed a POD-Galerkin reduced order methods for CFD using finite volume discretization. This was performed as a consequence of the projection of the Navier–Stokes equations onto different reduced basis space for velocity and pressure, respectively. Stabile et al. [10], developed finite volume POD-Galerkin stabilized reduced order methods for the incompressible Navier–Stokes equations. These methods are based on the pressure Poisson equation and supremizer enrichment which ensures that a reduced version of the inf-sup condition is satisfied.

Among the work concerning the stabilisation of the POD-based ROM for turbulent fluids dynamics, Amsallem et al. [11] have proposed to perform the POD-based model reduction using the descriptor form of the discretized Euler or Navier–Stokes equations, i.e., the natural variables of these equations, whereas many computations are performed in CFD codes using the non-descriptor form of these equations. In [12], the authors proposed to use the stability of the reduced-order Galerkin models in incompressible flows in order to study the limit cycle of the hydrodynamic vortex acting on a circular cylinder. Amsallem et al. [13] have proposed a stabilization of the projection-based linear reduced order models because of a convex optimization problem that operates directly on available reduced order operators. This method was tested for computational fluid dynamics-based model of a linearized unsteady supersonic flow, the reduction of a computational structural dynamic system, and the stabilization of the reduction of a coupled computational fluid dynamics–computational structural dynamics model of a linearized aeroelastic system in the transonic flow regime.

Besides these reduced order techniques which are intrusive for the computational fluid dynamics physics, there are some new non-intrusive reduced order ones, as they rely only on the available snapshots data, without taking into account all the equations of physics as constraints, but rather some physical properties as the turbulent kinetic energy conservation by learning the orthogonal projection coefficients of the solutions over a POD basis using a metamodel or a neural network, see the work of Wang et al. [14]. These non-intrusive techniques would take into account also the parameters calibration of the closure terms and the turbulent sub-grid scale modeling in the large eddy simulation models or the reynolds average Navier–Stokes equations ones, see the work of Lapeyre et al. [15]. There are also some research directions towards combining the machine learning non-intrusive reduced order techniques with the physics based reduced order ones in order to improve the quality of these latters, see the work of Xie et al. [16].

In this paper, we are interested in the stabilization of POD-Galerkin reduced order modeling for the turbulent and incompressible Navier–Stokes equations and we propose a new stabilization and purely physical approach for the POD-Galerkin approximation of the turbulent and incompressible Navier–Stokes equations. More precisely, a simplified POD-Galerkin projection of the complete Navier–Stokes equations is performed within an extended and minimal reduced order subspace, which reproduces accurately all the scales contained in the different terms of the Navier–Stokes equations, in order to recover a proper evolution of the fluctuating TKE. The proposed approach is based on the POD representation of the velocity gradient. A solution for the issue due to the combination of POD velocity and POD gradient modes is proposed. Moreover, we point out that the proposed stabilization is based on the a priori analysis of the different velocity modes and gradient velocity modes, in order to enforce the energetic dissipative modes within the first vectors of the new reduced order basis. This a priori enrichment by scale seperation is a key point to our proposed approach, and this will lead to the desired time stability of the reduced order model. We also point out that our proposed strategy is different from the ones that propose a scalar product change while defining the correlations matrix of the singular value decomposition (SVD) step of the POD method, in order to take into account the gradient scales within the scalar product computations, typically as in [17].

The manuscript is organized as follows: the proposed theoretical framework for the construction of the enhanced reduced order basis is detailed in Section 2. In this section, we motivate the use of an a priori enhancement of the classical POD basis by POD modes associated with the gradient velocity. All the numerical results are detailed in Section 3 for a benchmark problem of a typical aeronautical injection system at $Re = 45{,}000$ and 14 millions mesh elements. The flow solver of the High Fidelity Navier–Stokes equations is first exposed. Reduced order modeling via the enriched POD is presented and analyzed. In Section 4, we show the dynamic temporal coefficients obtained by running the enhanced fluid dynamics reduced order model for very long time integrations, even longer than the one of the high-fidelity solutions that generated the enhanced reduced order basis. In Section 5, we give some conclusions and prospects to this work. More precisely, we introduce our future work concerning the use of this stabilization technique for the extrapolation of the reduced order model in time so that, we can study efficiently the limit cycle of the PVC in an aeronautical injector [18]. It is well known that the Q-criterion of the smallest vortices is larger than the one of the large coherent structures, then the PVC is masked by small scales surrounding turbulence in the LES simulation. The enhanced ROM enables us to filter the PVC throughout the reduced order simulation even for large values of Q-criterion and for very long integration times.

2. Theoretical Framework

2.1. POD-Galerkin Reduced Order Modeling Applied to the Unsteady and Incompressible Navier–Stokes Equations

We denote by $X = [L^2(\Omega)]^3$ the functional Hilbert space of the squared integrable functions over a bounded $3D-$open set Ω. The corresponding inner product is the kinetic energy-based one associated with the X-functional norm. They will be denoted respectively by $(.,.)$ and $\|.\|$. Consider $v(t) \in X$ the velocity field of an unsteady incompressible flow. Denote $\bar{v}(t)$ the filtered field obtained by a given LES model. A reduced order POD subspace is obtained by the snapshots method [19]. More precisely, if we discretize the time interval to M points, then the snapshots set is given as follows: $\mathcal{S} = \{\bar{v}(t_i) \quad i = 1, ..., M\}$. The associated POD eigenmodes Φ_n, $n = 1, ..., M$ are solutions of the following eigenvalues problems given the temporal correlations matrix:

$$C_{ij} = (\bar{v}(t_i), \bar{v}(t_j)), \tag{1}$$

of size $M \times M$. We denote by $A_n = (A_{i,n})_{1 \le i \le M}$ for $n = 1, ..., M$, a set of orthonormal eigenvectors of the matrix C. Then, the POD-eigenmodes associated with $\bar{v}$, are given by:

$$\Phi_n(x) = \frac{1}{\sqrt{\lambda_n M}} \sum_{i=1}^{M} A_{i,n} \bar{v}(t_i, x), \quad \forall x \in \Omega \ \forall n = 1, ..., M, \tag{2}$$

where $(\lambda_n)_{n=1,...,M}$ is the decreasing sequence of the positive eigenvalues of the correlations matrix C.

To achieve the POD reduced order modeling of the filtered incompressible Navier–Stokes equations, the approximated velocity field is expressed in the reduced order POD subspace:

$$\bar{v}(t, x) = \sum_{n=1}^{N} a_n(t) \Phi_n(x), \quad \forall x \in \Omega, \tag{3}$$

where $N << M$ denotes the number of retained high energetic POD modes, and $a_1(t), a_2(t), ..., a_N(t)$ are the temporal weights which are solutions of the following coupled dynamical system:

$$\begin{cases} \dfrac{da_n}{dt} + (div(\bar{v}(t) \otimes \bar{v}(t)), \Phi_n) = \nu \, (\Delta \bar{v}(t), \Phi_n) - \dfrac{1}{\rho} (\nabla p(t), \Phi_n) \\ (q, div(\bar{v}(t)))_{H^0} = 0 \ \forall q \in H^0 \\ a_n(0) = (v_0, \Phi_n) \end{cases}, \tag{4}$$

where div denotes the divergence operator, $p(t)$ is the pressure field, ρ the density, ν denotes the kinematic viscosity, v_0 is the initial condition of the velocity field and H^0 is the subspace of the divergence free X-functions.

We point out the fact that the equations upon which we perform the POD-Galerkin projection are the continuous high-fidelity incompressible Navier–Stokes equations without any turbulence model taken into account. So, our reduced order modeling formulation is the one associated with the continuous Navier–Stokes equations. However, it is clear that the POD computation is associated with High-Fidelity snapshots $\bar{v}(t)$ which are usually obtained from LES of the Navier–Stokes equations.

In general, the first POD mode $\Phi_1(x)$ which describes the mean topology of the fluid flow is not kept and a ROM of the fluid dynamics equations represents only the fluctuating part. However in our case the first POD mode $\Phi_1(x)$ is kept within the ROM. This could be very valuable when we are interested in using the reduced order modeling in order to predict the flow with respect to parametric variations, or even for new geometries [20]. This enables the ROM to consider naturally the influence of the velocity fluctuations on the velocity mean.

We point out the following two remarks concerning our formulation of the reduced order modeling:

Remark 1. *The POD modes contain only the energetic scales of the flow. The dissipative scales at the Taylor macro-scale are not present in the basis.*

Remark 2. *The flow rate in the flow domain is not guaranteed except if penalization is added in the pressure term to take into account the pressure difference between inlet and outlet.*

We propose to tackle these remarks on account of a physical stabilization by satisfying the kinetic energy budget.

2.2. Physical Stabilization by Satisfying the Kinetic Energy Budget

2.2.1. Enrichment of the POD-Galerkin ROM with the Flow Rate Driving Forces

If we integrate by part the pressure term in the reduced order model (4), then we get the following equality:

$$(\nabla p(t), \Phi_n) = -(p(t), div(\Phi_n)) + \int_{\delta\Omega} p(t)(\Phi_n(x), \vec{n}(x))_{\mathbb{R}^3} d\delta\Omega, \tag{5}$$

where $\vec{n}$ is the normal vector to the domain boundaries $\delta\Omega$.

Using the fact that the incompressibility constraint is also verified by the velocity POD modes, the pressure term could be written:

$$(\nabla p(t), \Phi_n) = \int_{\delta\Omega} p(t)(\Phi_n(x), \vec{n}(x))_{\mathbb{R}^3} d\delta\Omega. \tag{6}$$

We propose to model the pressure difference between the inlet Γ_{in} and the outlet, because of a penalization for the flow rate. This could be written mathematically as follows:

$$\sum_{m=1}^{N} a_m(t) \int_{\Gamma_{in}} (\Phi_n(x), \Phi_m(x))_{\mathbb{R}^3} \, d\Gamma_{in} = \int_{\Gamma_{in}} (\Phi_n(x), v_{\Gamma_{in}}(x))_{\mathbb{R}^3} \, d\Gamma_{in}, \forall n = 1, ..., N, \tag{7}$$

where $v_{\Gamma_{in}}$ is the inlet boundary condition of the corresponding High-Fidelity Navier–Stokes equations.

The reduced order model satisfying the flow rate production forces is now written as follows: $\forall\, n = 1, ..., N$,

$$\left\{ \begin{array}{l} \dfrac{da_n}{dt} + (div(\tilde{v}(t) \otimes \tilde{v}(t)), \Phi_n) = \nu\,(\Delta\tilde{v}(t), \Phi_n) \\ +\tau \left[\sum\limits_{m=1}^{N} a_m(t) \displaystyle\int_{\Gamma_{in}} (\Phi_n(x), \Phi_m(x))_{\mathbb{R}^3} \, d\Gamma_{in} - \int_{\Gamma_{in}} (\Phi_n(x), v_{\Gamma_{in}}(x))_{\mathbb{R}^3} \, d\Gamma_{in} \right], \\ a_n(0) = (\tilde{v}(0), \Phi_n) \end{array} \right. \tag{8}$$

where τ is the penalization coefficient and which has been chosen equal to 10,000 in the online resolution of the reduced order modeling (8).

The flow rate penalization will enforce the following equality:

$$a_1(t)\Phi_1(x) = V(x), \forall x \in \Omega \text{ and } \forall t,$$

where V is a steady velocity field, that should represents the mean motion. Denote by $v^{reduced}(t) = \sum_{n=2}^{N} a_n(t)\Phi_n$ the fluctuating reduced order velocity, then the evolution of the turbulent kinetic energy within the ROM (8) is given by:

$$\frac{1}{2}\frac{d}{dt}\sum_{n=2}^{N} a_n^2 = \nu\,(\Delta\tilde{v}(t)),\tilde{v}(t)) \tag{9}$$

$$-\left(div(v^{reduced}(t)\otimes V),v^{reduced}(t)\right)-\left(div(V\otimes V),v^{reduced}(t)\right) \tag{10}$$

$$-\left(div(V\otimes v^{reduced}(t)),V\right)-\left(div(v^{reduced}(t)\otimes v^{reduced}(t)),V\right) \tag{11}$$

$$= \nu\,(\Delta\tilde{v}(t)),\tilde{v}(t)) \tag{12}$$

$$-2\left(div(v^{reduced}(t)\otimes V),v^{reduced}(t)\right)-2\left(div(V\otimes V),v^{reduced}(t)\right) \tag{13}$$

The terms (13) in the assessment of the kinetic energy represents the production rate of the kinetic energy.

2.2.2. Enrichment of the POD-Galerkin ROM with the Most Dissipative Scales Based on the Velocity Gradient

To recover the dissipation rate of the fluctuating TKE in (8), we propose the following numerical algorithm, based on the enrichment of velocity-based POD modes by gradient velocity-based POD modes following a new a priori approach.

The proposed enrichment algorithm is the following:

- Compute the POD velocity modes $\Phi_n = \frac{1}{\sqrt{\lambda_n M}}\sum_{i=1}^{M} A_{i,n}\tilde{v}(t_i)$, $n = 1,...,M$ and truncate at $N << M$ these POD modes. We note that N is intentionally chosen to be less than the needed number of the POD modes to represent all the features of the coherent energetic scales of the kinetic energy.
- Compute the fluctuating POD gradient modes $\Psi_n = \frac{1}{\sqrt{\beta_n M}}\sum_{i=1}^{M} B_{i,n}\nabla\tilde{v}(t_i)$, $n = 1,...,M$ and truncate at $N' << M$. Where $B_n = (B_{i,n})_{1\leq i\leq M}$ for $n = 1,...,M$, a set of orthonormal eigenvectors of the temporal correlations matrix on the fluctuating velocity gradient: $\left(\nabla\tilde{v}(t_i) - W, \nabla\tilde{v}(t_j) - W\right)_{[L^2(\Omega)]^9}$, $i,j = 1,...,M$ (W the mean velocity gradient being removed from these correlations), and $(\beta_n)_{n=1,...,M}$ is the sequence of the eigenvalues of this latter matrix.
- Compute the following velocity basis functions: $\Phi_n^E = \frac{1}{\sqrt{\beta_n M}}\sum_{i=1}^{M} B_{i,n}\tilde{v}(t_i)$, $n = 1,...,N'$.
- Perform the Gram–Schmidt orthonormalization process for the enriched set $\{\Phi_1,...,\Phi_N,\Phi_1^E,...,\Phi_{N'}^E\}$ with respect to the energy-based inner product $(.,.)$. This step is the key of the enforcement of dissipative energy modes with high singular values $(\beta_n)_{n=1,...,N'}$ in early ranks of the reduced order basis, which is the opposite case when considering only the classical velocity-based POD modes (dissipative energy modes are classified respectively with very small singular values).

We will show that the new reduced order basis features new modes that represent larger ranges of spatial scales than the ones encountered in the energetic classical velocity POD modes. We point out also that the intentional a priori enforcement of these new modes within the first ranks of the reduced order basis has a major role on the quality of the resulting reduced order model. We will show that it ensures the stability of the reduced order model in an efficient fashion as a result of the availability of the driving forces and the dissipative ones within a reduced number of velocity modes.

3. Application of the Stabilization Approach to a Typical Aeronautical Injector

3.1. Flow Solver

For the present simulations, the low-Mach number solver YALES2 [21] for unstructured grids is retained. This flow solver has been specifically tailored for the direct numerical simulation and large-eddy simulation of turbulent reacting flows on large meshes counting several billion cells using massively parallel super-computers [22,23]. It features a central fourth-order scheme for spatial discretization while time integration of convective terms is performed with an explicit fourth-order temporal scheme. The Poisson equation that arises from the low-Mach formulation of the Navier–Stokes equations is solved with a highly efficient Deflated Preconditioned conjugated gradient method [23].

3.2. Typical Aeronautical Injector of Re = 45,000 Lean Preccinsta Burner

3.2.1. Test Case Presentation

In what follows, we apply our new approach for a 3D unsteady, turbulent and incompressible fluid flow in a fuel injection system. The main objective is to be able to have an efficient strategy in order to compute precisely the aerodynamic field in the primary zone of the combustion chamber. The so-called PRECCINSTA test case [24,25] is presented in Figure 1. This lean-premixed burner has been widely studied in the combustion community to validate large-eddy simulation models [22,26–31].

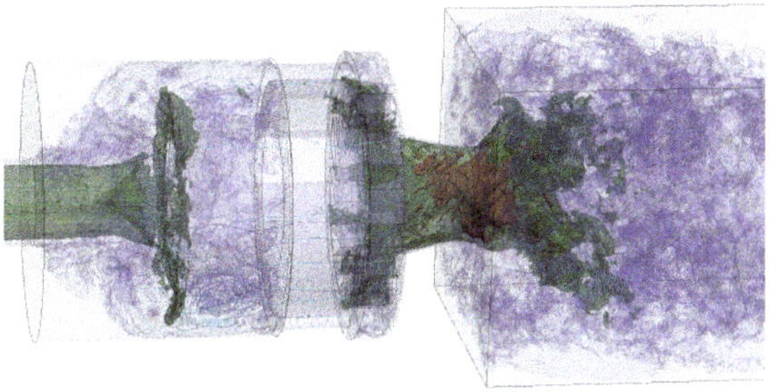

Figure 1. The 3D unsteady turbulent and incompressible flow in a fuel injection system and in the primary zone of the combustion chamber, given a constant inlet velocity, an outlet boundary condition on the channel outlet and a wall boundary condition on the upper and lower walls of the channel.

The 3D turbulent flow in the complex configuration presented in Figure 1 is considered. The kinematic viscosity $\nu = 10^{-5}\,\mathrm{m}^2/\mathrm{s}$ yields a Reynolds number 45,000 based on the inlet velocity and the length of the duct. The present High Fidelity simulation runs over 512 cores during 5 days in order to obtain a physical simulation time equal to 250 ms. A velocity-based POD-Galerkin reduced order modeling is performed, and an evaluation of its accuracy and efficiency is done before and after applying the a priori enrichment by the dissipative modes. By efficiency, we mean the online time needed to solve the mesh-independent ordinary differential Equation (8). In order to build the reduced basis, 2500 snapshots of the solution and its gradient are taken, extracted at each time step of the original HF simulation. We point out the fact that these 2500 snapshots are taken from 6644 time steps of the high fidelity simulation corresponding to the final 25 ms of its total physical time. We precise that these 25 ms represent two times the flow through time (FTT) of this test case.

3.2.2. POD Modes Computation for the Preccinsta

The velocity-based and gradient velocity-based POD modes were computed through a distributed snapshots POD performed in the YALES2 code. The CPU ressources needed for this computation are 768 cores (24 nodes), to guarantee a memory availability to read the 2500 time snapshots. The computation runs during 6 hours for the velocity-based POD modes and 9 hours for the gradient velocity-based POD modes. These POD modes for the velocity and gradient velocity fields are shown respectively starting from Figure 2–17.

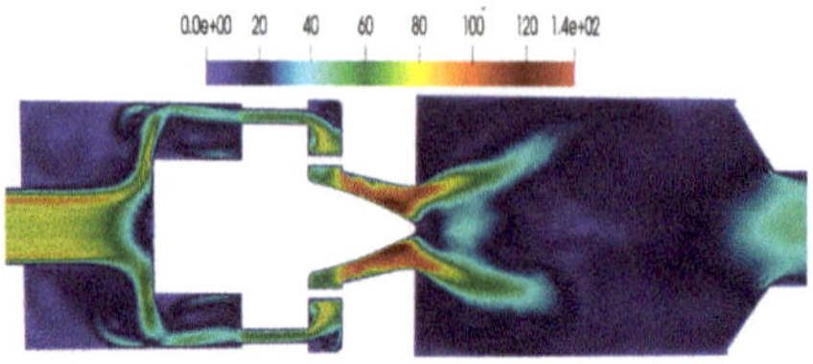

Figure 2. Velocity proper orthogonal decomposition (POD) mode Φ_1.

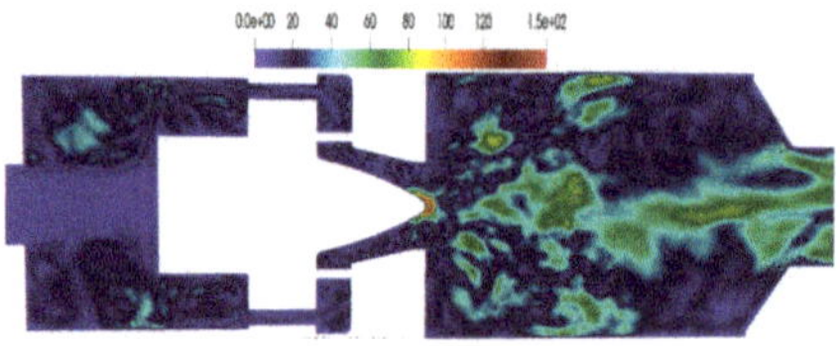

Figure 3. Velocity POD mode Φ_2.

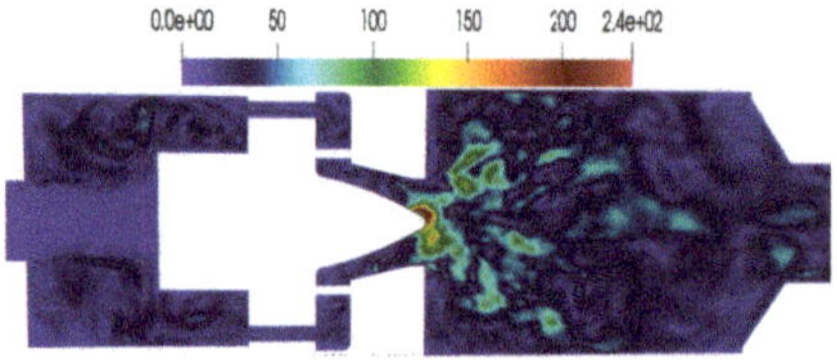

Figure 4. Velocity POD mode Φ_3.

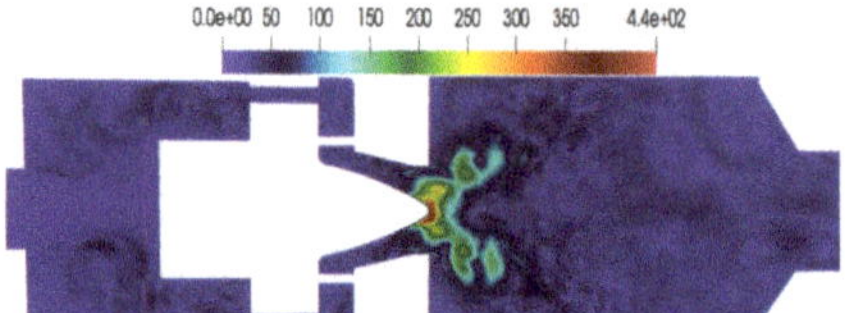

Figure 5. Velocity POD mode Φ_4.

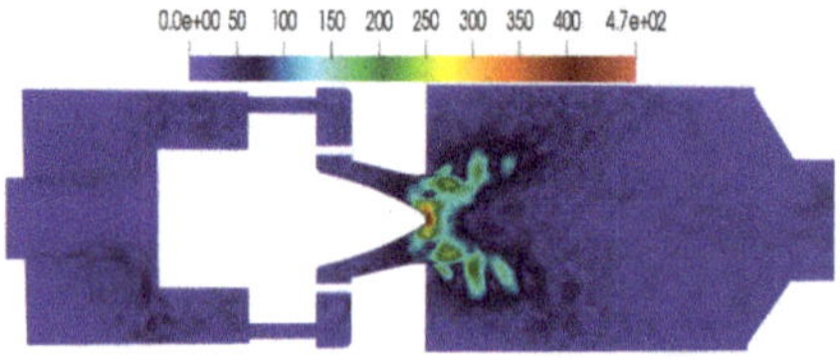

Figure 6. Velocity POD mode Φ_5.

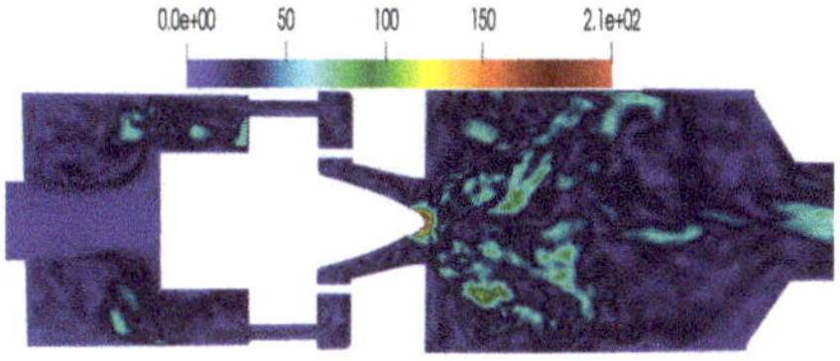

Figure 7. Velocity POD mode Φ_6.

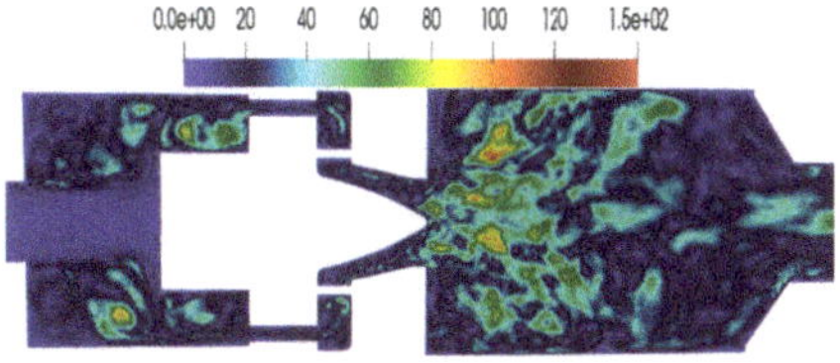

Figure 8. Velocity POD mode Φ_7.

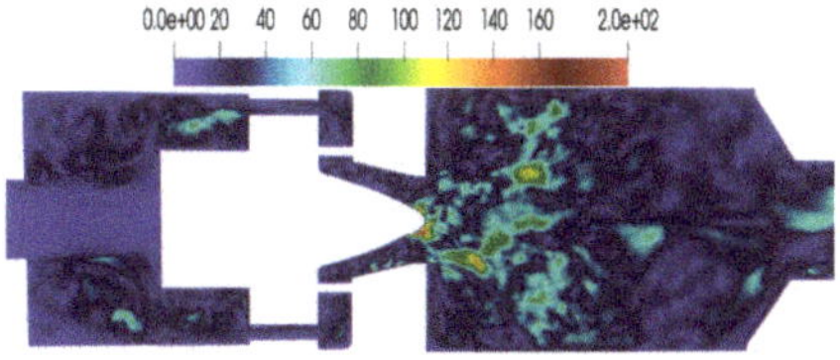

Figure 9. Velocity POD mode Φ_8.

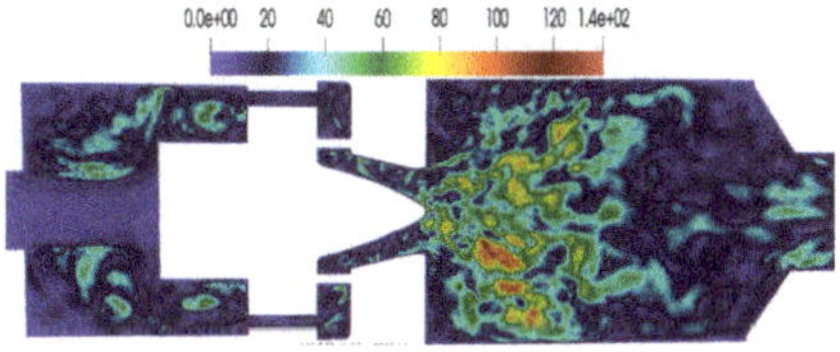

Figure 10. Velocity POD mode Φ_9.

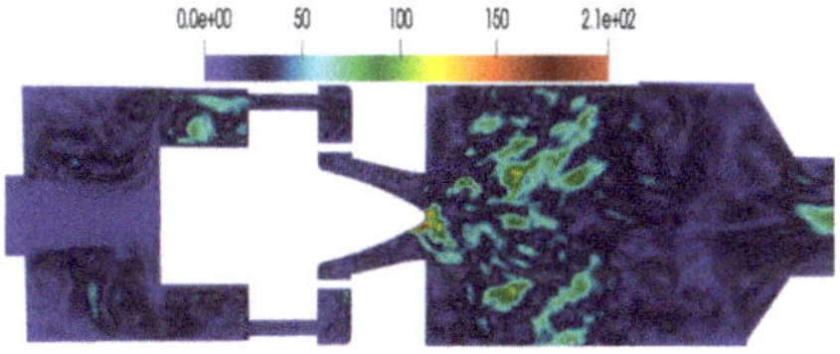

Figure 11. Velocity POD mode Φ_{10}.

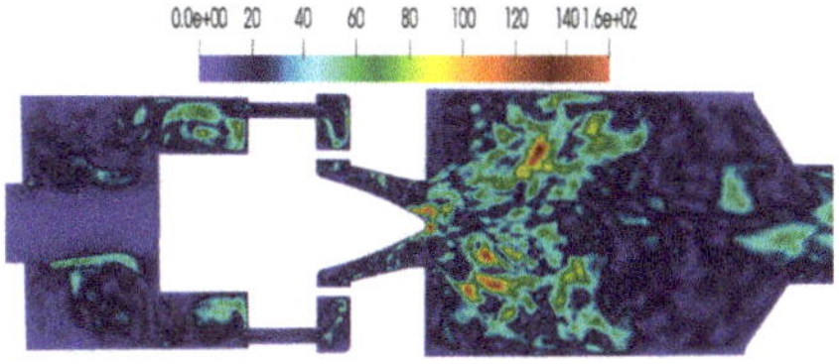

Figure 12. Velocity POD mode Φ_{11}.

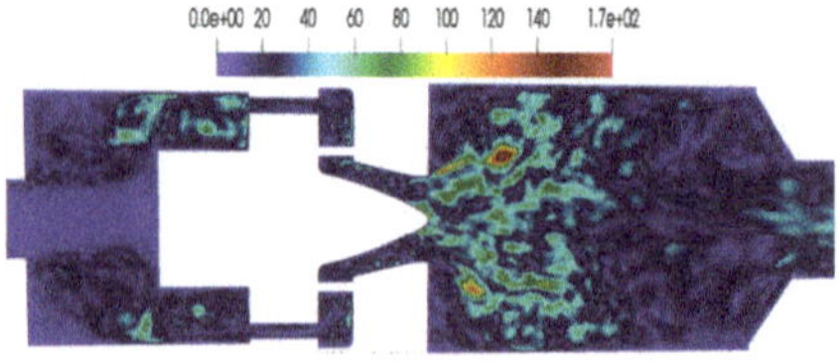

Figure 13. Velocity POD mode Φ_{12}.

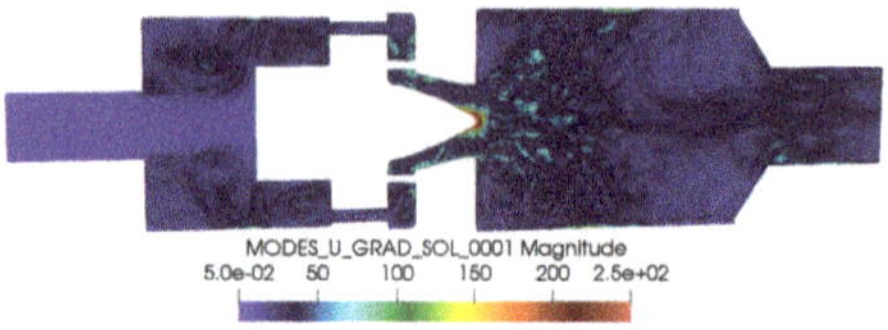

Figure 14. Fluctuating gradient velocity POD mode Ψ_1.

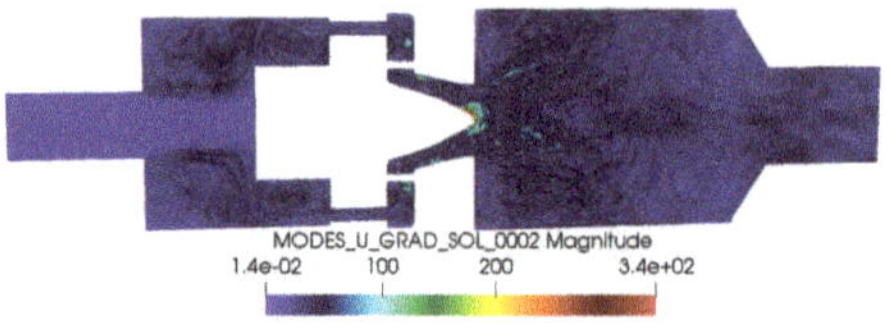

Figure 15. Fluctuating gradient velocity POD mode Ψ_2.

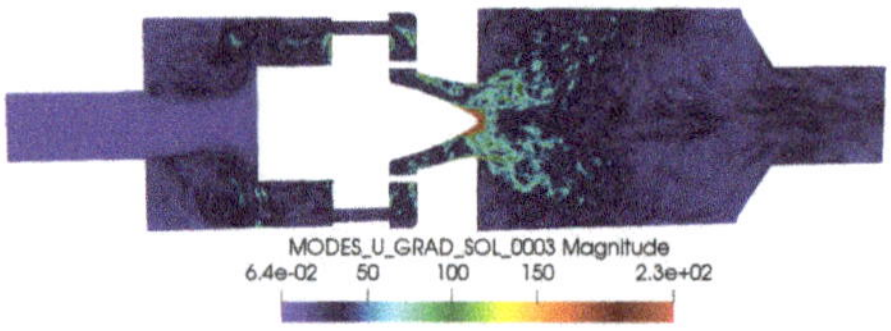

Figure 16. Fluctuating gradient velocity POD mode Ψ_3.

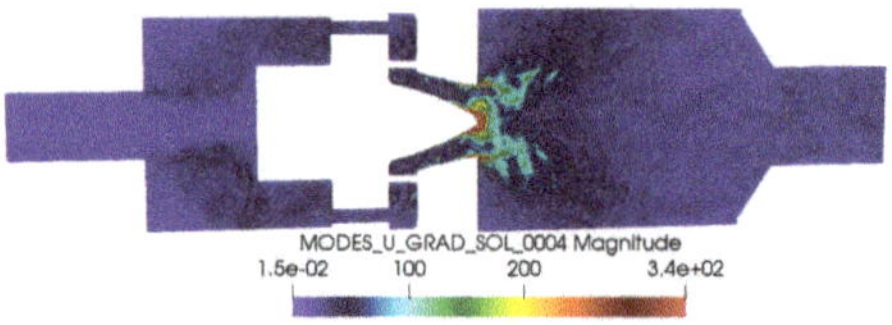

Figure 17. Fluctuating gradient velocity POD mode Ψ_4.

We can see that the velocity-based POD modes contain the high-scales of the principal coherent structures of the flow.

Interestingly, compared to the velocity-based POD modes, the velocity gradient-based POD ones feature high-scales in the dissipative regions such as in the wake of the two channels of the swirler and in the wake of the combustion chamber.

Moreover, if we compare the cumulative kinetic energies (Figures 18 and 19) associated respectively with the velocity-based POD modes and the gradient velocity-based ones, we can see that fewer than 10 velocity-based POD modes are sufficient to reproduce 90% of the high-scales TKE, however we need a larger number of velocity gradient-based POD modes in order to reproduce 90% of the small and dissipatives scales of the TKE. Then, it is clear that the dissipative scales are not considered in the velocity-based POD modes and should be added in order to preserve the energy conservation within the ROM.

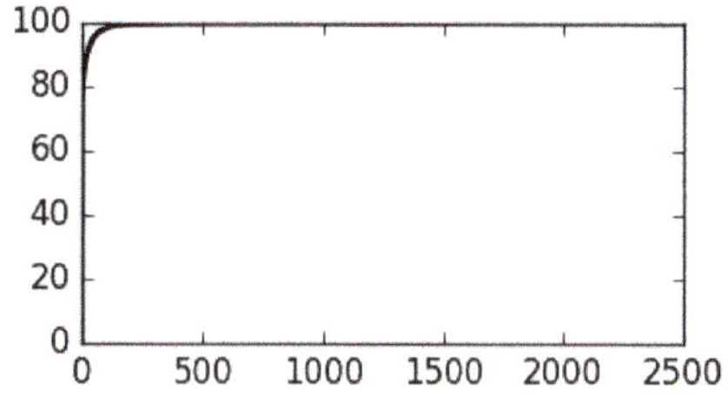

Figure 18. Cumulative kinetic energy of the velocity-based POD modes.

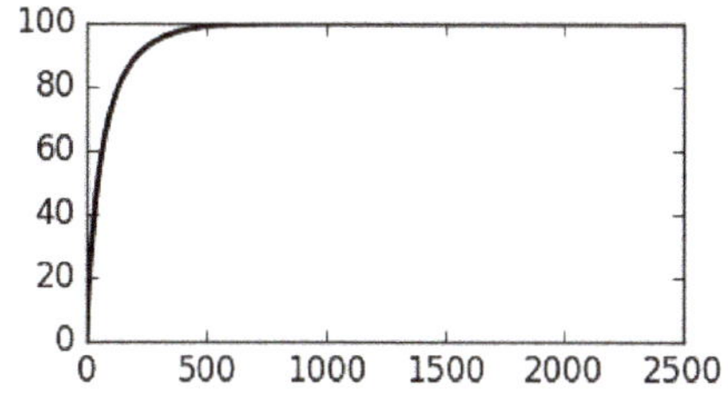

Figure 19. Cumulative kinetic energy of the velocity gradient-based POD modes.

3.2.3. The Enhanced Reduced Order Basis

We apply our a priori enforcement of the dissipative velocity modes defined previously by our new approach in the following fashion:

- We choose $N = 4$ and start the enforcement by the new velocity modes from the 5th rank. This choice is made because we want to limit the number of classical global POD modes which do not exhibit at the end very large features of spatial scales, as we can see on the modes Φ_5, Φ_6, Φ_7, Φ_8 and Φ_9.
- We choose $N' = 50$ because, as already discussed, we need a large number of velocity gradient-based POD modes in order to reproduce 90% of the small and dissipatives scales of the TKE as shown on Figure 19.
- We perform the Gram–Schmidt orthonormalization process for the enriched set $\{\Phi_1, ..., \Phi_N, \Phi_1^E, ..., \Phi_{N'}^E\}$ with respect to the energy-based inner product $(.,.)$.

By applying our proposed algorithm with the preceding choices, we get a new velocity-based reduced order basis as shown from Figures 20–31.

We give some further remarks in what follows:

- We recall that the choice $N = 4$ is done intentionally in order to retrieve some dissipative modes at earlier stages than in the classical POD technique where we can see that even after 12 modes we do not have any modes of large scale's features.
- The fact that the dissipative energy modes appear at late stages in the classical POD technique with very small singular values is the reason why we are not able to exploit their physical significance even if we increase the dimension of the classical POD reduced order model.
- We add starting $n = 5$ velocity-based modes of high singular values and large features of scales.
- This enrichment by small scale enforcement and separation is the key to multi-scale reproduction within the reduced order modeling. We precise once more that this approach is very different than the ones based on the change of the inner product that defines the matrix of the correlations between the instanteneous snapshots, typically the approach where the H^1 inner product is considered instead of the L^2 inner product. By our approach we enable scale separation, then

small scale's enforcement, which is very hard to distinguish when performing a H^1 correlations matrix and then retrieving a complete POD basis: the small scales will remain dominated by the L^2 correlated large scales even if we perform this inner product change. Some authors use mathematical calibration in order to retrieve the small scales [17].

The new velocity-based modes Φ_5^E, Φ_6^E, Φ_7^E, Φ_8^E and Φ_9^E show very large features of spatial scales which was not observed within the classical global POD modes Φ_5, Φ_6, Φ_7, Φ_8 and Φ_9. The margin of variation of these large features, as we can see on Figures 24–28, ranges from 0 to 380 (see Figures 25, 27 and 28). Moreover, the largest scales exhibit local structures in the fluid domain which are the small vortices carrying out the dissipative energy by analogy with the gradient velocity-based POD modes (see Figures 14–17).

In what follows, we call "dissipative ROM" the reduced order model computed using the proposed enriched basis $\left(\Phi_n^E\right)_n$, whereas "non-dissipative ROM" refers to the ROM using the classical POD basis $(\Phi_n)_n$.

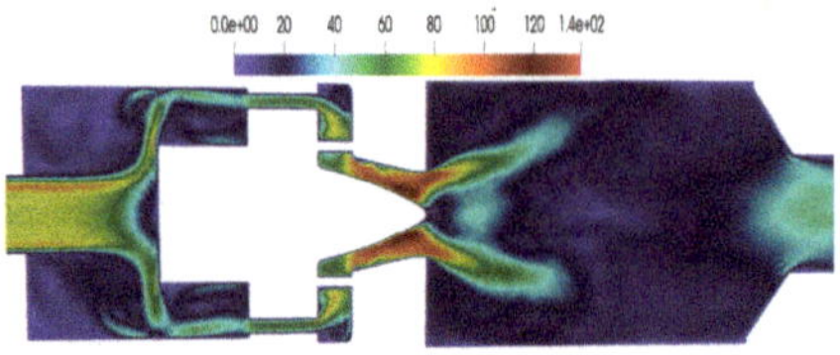

Figure 20. Velocity mode $\Phi_1^E = \Phi_1$.

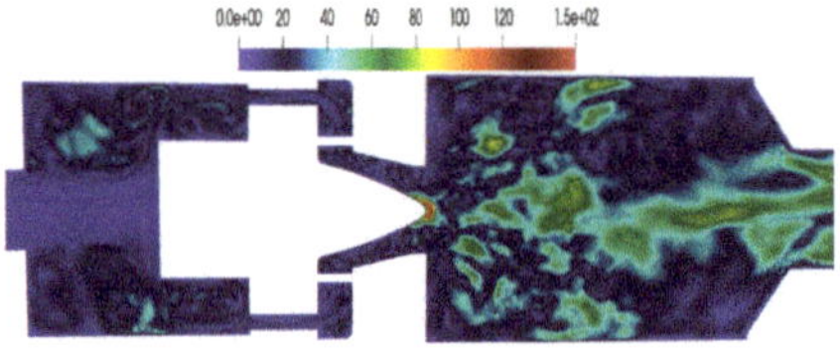

Figure 21. Velocity mode $\Phi_2^E = \Phi_2$.

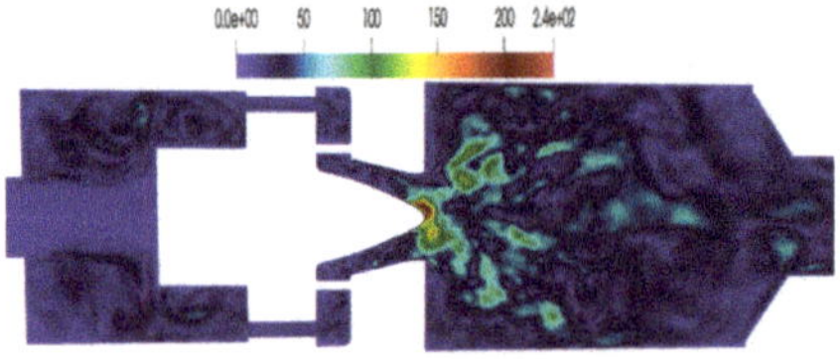

Figure 22. Velocity mode $\Phi_3^E = \Phi_3$.

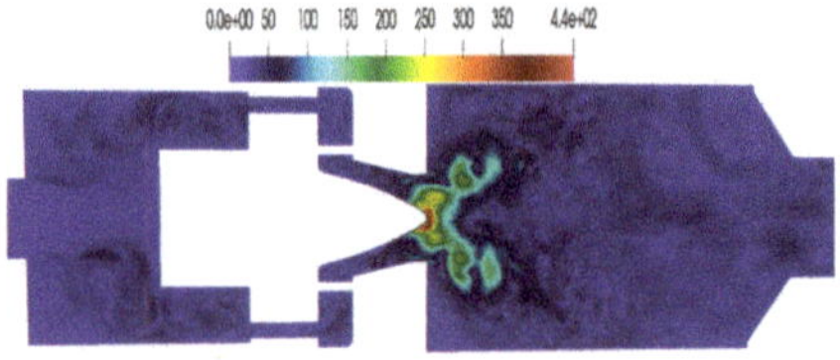

Figure 23. Velocity mode $\Phi_4^E = \Phi_4$.

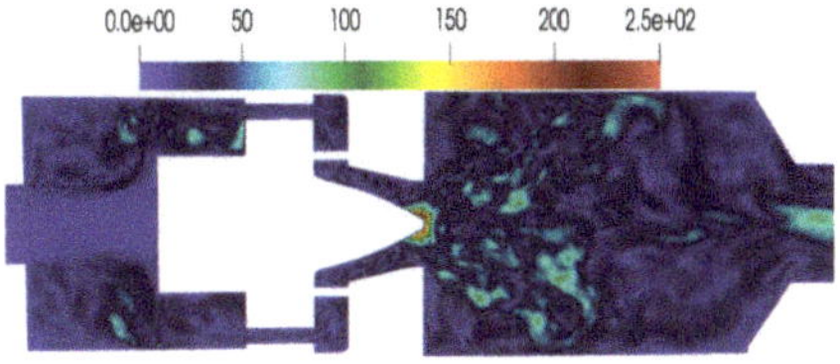

Figure 24. Velocity mode Φ_5^E.

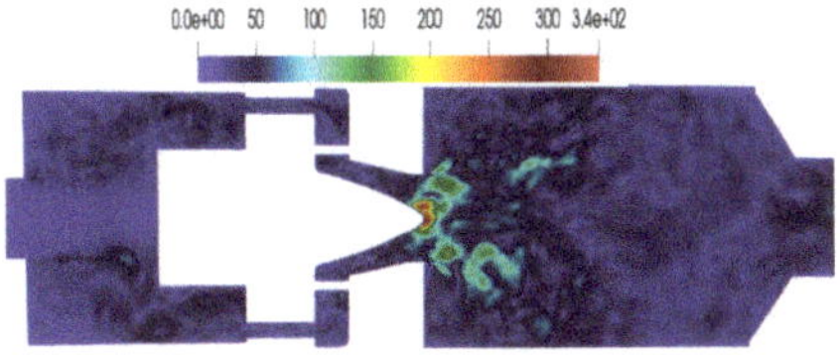

Figure 25. Velocity mode Φ_6^E.

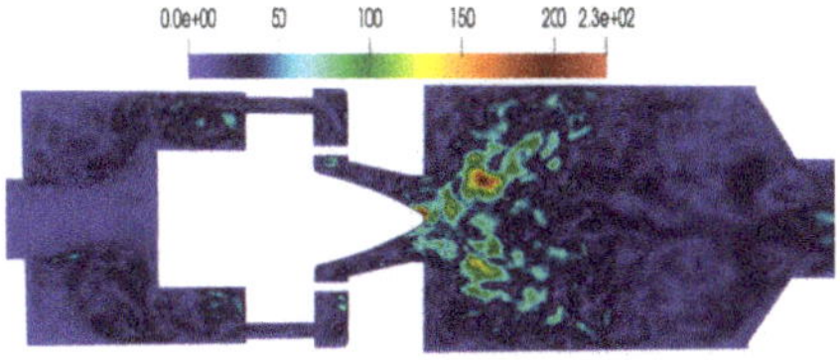

Figure 26. Velocity mode Φ_7^E.

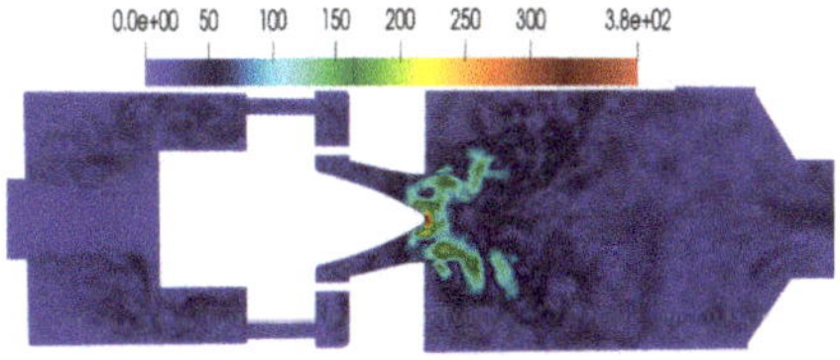

Figure 27. Velocity mode Φ_8^E.

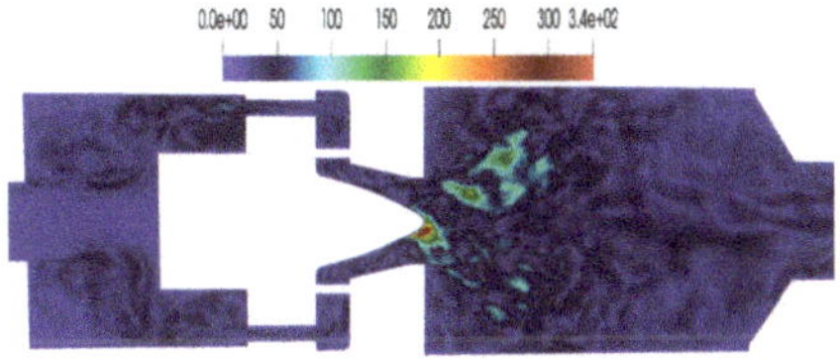

Figure 28. Velocity mode Φ_9^E.

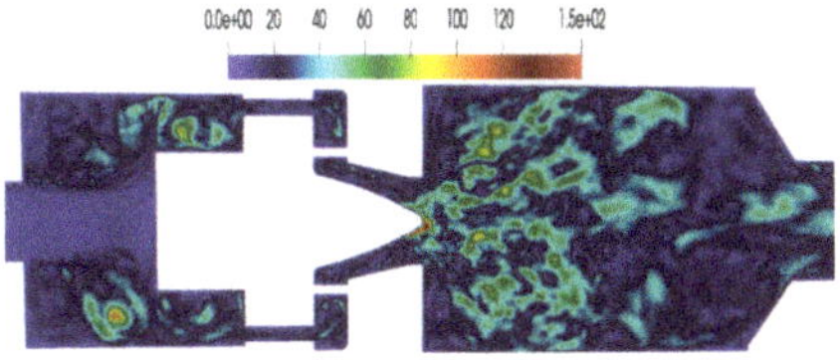

Figure 29. Velocity POD mode Φ_{10}^E.

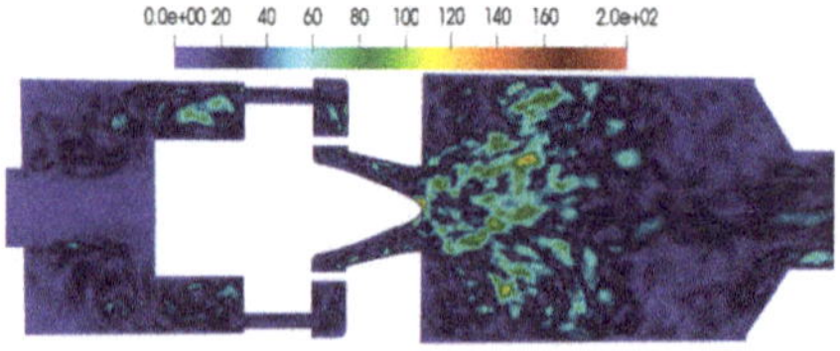

Figure 30. Velocity mode Φ^E_{11}.

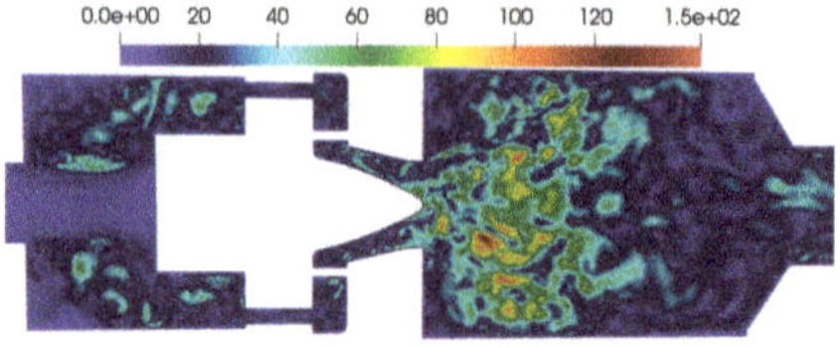

Figure 31. Velocity mode Φ^E_{12}.

3.2.4. The Temporal Coefficients and Kinetic Energy of the Enriched Reduced Order Model and the Comparaison with the Classical POD-Galerkin Reduced Order Model

Figures 32–36 show the time history of the stabilized ROM amplitudes, when the stabilization algorithm is performed by enrichment of $N = 4$ POD velocity modes with $N' = 50$ dissipative modes. We can see that these temporal coefficients obtained from the resolution of the reduced order model for a time interval equal to 25 ms which corresponds to the total time from which our data set was extracted in order to compute the POD modes for the velocity and the gradient velocity fields, tend to stabilize at the end of this resolution (from time step 2000). This could be explained by the order reduction using a limited number of modes, which means that the ROM needs to retrieve its equilibrium before the conservation of the kinetic energy. The ROM exhibits periodic fluctuations with a period of 2.2 ms which is the time scale of the precessing vortex core (PVC) associated with this test case.

This proves that the dissipative modes play a major role in the evolution of the Turbulent Kinetic Energy. Their introduction in the set of POD modes of the Galerkin ROM enables us to recover a better time evolution of the TKE in the system with fewer modes, see Figure 37. If we compare on this plot the kinetic energy evolutions respectively for the dissipative ROM and the non dissipative ROM, we can see that in the non dissipative case the plot of the kinetic energy is far from stabilization on the same time interval which is 25 ms.

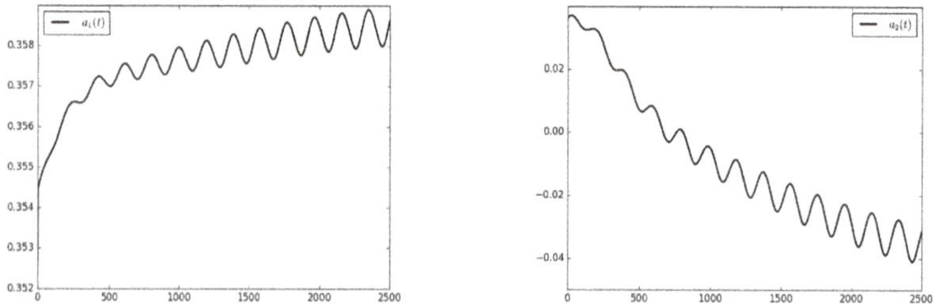

Figure 32. Time histories of the modal weights $a_1(t)$ and $a_2(t)$ for 25 ms time resolution of stabilized reduced order modeling (ROM)-POD.

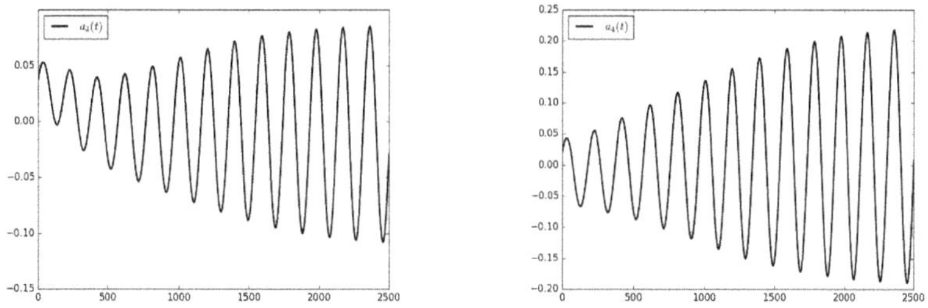

Figure 33. Time histories of the modal weights $a_3(t)$ and $a_4(t)$ for 25 ms time resolution of stabilized ROM-POD.

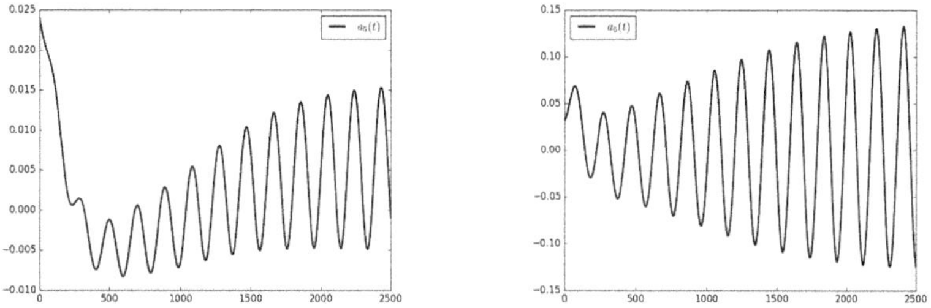

Figure 34. Time histories of the modal weights $a_5(t)$ and $a_6(t)$ for 25 ms time resolution of stabilized ROM-POD.

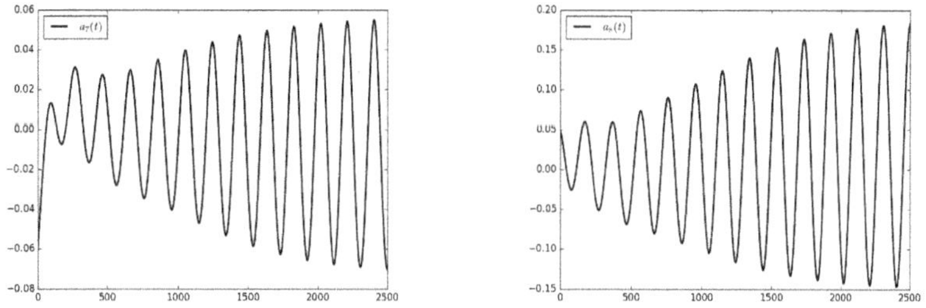

Figure 35. Time histories of the modal weights $a_7(t)$ and $a_8(t)$ for 25 ms time resolution of stabilized ROM-POD.

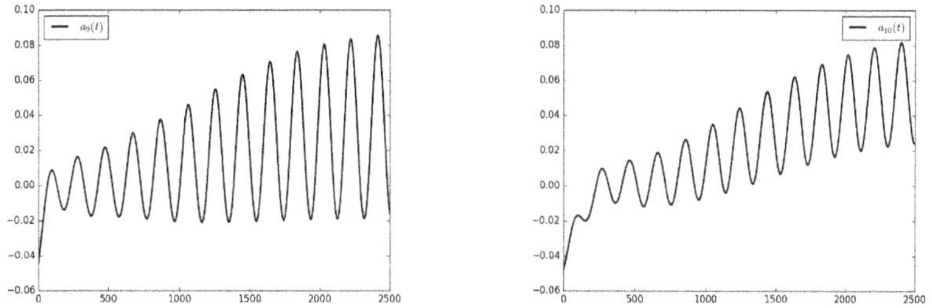

Figure 36. Time histories of the modal weights $a_9(t)$ and $a_{10}(t)$ for 25 ms time resolution of stabilized ROM-POD.

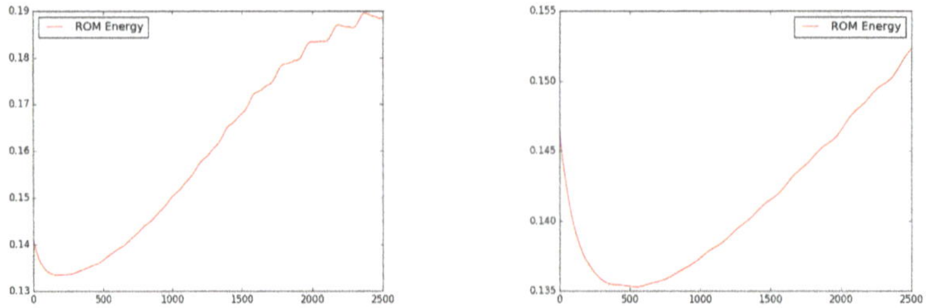

Figure 37. On the left: evolution the turbulent kinetic energy (TKE) in the dissipative ROM for 25 ms time resolution. On the right: evolution the TKE in the non-dissipative ROM for 25 ms time resolution.

These results are very encouraging to test the dynamic extrapolation of the stabilized reduced order model, so that we could access in real time (without any further offline operations) the evolution of the turbulent and incompressible flow outside the original snapshots data set. The first results of the dynamic extrapolation are shown in Section 4 in what follows.

3.2.5. 3D Time Fields Obtained by the ROM and the High-Fidelity Model

In what follows we show in Figure 38 plots of the 3*D* reduced order velocity fields when the stabilized ROM is applied, compared to the 3*D* high fidelity velocity fields obtained by LES. Large scale features of the flow are clearly reproduced by the ROM even for very long time integrations.

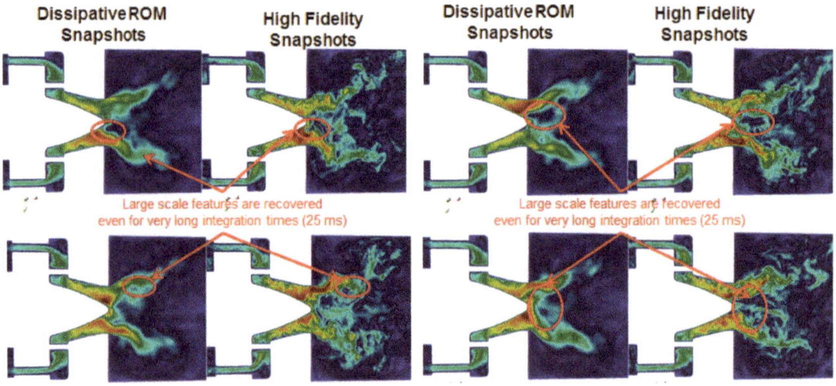

Figure 38. *X*-magnitude of the high-fidelity simulation against reduced order velocity fields.

3.2.6. CPU Time for Offline and Online Computation

In Table 1, we evaluate the efficiency of the stabilized reduced order modeling with respect to the High Fidelity simulation. Furthermore, we evaluate approximately the cost of the offline phase (including the snapshots POD, the Galerkin projection and the stabilization when applied) and the online ROM phase.

We precise that the speed-up is defined by the ratio of the ROM return time and the YALES2 return time. As a consequence of the proposed strategy we are able to enhance the accuracy of the reduced order modeling with a very good efficiency, regarding the online resolution. Furthermore, the offline effort associated with the additional stabilization algorithm scales with the high fidelity YALES2 return time.

It is important to note that the steps which are the most CPU consuming in the offline stage are the velocity-based POD and the gradient velocity-based one computations, followed by the stabilization by Gram–Schmidt. This took 18 h over 768 cores (24 nodes are required), because of the memory cost needed to read all of the 2500 time snapshots. This operation was not well distributed over the 768 cores due to the following issue: a temporal snapshot was not post-processed as one file per subdomain, i.e. the number of solution files per time step was less than the number of mesh partitions. This means that the running cores are working the available memory on the saved nodes in order to read lots of data per process. However, the Galerkin projection of the Navier–Stokes equations' operators took only three minutes over 768 cores. This is a consequence of the fact that we do not need to read any snapshots but, we read only the reduced number of the enhanced reduced order vectors and, we perform distributed scalar product and classical differentiation operations which scale with the mesh complexity but are very well parallelized due to distributed tasks on the mesh parts.

Table 1. Offline and online computational cost.

Operation	Wall Clock Time
High-fidelity YALES2 solver (512 cores)	5 days
Velocity-based POD + Disipative modes computation (768 cores)	15 h
Stabilization by Gram–Schmidt (768 cores)	3 h
Galerkin projection (768 cores)	3 min
Time python ROM-POD solver (1 core)	3.7 s
Speed up factor	10^8

4. Temporal Extrapolation of the Dissipative ROM

Running the reduced order model for 250 ms (i.e., 10 times longer than 25 ms that is the time interval over which the POD basis has been performed), yields the dynamic weight coefficients (Figures 39–43) and the evolution of the turbulent kinetic energy represented on Figure 44. These coefficients were obtained as a consequence of the run of the stable ROM over 1 core. In this case, we can legitimately state that the speed up of the reduced order modeling is 10^8, due to the fact that we are accessing physical solutions that were not seen by the offline phase and the learning phase of the ROM.

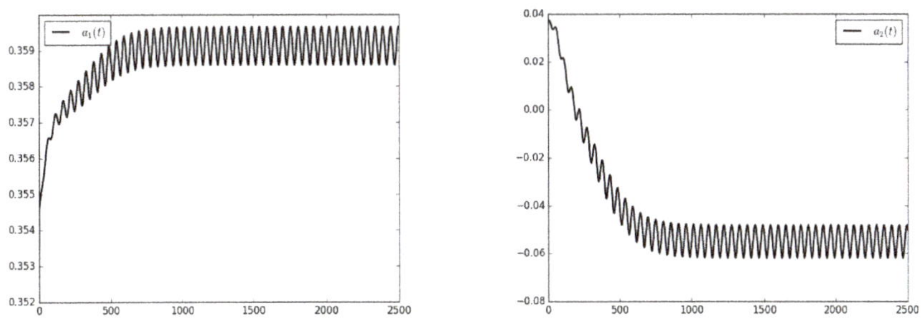

Figure 39. Time histories of the modal weights $a_1(t)$ and $a_2(t)$ for 250 ms time resolution of stabilized ROM-POD (ten times further than the total time of the snapshots data set).

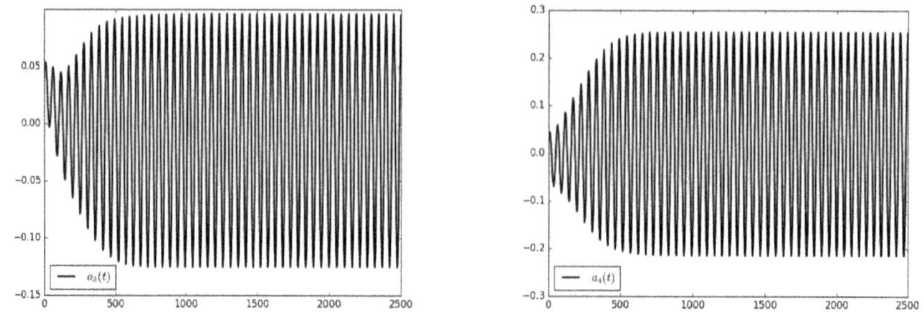

Figure 40. Time histories of the modal weights $a_3(t)$ and $a_4(t)$ for 250 ms time resolution of stabilized ROM-POD (ten times further than the total time of the snapshots data set).

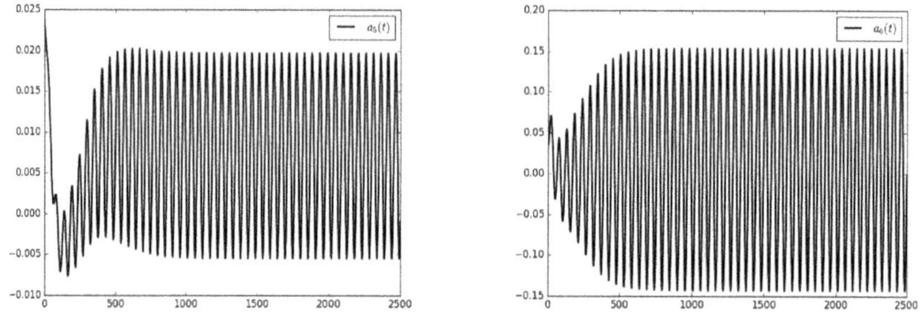

Figure 41. Time histories of the modal weights $a_5(t)$ and $a_6(t)$ for 250 ms time resolution of stabilized ROM-POD (ten times further than the total time of the snapshots data set).

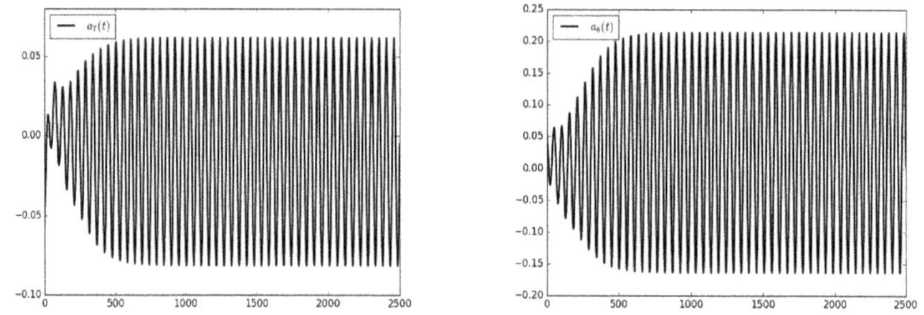

Figure 42. Time histories of the modal weights $a_7(t)$ and $a_8(t)$ for 250 ms time resolution of stabilized ROM-POD (ten times further than the total time of the snapshots data set).

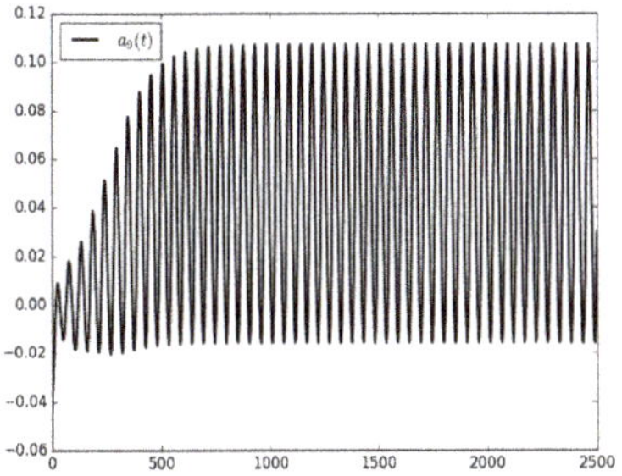

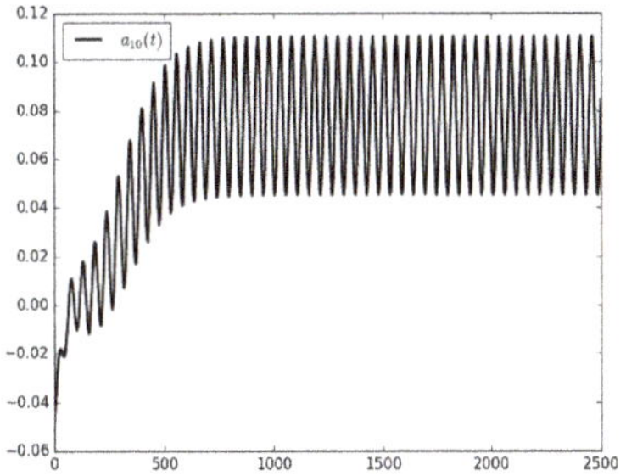

Figure 43. Time histories of the modal weights $a_9(t)$ and $a_{10}(t)$ for 250 ms time resolution of stabilized ROM-POD (ten times further than the total time of the snapshots data set).

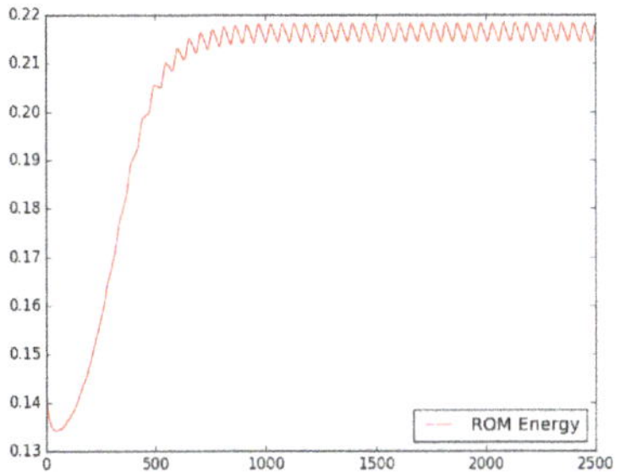

Figure 44. Time history of the TKE evolution in the dissipative ROM for 250 ms (ten times further than the total time of the snapshots data set).

5. Conclusions and Prospects

A new methodology is proposed for the stabilisation of Galerkin reduced order models by POD for the turbulent and incompressible 3D Navier–Stokes equations. The method is based on adding the necessary physics in the new reduced order space, so that all the scales modeled in the high-fidelity Navier–Stokes equations are taken into account by the reduced order model. The only ingredient which is not represented by the retained POD modes for the reduction process in the classical methodology, is the small rank scales which are responsible for the dissipation of the turbulent kinetic energy. This ingredient is added as a result of an a priori enrichment strategy and an enforcement to the velocity-based POD modes, by a minimal number of new velocity modes which contain the low dissipative energy in the new reduced order basis. This strategy shows a very good performance when applied to an unsteady turbulent flow of Reynolds 45,000 in a typical aeronautical injection system.

The prospects of this work are the use of the proposed stable reduced model in order to perform time extrapolation, that could be a way to study the limit cycle of the Precessing Vortex Core of an aeronautical injection system.

Author Contributions: Conceptualization, N.A.; methodology, N.A., F.C. and V.M.; software, N.A.; validation, N.A. and F.C.; formal analysis, N.A.; investigation, N.A.; data curation, N.A.; writing—original draft preparation, N.A.; writing—review and editing, F.C. and V.M.; visualization, N.A.; supervision, V.M.

Conflicts of Interest: The authors declare no conflict of interest.

Abbreviations

The following abbreviations are used in this manuscript:

ROM	Reduced order modeling
POD	Proper orthogonal decomposition
PVC	Precessing vortex core
SVD	Singular value decomposition
HF	High-fidelity
LES	Large eddy simulation
FTT	Flow through time
TKE	Turbulent kinetic energy

References

1. Couplet, M.; Sagaut, P.; Basdevant, C. Intermodal energy transfers in a proper orthogonal decomposition-Galerkin representation of a turbulent separated flow. *J. Fluid Mech.* **2003**, *491*, 275–284. [CrossRef]
2. Rowley, C.; Colonius, T.; Murray, R. Model Reduction for compressible flows using POD and Galerkin projection. *Phys. D Nonlinear Phenom.* **2004**, *189*, 115–129. [CrossRef]
3. Baiges, J.; Codina, R.; Idelsohn, S. Reduced-order subscales for POD models. *Comput. Methods Appl. Mech. Eng.* **2015**, *291*, 173–196. [CrossRef]
4. Balajewicz, M.; Tezaur, I.; Dowell, E. Minimal subspace rotation on the Stiefel manifold for stabilization and enhancement of projection-based reduced order models for the incompressible Navier–Stokes equations. *J. Comput. Phys.* **2016**. [CrossRef]
5. Xie, X.; Wells, D.; Wang, Z.; Iliescu, T. Approximate Deconvolution Reduced Order Modeling. *Comput. Methods Appl. Mech. Eng.* **2017**, *313*, 512–534. [CrossRef]
6. Mohebujjaman, M.; Rebholz, L.G.; Xie, X.; Iliescu, T. Energy balance and mass conservation in reduced order models of fluid flows. *J. Comput. Phys.* **2017**, *346*, 262–277. [CrossRef]
7. McLaughlen, B.; Peterson, J.; Ye, M. Stabilized reduced order models for the advection-diffusion-reaction equation using operator splitting. *Comput. Math. Appl.* **2016**, *71*, 2407–2420. [CrossRef]
8. Carlberg, K.; Choi, Y.; Sargsyan, S. Conservative model reduction for finite-volume models. *J. Comput. Phys.* **2017**, *371*, 280–314. [CrossRef]
9. Stabile, G.; Hijazi, S.; Mola, A.; Lorenzi, S.; Rozza, G. POD-Galerkin reduced order methods for CFD using Finite Volume Discretisation: Vortex shedding around a circular cylinder. *Commun. Appl. Ind. Math.* **2017**, *8*, 210–236. [CrossRef]
10. Stabile, G.; Rozza, G. Finite volume POD-Galerkin stabilized reduced order methods for the parametrised incompressible Navier–Stokes equations. *Comput. Fluids* **2018**, *173*, 273–284. [CrossRef]
11. Amsallem, D.; Farhat, C. On the Stability of Reduced-Order Linearized Computational Fluid Dynamics Models Based on POD and Galerkin Projection: Descriptor vs Non-Descriptor Forms. In *Reduced Order Methods for Modeling and Computational Reduction*; Quarteroni, A., Rozza, G., Eds.; Springer International Publishing: Cham, Switzerland, 2014; pp. 215–233.
12. Akhtar, I.; Nayfeh, A.H.; Ribbens, C.J. On the stability and extension of reduced-order Galerkin models in incompressible flows. *Theor. Comput. Fluid Dyn.* **2009**, *23*, 213–237. [CrossRef]
13. Amsallem, D.; Farhat, C. Stabilization of projection-based reduced-order models. *Int. J. Numer. Methods Eng.* **2012**, *91*, 358–377. [CrossRef]
14. Wang, Q.; Jan, S.H.; Ray, D. Non-intrusive reduced order modeling of unsteady flows using artificial neural networks with application to a combustion problem. *J. Comput. Phys.* **2019**, *384*, 289–307. [CrossRef]
15. Lapeyre, C.J.; Misdariis, A.; Casard, N.; Veynant, D.; Poinsot, T. Training convolutional neural networks to estimate turbulent sub-grid scale reaction rates. *Combust. Flame* **2019**, *203*, 255–264. [CrossRef]
16. Xie, X.; Zhang, G.; Webster, C.G. Data Driven Reduced Order Modeling of Fluid Dynamics Using Linear Multistep Network. *arXiv* **2018**, doi:arXiv:1809.07820. [CrossRef]
17. Iollo, A.; Lanteri, S.; Desideri, J. Stability properties of POD–Galerkin approximations for the compressible Navier–Stokes equations. *Theor. Comput. Fluid Dyn.* **2000**, *13*, 377–396. [CrossRef]

18. Guedot, L.; Lartigue, G.; Moureau, V. Numerical study of spray/precessing vortex core interaction in realistic swirling flows. In Proceedings of the 10th International ERCOFTAC Symposium on Engineering Turbulence Modelling and Measurements, Marbella, Spain, 17–19 September 2014.
19. Sirovich, L. Turbulence and the dynamics of coherent structures. III. Dynamics and scaling. *Q. Appl. Math.* **1987**, *45*, 583–590. [CrossRef]
20. Akkari, N.; Mercier, R.; Moureau, V. Geometrical Reduced Order Modeling (ROM) by Proper Orthogonal Decomposition (POD) for the incompressible Navier Stokes equations. In Proceedings of the 2018 AIAA Aerospace Sciences Meeting, Kissimmee, FL, USA, 8–12 January 2018.
21. Moureau, V.; Domingo, P.; Vervisch, L. Design of a massively parallel CFD code for complex geometries. *Comptes Rendus Mécanique* **2011**, *339*, 141–148. [CrossRef]
22. Moureau, V.; Domingo, P.; Vervisch, L. From Large-Eddy Simulation to Direct Numerical Simulation of a Lean Premixed Swirl Flame: Filtered Laminar Flame-PDF Modeling. *Combust. Flame* **2011**, *158*, 1340–1357. [CrossRef]
23. Malandain, M.; Maheu, N.; Moureau, V. Optimization of the deflated conjugate gradient algorithm for the solving of elliptic equations on massively parallel machines. *J. Comput. Phys.* **2013**, *238*, 32–47. [CrossRef]
24. Meier, W.; Weigand, P.; Duan, X.R.; Giezendanner-Thoben, R. Detailed characterization of the dynamics of thermoacoustic pulsations in a lean premixed swirl flame. *Combust. Flame* **2007**, *150*, 2–26. [CrossRef]
25. Weigand, P.; Duan, X.R.; Meier, W.; Meier, U.; Aigner, M.; Bérat, C. Experimental Investigations of an Oscillating Lean Premixed CH_4/Air Swirl Flame in a Gas Turbine Model Combustor. In Proceedings of the European Combustion Meeting, Louvain-la-Neuve, Belgium, 3–6 April 2005; p. 235.
26. Lartigue, G.; Meier, U.; Berat, C. Experimental and numerical investigation of self-excited combustion oscillations in a scaled gas turbine combustor. *Appl. Therm. Eng.* **2004**, *24*, 1583–1592. [CrossRef]
27. Roux, S.; Lartigue, G.; Poinsot, T.; Meier, U.; Bérat, C. Studies of mean and unsteady flow in a swirled combustor using experiments, acoustic analysis, and large eddy simulations. *Combust. Flame* **2005**, *141*, 40–54. [CrossRef]
28. Moureau, V.; Minot, P.; Pitsch, H.; Bérat, C. A ghost-fluid method for large-eddy simulations of premixed combustion in complex geometries. *J. Comput. Phys.* **2007**, *221*, 600–614. [CrossRef]
29. Fiorina, B.; Vicquelin, R.; Auzillon, P.; Darabiha, N.; Gicquel, O.; Veynante, D. A filtered tabulated chemistry model for LES of premixed combustion. *Combust. Flame* **2010**, *157*, 465–475. [CrossRef]
30. Franzelli, B.; Riber, E.; Gicquel, L.Y.M.; Poinsot, T. Large Eddy Simulation of Combustion Instabilities in a Lean Partially Premixed Swirled Flame. *Combust. Flame* **2012**, *159*, 621–637. [CrossRef]
31. Lourier, J.M.; Stöhr, M.; Noll, B.; Werner, S.; Fiolitakis, A. Scale Adaptive Simulation of a thermoacoustic instability in a partially premixed lean swirl combustor. *Combust. Flame* **2017**, *183*, 343–357. [CrossRef]

Mathematical and Computational Applications

MDPI

Article

Toward Optimality of Proper Generalised Decomposition Bases

Shadi Alameddin [1,*], Amélie Fau [1], David Néron [2], Pierre Ladevèze [2] and Udo Nackenhorst [1]

[1] IBNM, Leibniz Universität Hannover, Appelstraße 9a, 30167 Hannover, Germany; amelie.fau@ibnm.uni-hannover.de (A.F.); nackenhorst@ibnm.uni-hannover.de (U.N.)

[2] LMT, ENS Paris-Saclay, CNRS, Université Paris Saclay, 61 Avenue du Président Wilson, 94235 Cachan, France; neron@lmt.ens-cachan.fr (D.N.); pierre.ladeveze@lmt.ens-cachan.fr (P.L.)

* Correspondence: shadi.alameddin@ibnm.uni-hannover.de; Tel.: +49-511-762-2286

Received: 31 January 2019; Accepted: 1 March 2019; Published: 5 March 2019

Abstract: The solution of structural problems with nonlinear material behaviour in a model order reduction framework is investigated in this paper. In such a framework, greedy algorithms or adaptive strategies are interesting as they adjust the reduced order basis (ROB) to the problem of interest. However, these greedy strategies may lead to an excessive increase in the size of the ROB, i.e., the solution is no more represented in its optimal low-dimensional expansion. Here, an optimised strategy is proposed to maintain, at each step of the greedy algorithm, the lowest dimension of a Proper Generalized Decomposition (PGD) basis using a randomised Singular Value Decomposition (SVD) algorithm. Comparing to conventional approaches such as Gram–Schmidt orthonormalisation or deterministic SVD, it is shown to be very efficient both in terms of numerical cost and optimality of the ROB. Examples with different mesh densities are investigated to demonstrate the numerical efficiency of the presented method.

Keywords: model order reduction (MOR); low-rank approximation; proper generalised decomposition (PGD); PGD compression; randomised SVD; nonlinear material behaviour

1. Introduction

Numerical simulations appeal as an attractive augmentation to experiments to design and analyse mechanical structures. Despite the recent developments in computational resources that makes it feasible to solve systems with a substantial number of degrees of freedom efficiently, it is of common interest to reduce the numerical cost of numerical models throughout model order reduction (MOR) strategies [1]. The performance of MOR techniques has been shown in different fields such as their application to nonlinear problems [2,3], real-time computations [4] or for performing cyclic, parametric or probabilistic computations in which the information provided by some queries can be efficiently reused to respond to other queries that exhibit some similarities [5,6].

A posteriori model reduction techniques such as the Proper Orthogonal Decomposition (POD) is based on an offline training computations which extract a reduced order basis (ROB) from the solution of a high fidelity model. This optimal basis is practically built through a singular value decomposition (SVD) of a snapshot matrix. The singular vectors corresponding to the highest singular values are used to build the ROB [7]. Then, the problem of interest is confined to this ROB resulting in a drastic reduction in the numerical cost [1,8]. However, since the ROB has been defined as an optimal basis for the training stage, some advanced adaptive approaches are required to enrich the basis to tackle nonlinearities [9]. On the other hand, a priori MOR technique such as the Proper Generalized Decomposition (PGD) is based on the assumption that the quantities of interest can be written as a finite sum of products of separated functions, of generalised coordinates, which are sought in online computations [8,10]. No prior knowledge of the system is required in such a case and the ROB is directly adapted to the problem of interest by using a greedy algorithm, which enriches the basis when required [3,11]. However, an issue may be caused by the rapid growth of the ROB basis, whereas the primary interest of MOR is to benefit from a small sized ROB which provides a nondemanding temporal updating step. This step is equivalent to a POD step where the spatial modes are fixed and only the temporal ones are updated. It has been observed that the basis can increase to count some hundreds of modes for parametric studies of nonlinear cyclic loading [12], or some thousands for parametric computations [13]. In [5], some advanced strategies have been proposed to use an optimal parametric path allowing for controlling the basis expansion optimally.

In the context of reusing an ROB from a previous computation, a learning strategy has been proposed in [14,15] to extract an optimal basis from the reduced order model (ROM) through a Karhunen–Loève expansion. In a PGD framework, recompression based on SVD has been evaluated in [16]. However, the SVD step turns out to be numerically expensive prohibiting its implementation at each iteration. Therefore, it is common to let the basis increase and compress the results only at convergence to decrease their storage requirements. Therefore, it appears of interest to investigate probabilistic algorithms to compress the ROB on-the-fly without creating a bottleneck in the ROM. A detailed review of the most established algorithms to compute an SVD is provided in [17,18]. These algorithms are not limited to conventional deterministic methods such as truncated, incremental or iterative SVD but also randomised algorithms [19]. Different algorithms have been tested for POD applications in the case of dynamical problems in [18]. It has been noticed that randomised SVD algorithms can reduce drastically the numerical cost of the decomposition required after the training stage. Even if this step occurs only once in the offline stage of POD based ROM, the number of degrees of freedom and time steps can be vast for the high fidelity model so that the decomposition process can be a bottleneck.

Our goal here is to maintain the flexibility of the greedy algorithm through the usage of PGD while controlling the size of the ROB with a minimal numerical cost, by proposing to use a randomised SVD algorithm that provides a nondemanding compressive step after each enrichment of the basis. The numerical approach will be herein exemplified for the specific case of a fatigue computation based on continuum damage mechanics in a large time increment (LATIN) framework. However, the proposed numerical strategy can be generally used to optimise efficiently PGD basis for any application.

This paper is structured as follows. An overview of the LATIN-PGD scheme is provided in Section 2, followed by a discussion on the optimality of the PGD modes and the different algorithms to ensure that in Section 3. Lastly, in Section 4, different numerical examples are presented to illustrate the robustness and efficiency of the proposed algorithm.

Notation

The notation used in this paper is summarised in Table 1.

Table 1. Symbols and their representation.

Symbolic Representation	Verbal Representation
a, φ	scalars: lowercase letters
$\boldsymbol{u}, \boldsymbol{x}$	first-order tensors: lowercase boldface letters
$\boldsymbol{I}, \boldsymbol{N}$	second-order tensors: uppercase boldface letters
$\boldsymbol{\sigma}, \boldsymbol{\varepsilon}$	second-order tensors: Greek boldface letters
$\mathbb{C}, \mathbb{H}$	fourth-order tensor: blackboard bold letters
$\underline{a}$	column vector
$\underline{\underline{A}}$	matrix

2. An Overview of the LATIN-PGD Method

LATIN is a linearisation scheme that makes it easier to introduce PGD in nonlinear mechanical computations. A review of the LATIN-PGD method and some of its recent extensions to nonlinear solid mechanics problems can be found in [8,11].

The LATIN method is a fully discrete non-incremental solution scheme that inherits its efficiency for mechanical problems from incorporating an a priori model order reduction technique, namely PGD. It is shown in [20] that functions defined over space-time domain, with some regularities, may be approximated by PGD. However, it is vital that the number of modes (approximating functions) is small and the approximation error is low. A summary of the implemented framework is provided below.

For a generic structural problem defined over space-time domain $\Xi = \Omega \times \mathcal{I} = \Omega \times [0, T]$ in an infinitesimal strain and quasi-static framework, the strong form to be solved is represented in Figure 1 [21,22]. The equilibrium equation is linear in terms of the stress and the nonlinearity, in this case, is introduced through the constitutive model, i.e., in the stress–strain (σ, ε) relationship.

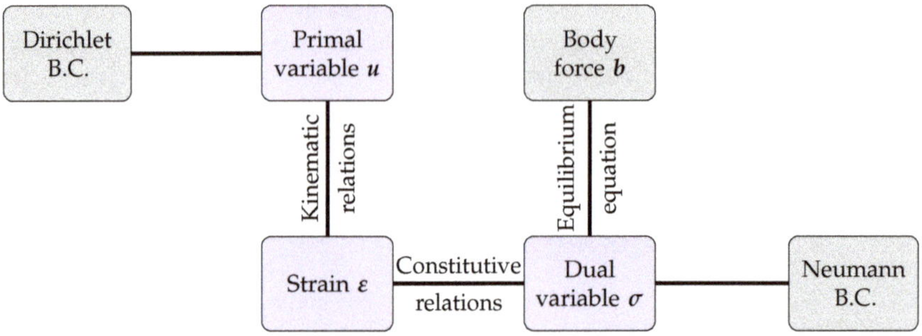

Figure 1. Graphical representation of the strong form of a structural problem (Tonti Diagram).

In a standard incremental Newton–Raphson scheme, the constitutive relations along with the kinematic relations are substituted into the balance equation resulting in a nonlinear problem in terms of the primal variable. However, a different linearisation strategy, termed LATIN, consists of solving the equilibrium equations along with kinematic relations in one step and solving the constitutive relations in the following step. Then, a solution that satisfies both of these systems is sought. In such a framework, two sets of equations are distinguished, the local equations described by constitutive relations (evolution and state laws [23]) and the global equilibrium equation along with the kinematic compatibility. Data flow between these two systems is required, i.e., to get statically admissible stress and kinematically permissible strain or displacement, a relation between the stress and the strain should be assumed. In the same manner, a decision should be taken on what data to pass back from the global system to the local one; these relations are referred to as search direction equations because they are affine equations in a 12-dimensional space hosting the stress and strain fields. The main advantage of the LATIN linearisation scheme is confining the computational cost to the solution of a global linear equation, which allows for introducing a model order reduction technique such as the PGD to reduce this numerical cost [8].

PGD is often used in many query context and quick response simulations where the solution is approximated by a finite sum of separated functions on each of the problem generalised coordinates, e.g., the displacement field may be approximated by a finite sum of globally spatial and temporal functions as

$$u(x,t) \approx \sum_{j=1}^{N} v_j(x) \circ \lambda_j(t), \qquad v_j(x) : \Omega \to \mathbb{R}^d,\ \lambda_j(t) : \mathcal{I} \to \mathbb{R}, \tag{1}$$

where $d \leq 3$ is the spatial dimension, $N \in \mathbb{N}$ and $\circ$ is the entry-wise Hadamard or Schur multiplication of vectors [8,24]. It is shown in [8] that a small number of pairs/modes is sufficient to approximate the solution of many problems with substantial savings in terms of CPU time and memory. In contrary to POD based techniques that include a preliminary learning phase, PGD defines the basis of the problem on-the-fly using a greedy algorithm such that additional pairs are added if necessary, i.e., the approximation error is controlled by the successive enrichment of the generated basis [25].

The LATIN solution algorithm starts with an elastic initialisation followed by a sequence of two stages, namely the local and the global ones. These two steps form one LATIN iteration, and they are repeated until convergence is reached. Note that, at every local and global step, the quantities of interest over all the space-time points are approximated. The space that belongs to the solution manifold of the constitutive relations is denoted by Γ while $\mathcal{A}$ represents the admissible space that satisfies the equilibrium equation (static admissibility) along with the kinematic relations (kinematic admissibility). Hence, the exact solution is defined as a set $s = \{\boldsymbol{X}, \dot{\boldsymbol{Y}}\} \in \Gamma \cap \mathcal{A}$, where $\boldsymbol{X}$ contains the dynamic conjugate variables and $\dot{\boldsymbol{Y}}$ represents the evolution of the internal variables. For discussions on the LATIN convergence behaviour, refer to [8,20,26].

The elastic solution $s_0 = \{\boldsymbol{X}_0\}$ takes all the boundary conditions into account, and the following solutions are computed in terms of corrections to s_0. Then, the constitutive model, consisting of the nonlinear evolution equations in addition to the state equations, is solved and integrated within the local stage at every space-time point. The outcome of this stage, at the ith iteration, is the solution $\hat{s}_i = \{\hat{\boldsymbol{X}}_i, \hat{\dot{\boldsymbol{Y}}}_i\}$, which is used in the following global stage to obtain s_{i+1}. The admissibility equations are the only ones left to be solved in the global stage. The kinematic admissibility is satisfied by deriving the strain as the symmetric gradient of the displacement field $\boldsymbol{\varepsilon} = \nabla^s \boldsymbol{u}$ with $\boldsymbol{u} = \bar{\boldsymbol{u}}$ on $\partial\Omega_D$ and the static admissibility condition is obtained from the equilibrium equation, which reads [27]

$$\nabla \cdot \boldsymbol{\sigma}(\boldsymbol{x}, t) + \boldsymbol{b} = \boldsymbol{0} \qquad \forall (\boldsymbol{x}, t) \in \Omega \times \mathcal{I}, \tag{2}$$

with $\boldsymbol{\sigma} \cdot \boldsymbol{n} = \bar{\boldsymbol{t}}$ on $\partial\Omega_N$, $\boldsymbol{\sigma}$ is Cauchy stress and $\boldsymbol{b}$ is the body force in the spatial domain Ω. The use of the Hamilton's law of varying action, which is the principle of virtual work integrated over time [28], leads to the following weak form

$$\int\limits_{\Omega \times \mathcal{I}} \boldsymbol{\sigma} : \boldsymbol{\varepsilon}(\boldsymbol{u}^*)\, \mathrm{d}\Omega\, \mathrm{d}t = \int\limits_{\Omega \times \mathcal{I}} \boldsymbol{b} \cdot \boldsymbol{u}^*\, \mathrm{d}\Omega\, \mathrm{d}t + \int\limits_{\partial\Omega_N \times \mathcal{I}} \bar{\boldsymbol{t}} \cdot \boldsymbol{u}^*\, \mathrm{d}S\, \mathrm{d}t \qquad \forall \boldsymbol{u}^* \in \mathcal{U}_0, \tag{3}$$

where $\mathcal{U}_0 = \{\boldsymbol{u}(\boldsymbol{x}, t) \mid \boldsymbol{u}(\boldsymbol{x}, t) \in \mathcal{H}_0^1(\Omega) \otimes \mathcal{C}^0(\mathcal{I}),\ \boldsymbol{u} = \boldsymbol{0} \text{ on } \partial\Omega_D \times \mathcal{I}\}$. As long as the boundary conditions are satisfied by the elastic initialisation, the corrections in each iteration, in terms of displacement, are defined as $\Delta \boldsymbol{u}_{i+1} = \boldsymbol{u}_{i+1} - \boldsymbol{u}_i$, where the i and $i+1$ subscripts refer to the previous and the current global stage, respectively. The solution of Equation (3) is computationally expensive due to the integration over the spatial domain. Therefore, the kinematically and statically admissible fields are computed for the whole space-time domain with the help of PGD, where a separate representation of the displacement and consequently the strain corrections is introduced as

$$\Delta \boldsymbol{u} = \boldsymbol{v}(\boldsymbol{x}) \circ \boldsymbol{\lambda}(t), \qquad \Delta \boldsymbol{\varepsilon} = \nabla \boldsymbol{v}(\boldsymbol{x}) \circ \boldsymbol{\lambda}(t). \tag{4}$$

Note that the subscript $i+1$ is dropped to simplify the notations, and it is assumed that only one PGD term/pair is generated within one LATIN iteration. Following the derivations in [3,29] by writing Equation (3) in terms of corrections and introducing the aforementioned PGD scheme results in a spatial and a temporal problem. These two problems are solved iteratively in a staggered manner using a fixed-point, alternated directions algorithm [8]. After introducing a Galerkin finite element discretisation, for the spatial and the temporal domains, this algorithm renders a space problem, with homogeneous boundary conditions,

$$\gamma\, \underline{\underline{K}}\, \underline{v} = \underline{f} \qquad \gamma \in \mathbb{R} \qquad \underline{\underline{K}} \in \mathbb{R}^{n \times n} \qquad \underline{v}, \underline{f} \in \mathbb{R}^n, \tag{5}$$

and a temporal problem, with zero initial conditions,

$$a\,\underline{\lambda} = \underline{b} \qquad a \in \mathbb{R} \qquad \underline{\lambda}, \underline{b} \in \mathbb{R}^{n_t}, \tag{6}$$

where (n, n_t) are the spatial and temporal degrees of freedom and $(\underline{v}, \underline{\lambda})$ are the spatial and temporal functions. The stiffness matrix is defined as $\underline{\underline{K}} = \int_\Omega \underline{\underline{B}}^\mathsf{T} \underline{\underline{C}}\, \underline{\underline{B}}\, \mathrm{d}\Omega$, where $\underline{\underline{B}}$ is a globally assembled matrix containing the derivatives of the shape functions and $\underline{\underline{C}}$ is a block diagonal matrix with 6×6 diagonal blocks representing the elasticity tensor at each integration point. The scaling factor in front of the stiffness is defined as $\gamma = \int_\mathcal{I} \underline{\lambda}^\mathsf{T} \underline{\lambda}\, \mathrm{d}t$ and the right-hand side $\underline{f} = -\int_{\Omega \times \mathcal{I}} \underline{\underline{B}}^\mathsf{T}\, (\underline{\underline{\hat{f}}}\, \underline{\lambda})\ \mathrm{d}\Omega\ \mathrm{d}t$, where $\underline{\underline{\hat{f}}}$ is a residual obtained from the previous local stage. The temporal problem is defined by $a = \int_\Omega (\underline{\underline{B}}\, \underline{v})^\mathsf{T} \underline{\underline{C}}\, (\underline{\underline{B}}\, \underline{v})\ \mathrm{d}\Omega$ and $\underline{b} = -\int_\Omega \underline{\underline{\hat{f}}}^\mathsf{T}\ (\underline{\underline{B}}\, \underline{v})\ \mathrm{d}\Omega$. Using μ modes at iteration $i+1$, the displacement field is approximated by, its discrete counterpart,

$$\underline{\underline{u}}_{i+1} = \underline{\underline{u}}_0 + \sum_{j=1}^{\mu} \underline{v}_j\, \underline{\lambda}_j^\mathsf{T}, \tag{7}$$

where $\underline{\underline{u}}_0$ corresponds to the elastic solution and the Hadamard multiplication is replaced by an outer product of the discrete values of $\boldsymbol{v}(\boldsymbol{x})$ and $\boldsymbol{\lambda}(t)$. It is seen that the cost of the global stage is dominated by the computational cost of the spatial problem, Equation (5), that has an identical dimension to the linear elastic problem associated with the finite element discretisation. Thus, a trial, POD-like, step is introduced at the beginning of the global stage that consists of reusing the previously generated spatial modes while updating the temporal ones [30].

2.1. Temporal Modes Update

Starting with an ROB that consists of a certain number (μ) of previously generated PGD pairs, the displacement correction is written as

$$\Delta\underline{\underline{u}}_{i+1} = \sum_{j=1}^{\mu} \underbrace{\underline{v}_j}_{known}\ \Delta\underline{\lambda}_j^\mathsf{T}, \tag{8}$$

where $\Delta\underline{\lambda}_j(t)$ is the correction added to the temporal function $\underline{\lambda}_j(t)$. Introducing this assumption into the temporal problem, Equation (6) leads to

$$\underline{\underline{\tilde{A}}}\, \underline{\underline{\tilde{\Lambda}}}^\mathsf{T} = \underline{\underline{\tilde{B}}} \qquad \underline{\underline{\tilde{A}}} \in \mathbb{R}^{\mu \times \mu} \qquad \underline{\underline{\tilde{B}}} \in \mathbb{R}^{\mu \times n_t}, \tag{9}$$

where

$$\tilde{A}_{kl} = \int_\Omega (\underline{\underline{B}}\, \underline{v}_k)^\mathsf{T} \underline{\underline{C}}\, (\underline{\underline{B}}\, \underline{v}_l)\ \mathrm{d}\Omega, \qquad \underline{\underline{\tilde{\Lambda}}} = [\Delta\underline{\lambda}_1, \cdots, \Delta\underline{\lambda}_\mu], \qquad \tilde{B}_{kl} = \int_\Omega (\underline{\underline{B}}\, \underline{v}_k)^\mathsf{T} \underline{\hat{f}}_{t_l}\ \mathrm{d}\Omega. \tag{10}$$

The cost of this step depends only on the temporal discretisation n_t and the number of already generated modes μ. If the computed approximation introduces a significant change to the original temporal modes, measured by $(\|\Delta\underline{\lambda}_j\| / \|\underline{\lambda}_j\|)$, then no further enrichment to the spatial modes is required and the algorithm continues to the next local stage. Otherwise, this update is ignored so as not to introduce unwanted numerical errors into the temporal functions and a new pair of temporal and spatial modes is generated.

3. Optimality of the Generated ROB

Recall that the correction/solution at the ith iteration of the LATIN algorithm, in matrix notation, reads

$$\underline{\underline{\tilde{U}}} = \sum_{j=1}^{\mu} \underline{v}_j\, \underline{\lambda}_j^{\mathsf{T}} = \underline{\underline{V}}\, \underline{\underline{\Lambda}}^{\mathsf{T}} \in \mathbb{R}^{n \times n_t}, \tag{11}$$

where $\underline{\underline{V}} = [\underline{v}_1, \cdots, \underline{v}_\mu] \in \mathbb{R}^{n\times\mu}$ and $\underline{\underline{\Lambda}} = [\underline{\lambda}_1, \cdots, \underline{\lambda}_\mu] \in \mathbb{R}^{n_t\times\mu}$. The representation in Equation (11) is referred to as an outer-product form [31], and such a form requires the storage of $\mu(n + n_t)$ entries only to represent $\underline{\underline{\tilde{U}}}$ with nn_t entries. It is practical to orthonormalise the spatial functions $\underline{v}_j$ before generating the temporal ones in order to limit the ROB size, i.e., the PGD expansion. This is traditionally done via a Gram–Schmidt (GS) procedure [3]. An orthonormalisation scheme based on a GS procedure is summarised in Algorithm 1, where $\underline{v}_l^{\mathsf{T}}\, \underline{v}_m = \delta_{lm}$ is the inner product between the spatial modes, δ_{lm} is the Kronecker delta and $\|\underline{v}_j\|_2^2 = \underline{v}_j^{\mathsf{T}}\, \underline{v}_j$.

Algorithm 1: Gram–Schmidt based orthonormalisation procedure

Data: Previously generated modes $\{\underline{v}_j, \underline{\lambda}_j\}$ $(j = 1, \cdots, \mu)$ with $\underline{v}_l^{\mathsf{T}}\, \underline{v}_m = \delta_{lm}$
New pair of modes $\{\underline{v}_{\mu+1}, \underline{\lambda}_{\mu+1}\}$

Result: Enriched basis $\{\underline{v}_j, \underline{\lambda}_j\}$ $(j = 1, \cdots, \mu + 1)$ with $\underline{v}_l^{\mathsf{T}}\, \underline{v}_m = \delta_{lm}$

for $j \leftarrow 1$ **to** μ **do**
  Calculate the inner product of $\underline{v}_{\mu+1}$ and an existing mode via $p = \underline{v}_j^{\mathsf{T}}\, \underline{v}_{\mu+1}$
  Subtract the projection from the new mode via $\underline{v}_{\mu+1} \leftarrow \underline{v}_{\mu+1} - p\, \underline{v}_j$
  Update existing temporal mode $\underline{\lambda}_j = \underline{\lambda}_j + p\, \underline{\lambda}_{j+1}$
end
Normalise the new spatial mode $v_{\mu+1} \leftarrow v_{\mu+1} / \|\underline{v}_{\mu+1}\|_2$
Update the new temporal mode $\lambda_{\mu+1} \leftarrow \lambda_{\mu+1}\, \|\underline{v}_{\mu+1}\|_2$

While experimenting on the LATIN-PGD scheme in a three-dimensional finite element framework, it has been noticed that reaching a small error required generating many modes, further discussion about the computational cost is provided in Section 4. This confirms the findings in [2] that orthonormality of the spatial modes is not enough to confine the PGD expansion, i.e., compressing the spatial modes only, leaves the temporal ones susceptible to redundancy.

3.1. SVD Compression of PGD

As long as PGD is not a unique decomposition and does not ensure the optimality of the generated modes in terms of a minimal expansion, an optimal decomposition can be obtained via an SVD of the full solution [32]. An SVD of the solution provides a straightforward scheme to compress both spatial and temporal information into a minimal set of modes, following Algorithm 2. This is similar to compressing information from different spatial directions into a single spatial mode.

It is known via the Schmidt–Eckart–Young theorem that the solution $\underline{\underline{\tilde{U}}}$ has an optimal approximation of rank $k \leq \mu + 1$ with respect to the Frobenius norm that satisfies [31]

$$\|\underline{\underline{\tilde{U}}} - \underline{\underline{\tilde{U}}}^{(k)}\|_{\mathrm{F}} = \sum_{j=k+1}^{\mu+1} \tilde{s}_j^2. \tag{12}$$

The corresponding approximation error in terms of the spectral norm reads

$$\|\underline{\underline{\tilde{U}}} - \underline{\underline{\tilde{U}}}^{(k)}\|_2 = \tilde{s}_{k+1}. \tag{13}$$

Hence, the PGD expansion may be restricted to a maximum number of modes and Equation (12) will give a measure of the approximation error due to this enforced truncation. Another way is to prescribe a subjectively acceptable tolerance ϵ_{tol} that the approximation error should not exceed, e.g., in the spectral norm this renders to

$$\frac{\|\underline{\underline{\tilde{U}}} - \underline{\underline{\tilde{U}}}^{(k)}\|_2}{\|\underline{\underline{\tilde{U}}}\|_2} = \frac{\tilde{s}_{k+1}}{\tilde{s}_1} < \epsilon_{\text{tol}}. \tag{14}$$

Algorithm 2: SVD compression of a PGD expansion

	Previously generated modes	$\{\underline{v}_j, \underline{\lambda}_j\}\ (j = 1, \cdots, \mu)$
Data:	New pair of modes	$\{\underline{v}_{\mu+1}, \underline{\lambda}_{\mu+1}\}$
	Required number of modes / truncation threshold	$k \leq \mu + 1,\ \epsilon_{\text{tol}}$

Result: Enriched basis $\{\underline{v}_j, \underline{\lambda}_j\}\ (j = 1, \cdots, k)$ with $\underline{v}_l^{\mathsf{T}}\ \underline{v}_m = \delta_{lm}$

Compute the full solution $\underline{\underline{\tilde{U}}} = \underline{\underline{V}}\ \underline{\underline{\Lambda}}^{\mathsf{T}}$

Compute a thin/truncated SVD of the solution $\underline{\underline{\tilde{U}}}^{(\mu+1)} = \sum_{j=1}^{\mu+1} \tilde{s}_j\ \underline{\tilde{v}}_j\ \underline{\tilde{\lambda}}_j^{\mathsf{T}}$

Truncate the decomposition based on $\tilde{s}_{k+1}/\tilde{s}_1 < \epsilon_{\text{tol}}$ or directly using k

Recover the outer-product representation:

$\underline{\underline{V}} \leftarrow [\underline{\tilde{v}}_1, \cdots, \underline{\tilde{v}}_k] \in \mathbb{R}^{n \times k}$

$\underline{\underline{\Lambda}} \leftarrow [\tilde{s}_1\ \underline{\tilde{\lambda}}_1, \cdots, \tilde{s}_k\ \underline{\tilde{\lambda}}_k] \in \mathbb{R}^{n_t \times k}$

The computation of a full SVD, in case of $n > n_t$, requires $\mathcal{O}(nn_t^2)$ floating point operations (flops) while seeking a truncated SVD requires $\mathcal{O}(nn_tk)$ flops. Due to the high computational cost of applying an SVD at each enrichment step in a PGD context, a quasi-optimal iterative orthonormalisation scheme was proposed in [2,16]. However, another appealing straightforward approach to provide a direct compression of the PGD modes into a minimal set is utilised here. It consists of using a randomised SVD algorithm [19] to compress the PGD expansion.

3.2. Randomised SVD (RSVD) Compression of PGD

Low-rank matrix decompositions may be computed efficiently using randomised algorithms as illustrated in this section for an SVD case. Such methods are based on random sampling to approximate the range of the input matrix, i.e., a subspace that captures most of the matrix effect. Then, the matrix is restricted to this subspace, and the low-rank approximation of this reduced matrix is sought using classical deterministic schemes. If $\underline{\underline{\tilde{U}}}$ is a dense matrix, the required flops are reduced from $\mathcal{O}(nn_tk)$ to $\mathcal{O}(nn_t \log(k))$, where k is the number of the sought dominant singular values of an $n \times n_t$ matrix. It is worth mentioning that, even when randomised algorithms require a higher number of flops, they exploit modern multi-processors architecture more efficiently than standard deterministic schemes [19]. It has been shown in [18] that a randomised SVD algorithm can outperform a truncated SVD one with a speed-up factor over 50 when $k = 10$. An overview of the randomised SVD algorithm applied in a PGD context is briefed in Algorithm 3.

Algorithm 3 can be straightforwardly extended to sample the rows of $\underline{\underline{\tilde{U}}}$ when n_t is large. However, this is not the case in the current study. It is also possible to exploit the PGD decomposition of the solution when computing its SVD or RSVD [31]; see Algorithm 4 for details.

Algorithm 4 utilises a rank revealing QR-decomposition in order not to rebuild the full matrix $\underline{\underline{\tilde{U}}}$. Further algorithmic details of the presented deterministic and randomised algorithms may be found in [17–19]. However, the goal of this study is to investigate the behaviour, robustness and efficiency of the presented algorithms in a PGD framework.

Algorithm 3: RSVD compression of a PGD expansion

Data: Previously generated modes $\{\underline{v}_j, \underline{\lambda}_j\}$ $(j = 1, \cdots, \mu)$
New pair of modes $\{\underline{v}_{\mu+1}, \underline{\lambda}_{\mu+1}\}$

Result: Enriched basis $\{\underline{v}_j, \underline{\lambda}_j\}$ $(j = 1, \cdots, k)$ with $\underline{v}_l^{\mathsf{T}}\, \underline{v}_m = \delta_{lm}$

Compute the full solution $\underline{\underline{\tilde{U}}} = \underline{\underline{V}}\, \underline{\underline{\Lambda}}^{\mathsf{T}} \in \mathbb{R}^{n \times n_t}$

Approximate a basis $\underline{\underline{E}}$ of range$(\underline{\underline{\tilde{U}}})$ via $\underline{\underline{E}} \leftarrow \underline{\underline{\tilde{U}}}\, \underline{\underline{Q}} \in \mathbb{R}^{n \times \tilde{k}}$ with
$\underline{\underline{Q}} \in \mathbb{R}^{n_t \times \tilde{k}}$ is a random matrix, $\tilde{k} = k + p$ and p is an oversampling factor taken experimentally to be in the range of $5 \sim 10$ [19].

Orthonormalise the columns of $\underline{\underline{E}}$ such that $\underline{\underline{\tilde{U}}} \approx \underline{\underline{E}}\, \underline{\underline{E}}^{\mathsf{T}} \underline{\underline{\tilde{U}}}$.

Restrict $\underline{\underline{\tilde{U}}}$ to the span$\{$col$(\underline{\underline{E}})\}$ to get a small matrix $\underline{\underline{S}} = \underline{\underline{E}}^{\mathsf{T}} \underline{\underline{\tilde{U}}} \in \mathbb{R}^{\tilde{k} \times n_t}$

Compute a truncated SVD $\underline{\underline{S}} \approx \underline{\underline{S}}^{(k)} = \underline{\underline{\tilde{\tilde{V}}}}\, \underline{\underline{\tilde{S}}}\, \underline{\underline{\tilde{\Lambda}}}^{\mathsf{T}}$ with $k \leq \mu + 1$

Expand $\underline{\underline{S}}$ to span$\{$col$(\underline{\underline{\tilde{U}}})\}$, i.e., $\underline{\underline{\tilde{U}}} \approx \underline{\underline{E}}\, \underline{\underline{S}}^{(k)} = \underline{\underline{E}}\, \underline{\underline{\tilde{\tilde{V}}}}\, \underline{\underline{\tilde{S}}}\, \underline{\underline{\tilde{\Lambda}}}^{\mathsf{T}} = \underline{\underline{\tilde{V}}}\, \underline{\underline{\tilde{S}}}\, \underline{\underline{\tilde{\Lambda}}}^{\mathsf{T}}$

Recover the outer-product representation:

$\underline{\underline{V}} \leftarrow [\underline{\tilde{v}}_1, \cdots, \underline{\tilde{v}}_k] \in \mathbb{R}^{n \times k}$

$\underline{\underline{\Lambda}} \leftarrow [\tilde{s}_1\, \underline{\tilde{\lambda}}_1, \cdots, \tilde{s}_k\, \underline{\tilde{\lambda}}_k] \in \mathbb{R}^{n_t \times k}$

Algorithm 4: RSVD compression that exploits the PGD expansion (RSVD-PGD)

Data: Previously generated modes $\{\underline{v}_j, \underline{\lambda}_j\}$ $(j = 1, \cdots, \mu)$
New pair of modes $\{\underline{v}_{\mu+1}, \underline{\lambda}_{\mu+1}\}$
Required number of modes / truncation threshold $k \leq \mu + 1$, ϵ_{tol}

Result: Enriched basis $\{\underline{v}_j, \underline{\lambda}_j\}$ $(j = 1, \cdots, k)$ with $\underline{v}_l^{\mathsf{T}}\, \underline{v}_m = \delta_{lm}$

QR-decomposition:

$\underline{\underline{V}} = \underline{\underline{Q}}_v\, \underline{\underline{R}}_v \quad \sim \mathcal{O}((\mu+1)^2\, n)$

$\underline{\underline{\Lambda}} = \underline{\underline{Q}}_\lambda\, \underline{\underline{R}}_\lambda \quad \sim \mathcal{O}((\mu+1)^2\, n_t)$

Compute $\underline{\underline{R}}_v\, \underline{\underline{R}}_\lambda^{\mathsf{T}} \in \mathbb{R}^{(\mu+1) \times (\mu+1)}$

Apply Algorithm 3 to approximate $\underline{\underline{R}}_v\, \underline{\underline{R}}_\lambda^{\mathsf{T}}$ as $\sum_{j=1}^{k} \tilde{s}_j\, \underline{\tilde{v}}_j\, \underline{\tilde{\lambda}}_j^{\mathsf{T}}$

Recover the outer-product representation:

$\underline{\underline{V}} \leftarrow \underline{\underline{Q}}_v\, \underline{\underline{\tilde{V}}} \in \mathbb{R}^{n \times k}$

$\underline{\underline{\Lambda}} \leftarrow \underline{\underline{Q}}_\lambda\, \underline{\underline{\tilde{\Lambda}}}\, \underline{\underline{\tilde{S}}}^{\mathsf{T}} \in \mathbb{R}^{n_t \times k}$

4. Numerical Results

The different algorithms are tested in the case of a modified unified viscoplastic viscodamage model, in an infinitesimal strain settings, derived from [3,23,33,34]. The analysis is carried out on a three-dimensional plate made of Cr–Mo steel at 580 °C [35] with a central groove. One-eighth of the plate with symmetric boundary conditions is shown in Figure 2. The plate geometry is defined by its length, width and depth being $(40, 20, 2)$ mm while the length and width of the groove are $(10, 4)$ mm. This plate is subjected to a uniformly distributed displacement field of the form $U_d = U_0 \sin\left(\frac{2\pi}{T} t\right)$ with t and T being the time and the time period, respectively.

Three examples are discussed below. Firstly, the effect of the temporal functions update is investigated; see Section 2.1. Then, the PGD behaviour with different orthonormalisation schemes is analysed to illustrate the optimality of the ROB. Lastly, the computational requirements of the orthonormalisation schemes and their effect on the temporal functions update are discussed.

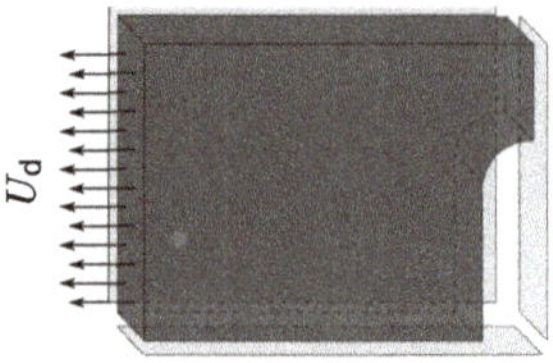

Figure 2. A plate with a central groove subjected to cyclic loading.

4.1. POD-Like Temporal Functions Update

The analysis of the plate, shown in Figure 2, is carried out on a mesh that consists of 387 hexahedron elements, with eight integration points in each element, resulting in 1884 spatial displacement degrees of freedom. The model is subjected to a uniformly distributed displacement field with an amplitude $U_0 = 0.00606\,\text{mm}$ and a time period $T = 10\,\text{s}$. The temporal discretisation is chosen such that the domain $[0, T]$ is discretised into 33 time steps. Since the whole time domain is computed at once, a total of 62,172 degrees of freedom are being sought. The commonly used GS scheme (Algorithm 1) is utilised in this example and the convergence criterion is considered to be 10^{-10}.

The purpose of this test case is to evaluate the importance of the updating step in a PGD approach. Hence, the number of generated modes along with the number of the LATIN iterations, with and without this POD-like step, is illustrated in Figure 3.

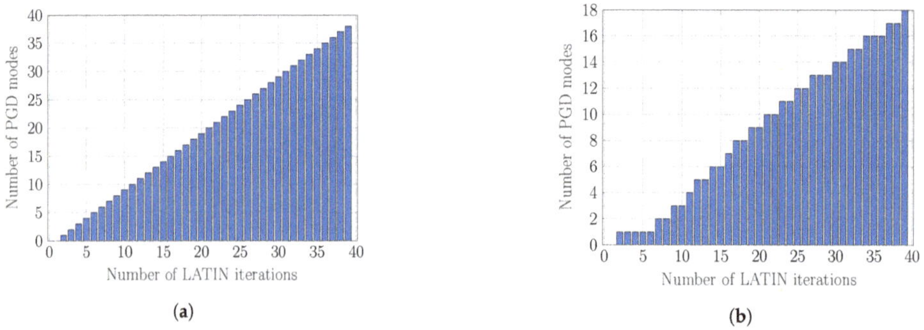

Figure 3. The size of the generated ROB. (**a**) ROB size without the updating step; (**b**) ROB size with the updating step.

It is seen in Figure 3 that the required number of LATIN iterations is not affected by this updating step, but the computational cost is sharply decreased. Moreover, such a step is crucial to limit the size of the PGD expansion. With the updating step, only half the number of modes were generated in comparison with the approach without any update. Due to this favourable nature, the updating step is implemented in the rest of the examples.

4.2. PGD Behaviour with Different Orthonormalisation Schemes

The previous example with the same spatial discretisation is subjected to 12 load cycles with different amplitudes, in the range of $[0.0033, 0.0066]$ mm. The temporal domain is divided into 12 intervals, each corresponding to one cycle, and the ROB generated within one cycle is reused in the following cycles. The convergence criterion is considered to be 10^{-4}.

The nonlinearity and the rapid damage evolution can be seen in Figure 4a where the damage value at the end of each cycle is plotted with respect to the number of cycles. The first PGD temporal function of each cycle, after convergence, is illustrated in Figure 4b.

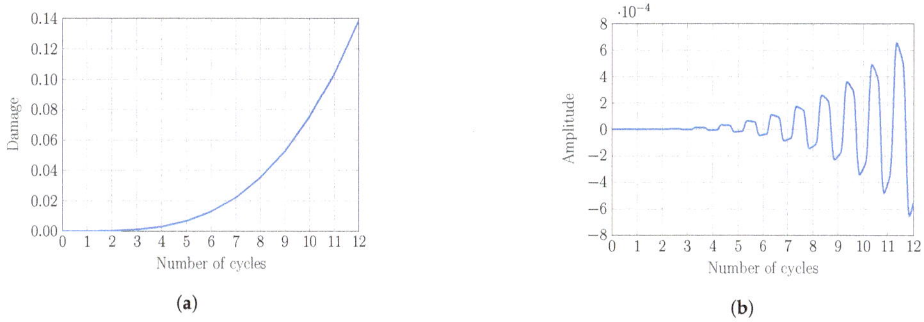

Figure 4. Damage evolution and the first PGD temporal mode in each cycle. (**a**) damage w.r.t. number of cycles; (**b**) first temporal function.

The simulation is carried out using Algorithms 1–4 and the resulting number of PGD modes with respect to the number of cycles is depicted in Figure 5. It is shown in Figure 5a that using Algorithm 1 resulted in an ROB with 18 modes while Algorithms 2–4 reduced this number to 11 modes by adding a maximum of one supplementary mode for each cycle. It is emphasised that Algorithms 2–4 provide the same ROB. However, their computational cost differs as illustrated in Section 4.3.

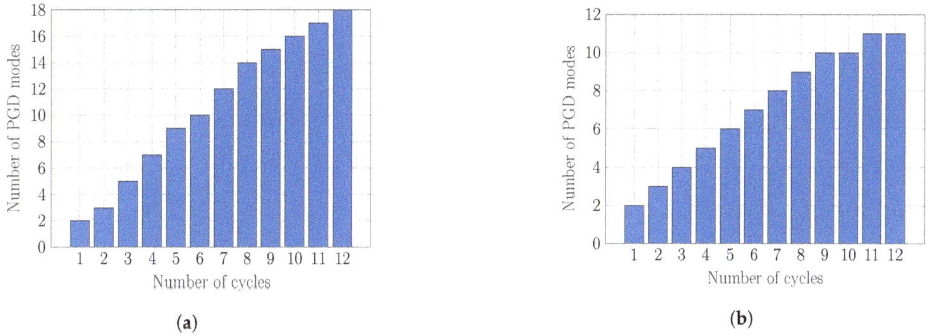

Figure 5. Number of PGD modes in each cycle using different orthonormalisation schemes. (**a**) number of PGD modes using GS; (**b**) number of PGD modes using (R)SVD.

It is observed that an SVD compression provides optimality of the ROB. It also has interesting properties such as not rejecting any mode in the current example. In other words, due to the optimality of the generated ROB, the POD-like step plays a noticeable role in convergence and there is no need for further enrichment of the ROB.

The inner product of the spatial modes in each case, after the last cycle, with their corresponding SVD of the acquired solution is shown in Figure 6. As expected, the GS modes are far from the optimal SVD ones while, trivially, Figure 6b depicts an almost diagonal matrix. The off-diagonal entries are caused by the temporal functions update at the final iterations.

It is of interest to point out that an excessive (R)SVD, termed e(R)SVD, step after the temporal functions update, which seems to be an unnecessary step, restricts the ROB to six modes only as illustrated in Figure 6c.

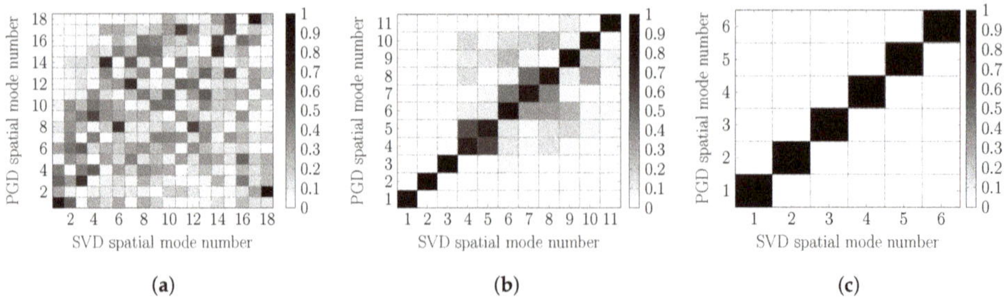

Figure 6. ROB optimality w.r.t. an SVD of the resulting solution. (**a**) GS; (**b**) (R)SVD; (**c**) e(R)SVD.

4.3. Relative Performance of the Different Orthonormalisation Schemes

The ensured optimality of the ROB is of interest when used with challenging examples such as in many-query context, due to the expected slow growth of the ROB. In order to investigate the robustness and the behaviour of the ROB, in a many-query context with a large number of degrees of freedom, the plate model is discretised into 13,812 hexahedron elements, with eight integration points in each element, resulting in 50,547 spatial displacement degrees of freedom. The temporal discretisation consists of 33 time steps in each cycle resulting in 1,668,051 degrees of freedom in each cycle. The plate is subjected to a uniformly distributed displacement field with a uniformly distributed random amplitudes in the range of $[18, 22] \times 10^{-5}$ mm and a time period $T = 10$ s. The convergence criterion is considered to be 10^{-4}.

The resulting number of PGD modes with respect to the number of cycles using GS and SVD algorithms is illustrated in Figures 7 and 8. It is seen that using a GS algorithm allows the ROB to grow to contain 126 pairs of modes while SVD algorithms confine this size to 21 modes, using a truncation threshold of 10^{-8}. Accepting a bigger approximation error with a truncation threshold of 10^{-5} reduces the number of modes to 11 pairs while an e(R)SVD scheme introduces further reduction to seven modes without any rejection or truncation due to the maintained optimality of the ROB.

It is worth noting that, in this example, the SVD orthonormalisation schemes, other than e(R)SVD, were invoked 53 times only compared to 125 times with the GS algorithm. Hence, this explains the low computational requirements of Algorithms 2–4 in comparison with Algorithm 1 as summarised in Figure 9b. The e(R)SVD scheme was invoked in each LATIN iteration. However, due to the small number of generated modes, the required time to update the temporal functions is drastically decreased in comparison with the other schemes; see Figure 9a for a profiler summary.

It is worth noting that the timing for each algorithm depends on the available computational resources. However, we expect their relative performance not to change. The RSVD algorithm is implemented in MATLAB® and uses its built-in SVD routine.

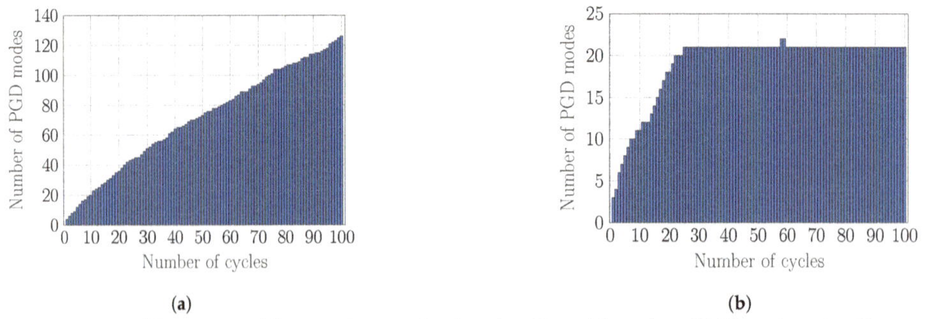

Figure 7. ROB size using different orthonormalisation algorithms. (**a**) number of PGD modes using GS; (**b**) number of PGD modes using (R)SVD ($\epsilon_{\text{tol}} = 10^{-8}$).

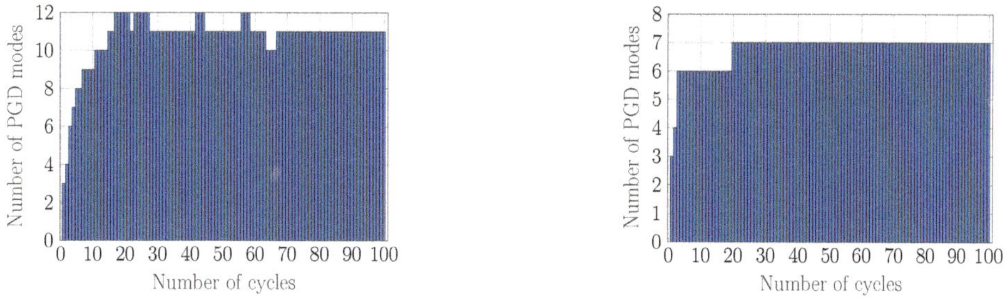

(**a**) (**b**)

Figure 8. ROB size using different orthonormalisation algorithms. (**a**) number of PGD modes using (R)SVD ($\epsilon_{\text{tol}} = 10^{-5}$); (**b**) number of PGD modes using e(R)SVD.

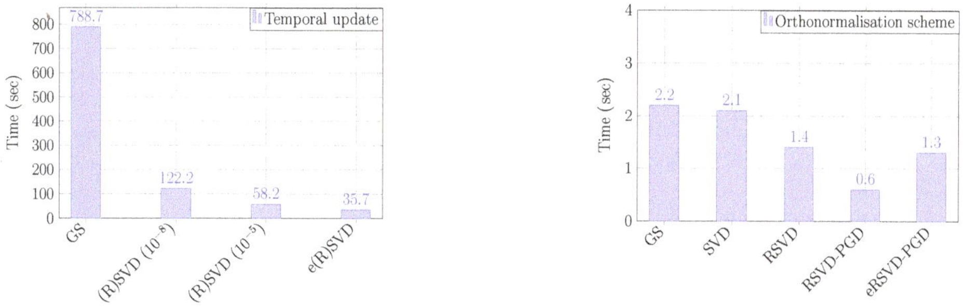

(**a**) (**b**)

Figure 9. The required time to perform the temporal update and the orthonormalisation steps. (**a**) timing of the temporal functions update; (**b**) timing of orthonormalisation schemes.

5. Conclusions and Further Research

Different orthonormalisation techniques were investigated to ensure the optimality of the PGD decomposition. These techniques and their effect on the PGD greedy algorithm are illustrated throughout examples with a varying number of degrees of freedom. It is found that a randomised SVD algorithm is a promising scheme to ensure the optimality of PGD expansions. It introduces beneficial time saving by limiting the number of modes compared to a Gram–Schmidt procedure and, at the same time, it shows a drastic speed-up compared to a deterministic SVD scheme. Another promising approach is proposed here where the randomised SVD scheme is invoked at each LATIN iteration, after the temporal update or the basis enrichment. This approach is referred to, in the current work, as e(R)SVD and it shows desired properties such as ensuring an optimal basis in each iteration and reducing the enrichment of basis functions to a minimum, i.e., no modes are rejected. The proposed numerical strategy, though it is presented in a LATIN-PGD framework, can be used to optimise PGD basis for any application.

Author Contributions: Conceptualization, S.A., A.F. and U.N.; Funding acquisition, U.N.; Investigation, S.A.; Methodology, S.A., D.N. and P.L.; Project administration, A.F. and U.N.; Resources, U.N.; Software, S.A.; Supervision, A.F., D.N., P.L. and U.N.; Validation, S.A.; Visualization, S.A.; Writing—original draft, S.A. and A.F.; Writing—review and editing, S.A.

Funding: This research was funded by the German Research Foundation/Deutsche Forschungsgemeinschaft (IRTG 1627).

Conflicts of Interest: The authors declare no conflict of interest.

Abbreviations

The following abbreviations are used in this manuscript:

flops	Floating point operations
GS	Gram–Schmidt
MOR	Model order reduction
ROM	Reduced order model
ROB	Reduced order basis
POD	Proper orthogonal decomposition
PGD	Proper generalised decomposition
SVD	Singular value decomposition
RSVD	Randomised singular value decomposition
e(R)SVD	Excessive SVD/RSVD applied at each iteration after ROB enrichment or temporal update
LATIN	Large time increment

References

1. Chinesta, F.; Huerta, A.; Rozza, G.; Willcox, K. Model Reduction Methods. In *Encyclopedia of Computational Mechanics*; John Wiley & Sons: Hoboken, NJ, USA, 2018; Volume 3.
2. Giacoma, A.; Dureisseix, D.; Gravouil, A.; Rochette, M. Toward an optimal a priori reduced basis strategy for frictional contact problems with LATIN solver. *Comput. Methods Appl. Mech. Eng.* **2015**, *283*, 1357–1381. [CrossRef]
3. Bhattacharyya, M.; Fau, A.; Nackenhorst, U.; Néron, D.; Ladevèze, P. A LATIN-based model reduction approach for the simulation of cycling damage. *Comput. Mech.* **2018**, *62*, 725–743. [CrossRef]
4. Niroomandi, S.; González, D.; Alfaro, I.; Bordeu, F.; Leygue, A.; Cueto, E.; Chinesta, F. Real-time simulation of biological soft tissues: A PGD approach. *Int. J. Numer. Methods Biomed. Eng.* **2013**, *29*, 586–600. [CrossRef] [PubMed]
5. Heyberger, C.; Boucard, P.A.; Néron, D. A rational strategy for the resolution of parametrized problems in the PGD framework. *Comput. Methods Appl. Mech. Eng.* **2013**, *259*, 40–49. [CrossRef]
6. Bhattacharyya, M.; Fau, A.; Nackenhorst, U.; Néron, D.; Ladevèze, P. A multi-temporal scale model reduction approach for the computation of fatigue damage. *Comput. Methods Appl. Mech. Eng.* **2018**, *340*, 630–656. [CrossRef]
7. Cline, A.; Dhillon, I. Computation of the Singular Value Decomposition. In *Handbook of Linear Algebra*; CRC Press: Boca Raton, FL, USA, 2013; pp. 1027–1039.
8. Chinesta, F.; Ladevèze, P. *Separated Representations and PGD-Based Model Reduction*; Springer: Vienna, Austria, 2014; Volume 554, p. 227.
9. Kerfriden, P.; Gosselet, P.; Adhikari, S.; Bordas, S. Bridging proper orthogonal decomposition methods and augmented Newton–Krylov algorithms: An adaptive model order reduction for highly nonlinear mechanical problems. *Comput. Methods Appl. Mech. Eng.* **2011**, *200*, 850–866. [CrossRef] [PubMed]
10. Ladevèze, P. Large time increment method for the analysis of structures with non-linear behavior caused by internal variables (La methode a grand increment de temps pour l'analyse de structures a comportement non lineaire decrit par variables internes). *Comptes Rendus de l'Académie des Sciences. Série 2, Mécanique, Physique, Chimie, Sciences de l'Univers, Sciences de la Terre* **1989**, *309*, 1095–1099.
11. Ladevèze, P. On reduced models in nonlinear solid mechanics. *Eur. J. Mech. A Solids* **2016**, *60*, 227–237. [CrossRef]
12. Nasri, M.A.; Robert, C.; Ammar, A.; El Arem, S.; Morel, F. Proper Generalized Decomposition (PGD) for the numerical simulation of polycrystalline aggregates under cyclic loading. *Comptes Rendus Mécanique* **2018**, *346*, 132–151. [CrossRef]
13. El Halabi, F.; González, D.; Sanz-Herrera, J.A.; Doblaré, M. A PGD-based multiscale formulation for non-linear solid mechanics under small deformations. *Comput. Methods Appl. Mech. Eng.* **2016**, *305*, 806–826. [CrossRef]
14. Ryckelynck, D. An a priori model reduction method for thermomechanical problems, Réduction a priori de modèles thermomécaniques. *Comptes Rendus Mécanique* **2002**, *330*, 499–505. [CrossRef]

15. Ryckelynck, D. A priori hyperreduction method: An adaptive approach. *J. Comput. Phys.* **2005**, *202*, 346–366. [CrossRef]
16. Giacoma, A.; Dureisseix, D.; Gravouil, A. An efficient quasi-optimal space-time PGD application to frictional contact mechanics. *Adv. Model. Simul. Eng. Sci.* **2016**, *3*, doi:10.1186/s40323-016-0067-7. [CrossRef]
17. Golub, G.H.; van Loan, C.F. *Matrix Computations*; The Johns Hopkins University Press: Baltimore, MD, USA, 2013.
18. Bach, C.; Ceglia, D.; Song, L.; Duddeck, F. Randomized low-rank approximation methods for projection-based model order reduction of large nonlinear dynamical problems. *Int. J. Numer. Methods Eng.* **2019**, 1–33. [CrossRef]
19. Halko, N.; Martinsson, P.G.; Tropp, J.A. Finding Structure with Randomness: Probabilistic Algorithms for Constructing Approximate Matrix Decompositions. *SIAM Rev.* **2011**, *53*, 217–288. [CrossRef]
20. Ladevèze, P. *Nonlinear Computational Structural Mechanics*; Springer: New York, NY, USA, 1999.
21. Steinke, P. *Finite-Elemente-Methode*; Springer: Berlin/Heidelberg, Germany, 2015; p. 391.
22. Felippa, C.A. Recent developments in parametrized variational principles for mechanics. *Comput. Mech.* **1996**, *18*, 159–174. [CrossRef]
23. Lemaitre, J. *A Course on Damage Mechanics*; Springer-Verlag: Berlin/Heidelberg, Germany, 1992.
24. Cueto, E.; González, D.; Alfaro, I. *Proper Generalized Decompositions*; Springer International Publishing: Cham, Switzerland, 2016.
25. Chinesta, F.; Keunings, R.; Leygue, A. *The Proper Generalized Decomposition for Advanced Numerical Simulations: A Primer*; Springer International Publishing: Cham, Switzerland, 2014.
26. Ladevèze, P.; Perego, U. Duality preserving discretization of the large time increment methods. *Comput. Methods Appl. Mech. Eng.* **2000**, *189*, 205–232. [CrossRef]
27. Wunderlich, W.; Pilkey, W.D. Mechanics of Structures. Variational and Computational Methods. *Meccanica* **2004**, *39*, 291–292. [CrossRef]
28. Gellin, S.; Pitarresi, J.M. Nonlinear analysis using temporal finite elements. *Eng. Anal.* **1988**, *5*, 126–132. [CrossRef]
29. Allix, O.; Ladevèze, P.; Gilletta, D.; Ohayon, R. A damage prediction method for composite structures. *Int. J. Numer. Methods Eng.* **1989**, *27*, 271–283. [CrossRef]
30. Bhattacharyya, M. A Model Reduction Technique in Space and Time for Fatigue Simulation. Ph.D. Thesis, Leibniz Universität Hannover, Hannover, Germany, 2018.
31. Bebendorf, M. *Hierarchical Matrices*; Springer-Verlag: Berlin/Heidelberg, Germany, 2008; Volumn 63, pp. 1–302.
32. Eckart, C.; Young, G. The approximation of one matrix by another of lower rank. *Psychometrika* **1936**, *1*, 211–218. [CrossRef]
33. Chaboche, J.L.; Nouailhas, D. A Unified Constitutive Model for Cyclic Viscoplasticity and Its Applications to Various Stainless Steels. *J. Eng. Mater. Technol.* **1989**, *111*, 424. [CrossRef]
34. De Souza Neto, E.A.; Peric, D.; Owen, D.R.J. *Computational Methods for Plasticity: Theory and Applications*; John Wiley & Sons: Hoboken, NJ, USA, 2011.
35. Lemaitre, J.; Desmorat, R. *Engineering Damage Mechanics*; Springer-Verlag: Berlin/Heidelberg, Germany, 2005.

Mathematical and Computational Applications

MDPI

Article

An Error Indicator-Based Adaptive Reduced Order Model for Nonlinear Structural Mechanics—Application to High-Pressure Turbine Blades

Fabien Casenave * and Nissrine Akkari

Safran Tech, Modelling and Simulation, Rue des Jeunes Bois, Châteaufort, 78114 Magny-Les-Hameaux, France; nissrine.akkari@safrangroup.com

* Correspondence: fabien.casenave@safrangroup.com

Received: 26 February 2019; Accepted: 15 April 2019; Published: 19 April 2019

Abstract: The industrial application motivating this work is the fatigue computation of aircraft engines' high-pressure turbine blades. The material model involves nonlinear elastoviscoplastic behavior laws, for which the parameters depend on the temperature. For this application, the temperature loading is not accurately known and can reach values relatively close to the creep temperature: important nonlinear effects occur and the solution strongly depends on the used thermal loading. We consider a nonlinear reduced order model able to compute, in the exploitation phase, the behavior of the blade for a new temperature field loading. The sensitivity of the solution to the temperature makes the classical unenriched proper orthogonal decomposition method fail. In this work, we propose a new error indicator, quantifying the error made by the reduced order model in computational complexity independent of the size of the high-fidelity reference model. In our framework, when the error indicator becomes larger than a given tolerance, the reduced order model is updated using one time step solution of the high-fidelity reference model. The approach is illustrated on a series of academic test cases and applied on a setting of industrial complexity involving five million degrees of freedom, where the whole procedure is computed in parallel with distributed memory.

Keywords: nonlinear reduced order model; elastoviscoplastic behavior; nonlinear structural mechanics; proper orthogonal decomposition; empirical cubature method; error indicator

1. Introduction

The application of interest for this work is the lifetime computation of aircraft engines' high-pressure turbine blades. Being located immediately downstream the combustion chamber, such parts undergo extreme thermal loading, with incoming fluid temperature higher than the material's melting temperature. These blades are responsible for a large part of the maintenance budget of the engine, with temperature creep rupture and high-cycle fatigue [1,2] as possible failure causes. Various technological efforts have been spent to increase the durability of these blades as much as possible, such as thermal barrier coatings [3], advanced superalloys [4] and complex internal cooling channels [5,6], see Figure 1 for a representation of a high-pressure turbine blade.

Computing lifetime predictions for high-pressure turbine blades is a challenging task: meshes involve large numbers of degrees of freedom to account for local structures such as the internal cooling channels, the behavior laws are strongly nonlinear with many internal variables, and a large number of cycles has to be computed. Besides, the temperature loading is poorly known in the outlet section of the combustion chamber. Our team has proposed in [7] a nonintrusive reduced order model

(ROM) strategy in parallel computation with distributed memory to mitigate the runtime issues: a domain decomposition method is used to compute the first cycle, and the reduced order model is used to speed up the computation of the following cycles, which can be considered as a reduced order model-based temporal extrapolation. As pointed out in [7], errors are accumulated during this temporal extrapolation. Moreover, quantifying the uncertainty on the lifetime with respect to some statistical description of the temperature loading using an already constructed reduced order model would introduce additional errors. In this context, error indicator-based enrichment of reduced order models is the topic of the present work.

Figure 1. Illustration of a high-pressure turbine blade [8]. The internal channels create a protective layer of cool air to protect the outer surface of the blade.

Error estimation for reduced model predictions is a topic that receives interest in the scientific literature. The reduced basis method [9,10] for parametrized problems is a reduced order modeling method that intrinsically relies on efficient a posteriori error bounds of the error between the reduced prediction and the reference high-fidelity (HF) solution. This method consists of a greedy enrichment of a current reduced order basis by the high-fidelity solution at the parametric value that maximizes the error bound on a rich sampling of the parametric space. Being intensively evaluated, the error bound must be computed in computational complexity independent of the number of degrees of freedom of the high-fidelity reference. Initially proposed for elliptic coercive partial differential equations [11], where the error bound is the dual norm of the residual divided by a lower bound of the stability constant, the method has been adapted to problems of increased difficulty, with the derivation of certified error bounds for the Boussinesq equation [12], the Burger's equation [13], and the Navier–Stokes equations [14]. Numerical stability of such error estimations with respect to round-off error can be an issue in nonlinear problems, which was investigated in [15–18].

Even if it is not a requirement for their execution, error estimation is a desired feature for all the other reduced order modeling methods. In proper generalized decomposition (PGD) methods [19], error estimation based on the constitutive relation error method is available [20–22]. In proper orthogonal decomposition (POD)-based reduced order modeling methods [23,24], error estimators have been developed for linear-quadratic optimal control problems [25], the approximation of mixte finite element problems [26], the optimal control of nonlinear parabolic partial differential equations [27], and for the reduction of magnetostatic problems [28] and Navier–Stokes equations [29]. To reduce nonlinear problems, the POD has been coupled with reduced integration strategies called hyperreduction, for which error estimates in constitutive relation have been proposed [30,31]. A priori sensitivity studies for POD approximations of quasi-nonlinear parabolic equations are also available [32].

The contribution of this work consists in the construction of a new error indicator, adapted to the model order reduction of nonlinear structural mechanics, where we are interested in the prediction of the dual quantities such as the cumulated plasticity or the stress tensor. These dual quantities need a reconstruction step to be represented on the complete structure of interest, usually done using a Gappy-POD algorithm based on the reduced solution. We illustrate that the ROM-Gappy-POD residual of the quantities of interest is highly correlated to the error in our cases. From this observation, we propose a calibration step, based on the data computed during the offline stage of the reduced order modeling, to construct an error indicator adapted to the considered problem and configuration. This error indicator is then used in enrichment strategies that improve the accuracy of the reduced order model prediction, when nonparametrized variations of the temperature field are considered in the online stage.

The problem of interest, the evolution of an elastoviscoplastic body under a time-dependent loading, in presented in Section 2. Then, the a posteriori reduced order modeling of this problem is detailed in Section 3. Section 4 presents the proposed error indicator, and the enrichment strategy based upon it. The performances of this error indicator and its ability to improve the quality of the reduced order model prediction via enrichment are illustrated in two numerical experiments involving elastoviscoplastic materials in Section 5. Finally, conclusions and prospects are given in Section 6.

2. High-Fidelity Elastoviscoplastic Model

We consider the model introduced in [7], which we briefly recall below for the sake of completeness. The structure of interest is noted Ω and its boundary $\partial\Omega$, where $\partial\Omega = \partial\Omega_D \cup \partial\Omega_N$ such that $\partial\Omega_D \cap \partial\Omega_N = \emptyset$, see Figure 2.

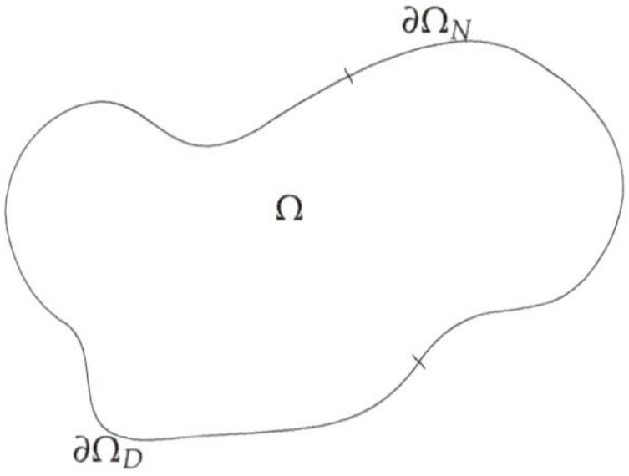

Figure 2. Schematics of the considered structure Ω.

Prescribed zero displacements are imposed on $\partial\Omega_D$, prescribed tractions T_N are imposed on $\partial\Omega_N$ and volumic forces are imposed to the structure Ω, in the form of a time-dependent loading. Assuming small deformations, the evolution of the structure Ω is governed by equations

$$\epsilon(u) = \frac{1}{2}\left(\nabla u + \nabla^T u\right) \quad \text{in } \Omega \times [0,\mathrm{T}] \quad \text{(compatibility)}, \tag{1a}$$
$$\mathrm{div}\,(\sigma) + f = 0 \quad \text{in } \Omega \times [0,\mathrm{T}] \quad \text{(equilibrium)}, \tag{1b}$$
$$\sigma = \sigma(\epsilon(u), y) \quad \text{in } \Omega \times [0,\mathrm{T}] \quad \text{(behavior law)}, \tag{1c}$$
$$u = 0 \quad \text{in } \partial\Omega_\mathrm{D} \times [0,\mathrm{T}] \quad \text{(prescribed zero displacement)}, \tag{1d}$$
$$\sigma \cdot n = T_N \quad \text{in } \partial\Omega_\mathrm{N} \times [0,\mathrm{T}] \quad \text{(prescribed traction)}, \tag{1e}$$
$$u = 0, y = 0 \quad \text{in } \Omega \text{ at } \mathrm{t} = 0 \quad \text{(initial condition)}, \tag{1f}$$

where σ is the Cauchy stress tensor, ϵ is the linear strain tensor, n is the exterior normal on $\partial\Omega$, y denotes the internal variables of the behavior law, and u is the displacement solution.

Consider $H^1_0(\Omega) = \{v \in L^2(\Omega) |\ \frac{\partial v}{\partial x_i} \in L^2(\Omega),\ 1 \leq i \leq 3 \text{ and } v|_{\partial\Omega_D} = 0\}$. We introduce a finite element basis $\{\varphi_i\}_{1\leq i\leq N}$, such that $\mathcal{V} := \mathrm{Span}\,(\varphi_i)_{1\leq i\leq N}$ is a conforming approximation of $\left[H^1_0(\Omega)\right]^3$.

In what follows, bold symbols are used to refer to vectors. Using the Galerkin method, problem (1a)–(1f) leads to a system of nonlinear equations, numerically solved using the following Newton algorithm:

$$\frac{D\mathcal{F}}{Du}\left(u^k\right)\left(\mathbf{u}^{k+1}-\mathbf{u}^k\right)=-\boldsymbol{\mathcal{F}}\left(u^k\right), \tag{2}$$

where $u^k \in \mathcal{V}$ is the k-th iteration of the discretized displacement field at the considered time-step and $\mathbf{u}^k = \left(u_i^k\right)_{1\leq i\leq N} \in \mathbb{R}^N$ is such that $u^k = \sum_{i=1}^{N} u_i^k \varphi_i$,

$$\frac{D\mathcal{F}}{Du}\left(u^k\right)_{ij}=\int_\Omega \epsilon\left(\varphi_j\right):\mathcal{K}\left(\epsilon(u^k),y\right):\epsilon\left(\varphi_i\right), 1\leq i,j\leq N, \tag{3}$$

where $\mathcal{K}\left(\epsilon(u^k),y\right)$ is the local tangent operator, and

$$\mathcal{F}_i\left(u^k\right)=\int_\Omega \sigma\left(\epsilon(u^k),y\right):\epsilon\left(\varphi_i\right)-\int_\Omega f\cdot\varphi_i-\int_{\partial\Omega_N} T_N\cdot\varphi_i, 1\leq i\leq N. \tag{4}$$

The Newton algorithm stops when the norm of the residual divided by the norm of the external forces vector is smaller than a user-provided tolerance, denoted $\epsilon^{\mathrm{HFM}}_{\mathrm{Newton}}$.

In Equation (2), f, T_N, u^k and y from Equation (4) are known quantities and contain the time-dependency of the solution. Notice that the computation of the functions $\left(u^k,y\right)\mapsto\sigma\left(\epsilon(u^k),y\right)$ and $\left(u^k,y\right)\mapsto\mathcal{K}\left(\epsilon(u^k),y\right)$ requires solving ordinary differential equations, whose complexity depends on the behavior law modeling the considered material.

In our application, the quantities of interest are not the displacement fields u, but rather the dual quantities stress tensor field σ and cumulated plasticity field, denoted p. The finite element software used to generate the high-fidelity solutions u is Zebulon, which contains a domain decomposition solver able to solve large scale problems, and the behavior laws are computed using Z-mat; both solvers belong to the Z-set suite [33].

3. Reduced Order Modeling

Reduced order modeling techniques are usually decomposed in two stages: the offline stage, where information from the high-fidelity model (HFM) is learned, and the online stage, where the reduced order model is constructed and exploited. In the offline stage, computationally demanding tasks occur, whereas the online stage is required to be efficient, in the sense that only operations in computational complexity independent of the number N of degrees of freedom of the high-fidelity model are allowed.

In what follows, we consider a posteriori reduced order modeling, which means that our reduced model involves an efficient Galerkin method no longer written in the finite element basis $(\varphi_i)_{1\leq i\leq N}$, but on a reduced order basis $(\psi_i)_{1\leq i\leq n}$, with $n \ll N$, adapted to the problem at hand. To generate this basis, the high-fidelity problem (1a)–(1f) is solved for given configurations. In the general case, the variations between the candidate configurations are quantified using a low-dimensional parametrization, leading to a parametrized reduced order model. In this work, we consider nonparametrized variations between the configurations of interest, which we call variability and denote μ. The variability contains the time step, as well as a nonparametrized description of the configuration, which in our case is the loading referred as a label. For instance, $\mu = \{t = 3, \text{"computation 1"}\}$, means that we consider the third time step of the configuration "computation 1", for which we have a description of the loading (center, axis and speed of rotation, temperature, and pressure fields in our applications). We denote by $\mathcal{P}_{\text{off.}}$ the set of variabilities encountered during the offline stage.

The reduced Newton algorithm reads

$$\frac{D\mathcal{F}_\mu}{Du}\left(\hat{u}_\mu^k\right)\left(\hat{\mathbf{u}}_\mu^{k+1}-\hat{\mathbf{u}}_\mu^k\right)=-\boldsymbol{\mathcal{F}}_\mu\left(\hat{u}_\mu^k\right), \tag{5}$$

where $\hat{u}_\mu^k \in \hat{\mathcal{V}} := \mathrm{Span}\left(\psi_i\right)_{1\leq i\leq n}$ is the k-th iteration of the reduced displacement field for the considered time-step and $\hat{\mathbf{u}}_\mu^k = \left(\hat{u}_{\mu,i}^k\right)_{1\leq i\leq n} \in \mathbb{R}^n$ is such $\hat{u}_\mu^k = \sum_{i=1}^{n}\hat{u}_{\mu,i}^k\psi_i$,

$$\frac{D\mathcal{F}_\mu}{Du}\left(\hat{u}_\mu^k\right)_{ij}=\int_\Omega \epsilon\left(\psi_j\right):\mathcal{K}\left(\epsilon(\hat{u}_\mu^k),y_\mu\right):\epsilon\left(\psi_i\right),1\leq i,j\leq n, \tag{6}$$

and

$$\mathcal{F}_{\mu,i}\left(\hat{u}^k\right)=\int_\Omega \sigma\left(\epsilon(\hat{u}_\mu^k),y_\mu\right):\epsilon\left(\psi_i\right)-\int_\Omega f_\mu\cdot\psi_i-\int_{\partial\Omega_N}T_{N,\mu}\cdot\psi_i,1\leq i\leq n. \tag{7}$$

The reduced Newton algorithm stops when the norm of the reduced residual divided by the norm of the reduced external forces vector is smaller than a user-provided tolerance, denoted $\epsilon_{\mathrm{Newton}}^{\mathrm{ROM}}$. In Equations (5)–(7), the online variability μ consists in the considered time step, the pressure field $T_{N,\mu}$, the centrifugal effects f_μ, and the temperature field in the internal variables y_μ.

Ensuring the efficiency of Equation (5) can be a complicated task, in particular for nonlinear problems, that requires methodologies recently proposed in the literature. For instance, the integrals in Equations (6) and (7) are computed in computational complexity dependent on N in the general case. We briefly present the choices made in our previous work [7]: the offline stage is composed of the following steps

- Data generation: this corresponds to the generation of the numerical approximation of the solutions to Equation (1a)–(1f), using the Newton algorithm (2). Multiple temporal solutions can be considered, for different loading conditions. The set of theses solutions $\{u_{\mu_i}\}_{1\leq i\leq N_c}$ is called the snapshots set.
- Data compression: this corresponds to the generation of the reduced order basis, usually obtained by looking for a hidden low-rank structure of the snapshots set. In this work, we consider the snapshot POD, see Algorithm 1 and [23,24].
- Operator compression: this step enables the efficient construction of (5), usually by replacing the computationally demanding integral evaluations by adapted approximation evaluated in computational complexity independent of N. In this work, we consider the empirical cubature method (ECM, see [34]), a method close to the energy conserving sampling and weighting (ECSW, see [35–37]) proposed earlier. Consider the vector of reduced internal forces appearing in (7):

$$\hat{F}_{\mu,i}^{\mathrm{int}} := \int_\Omega \sigma\left(\epsilon(\hat{u}_\mu),y_\mu\right)(x):\epsilon\left(\psi_i\right)(x)dx \approx \sum_{e\in E}\sum_{k=1}^{n_e}\omega_k\sigma\left(\epsilon(\hat{u}_\mu),y_\mu\right)(x_k):\epsilon\left(\psi_i\right)(x_k),1\leq i\leq n, \tag{8}$$

where the right-hand side is the high-fidelity quadrature formula used for numerical evaluation. In (8), the stress tensor $\sigma\left(\epsilon(\hat{u}_\mu),y_\mu\right)$ for the considered reduced solution $\hat{u}_\mu$ at variability μ and internal variables y_μ is seen as a function of space, and E denotes the set of elements of the mesh, n_e denotes the number of integration points for the element e, ω_k and x_k are the integration weights and points of the considered element. The ECM consists of replacing this high-fidelity quadrature (8) by an approximation adapted to the snapshots $\{u_{\mu_i}\}_{1\leq i\leq N_c}$ and the reduced order basis $\{\psi_i\}_{1\leq i\leq n}$, and involving a small number of integration points:

$$\hat{F}_{\mu,i}^{\mathrm{int}}(t) \approx \sum_{k'=1}^{d}\hat{\omega}_{k'}\sigma\left(\epsilon(\hat{u}_\mu),y_\mu\right)(\hat{x}_{k'}):\epsilon\left(\psi_i\right)(\hat{x}_{k'}),1\leq i\leq n, \tag{9}$$

where $d \ll \sum_{e\in E} n_e$, the reduced integration points $\hat{x}_{k'}$, $1\leq k'\leq d$, are taken among the integration points of the high-fidelity quadrature (8) and the reduced integration weights $\hat{\omega}_{k'}$ are positive.

We now briefly present how this reduced quadrature formula is obtained and we refer to [7,34] for more details. We denote $h_q := \sigma\left(\epsilon(u_{\mu_{(q//n)+1}}), y\right) : \epsilon\left(\psi_{(q\%n)+1}\right) \in L^2(\Omega)$, where $//$ and $\%$ are respectively the quotient and the remainder of the Euclidean division, $\mathcal{Z}$ is a subset of $[1; N_G]$ of size d, with N_G the number of integration points, and $J_{\mathcal{Z}} \in \mathbb{R}^{nN_c \times d}$ and $g \in \mathbb{N}^{nN_c}$ are such that for all $1 \leq q \leq nN_c$ and all $1 \leq k' \leq d$,

$$J_{\mathcal{Z}} = \left(h_q(x_{\mathcal{Z}_{k'}})\right)_{1\leq q\leq nN_c,\, 1\leq k'\leq d}, \qquad g = \left(\int_\Omega h_q\right)_{1\leq q\leq nN_c}, \tag{10}$$

where $\mathcal{Z}_{k'}$ denotes the k'-th element of $\mathcal{Z}$ and where we recall that n is the number of snapshot POD modes. Let $\hat{\omega} \in \mathbb{R}^{+d}$. From the introduced notation, $\left(J_{\mathcal{Z}}\hat{\omega}\right)_q = \sum_{k'=1}^{d} \hat{\omega}_{k'}\sigma\left(\epsilon(u_{\mu_{(q//n)+1}}), y\right)(x_{\mathcal{Z}_{k'}}) : \epsilon\left(\psi_{(q\%n)+1}\right)(x_{\mathcal{Z}_{k'}})$, $1 \leq q \leq nN_c$, which is a candidate approximation for $\int_\Omega \sigma\left(\epsilon(u_{\mu_{(q//n)+1}}), y\right) : \epsilon\left(\psi_{(q\%n)+1}\right) = g_q$, $1 \leq q \leq nN_c$. The best reduced quadrature formula of length d for the reduced internal forces vector is obtained as (c.f. [34], Equation (23))

$$(\hat{\omega}, \mathcal{Z}) = \arg \min_{\hat{\omega}'>0, \mathcal{Z}'\subset[1;N_G]} \left\|J_{\mathcal{Z}'}\hat{\omega}' - g\right\|_2, \tag{11}$$

where $\|\cdot\|_2$ stands for the Euclidean norm. Taking the length of the reduced quadrature formula in the objective function yields a NP-hard optimization problem, see ([35], Section 5.3), citing [38]. To produce a reduced quadrature formula in a controlled return time, we consider a nonnegative orthogonal matching pursuit algorithm, see ([39], Algorithm 1) and Algorithm 2 below, a variant of the matching pursuit algorithm [40] tailored to the nonnegative requirement.

A reduced quadrature is also used to accelerate the integral computation in (6). The remaining integral computations in (5) are $\int_\Omega f_\mu \cdot \psi_i$ and $\int_{\partial\Omega_N} T_{N,\mu} \cdot \psi_i$. They do not depend on the current solution, but only on the loading of the online variability μ, which is no longer efficient for nonparametrized variabilities. However, in our context of large scale nonlinear mechanics, these integrals are computed very fast with respect to the ones requiring behavior law resolutions, see Remark 1.

Algorithm 1: Data compression by snapshot proper orthogonal decomposition (POD).

Input: tolerance ϵ_{POD}, snapshots set $\{u_{\mu_i}\}_{1\leq i\leq N_c}$
Output: reduced order basis $\{\psi_i\}_{1\leq i\leq n}$

1 Compute the correlation matrix $C_{i,j} = \int_\Omega u_{\mu_i} \cdot u_{\mu_j}$, $1 \leq i,j \leq N_c$

2 Compute the n largest eigenvalues λ_i, $1 \leq i \leq n$, and associated orthonormal eigenvectors ξ_i, $1 \leq i \leq n$, of C such that $n = \max(n_1, n_2)$, where n_1 and n_2 are respectively the smallest integers such that $\sum_{i=1}^{n_1} \lambda_i \geq \left(1 - \epsilon_{\text{POD}}^2\right) \sum_{i=1}^{N_c} \lambda_i$ and $\lambda_{n_2} \leq \epsilon_{\text{POD}}^2 \lambda_0$

3 Compute the reduced order basis $\psi_i(x) = \frac{1}{\sqrt{\lambda_i N_c}} \sum_{j=1}^{N_c} u_{\mu_j}(x)\xi_{i,j}$, $1 \leq i \leq n$

Algorithm 2: Nonnegative orthogonal matching pursuit.

Input: J, b, tolerance $\epsilon_{\text{Op.comp.}}$

Output: $\hat{\omega}_k$, $\hat{x}_k$, $1 \leq k \leq d$

Initialization: $\mathcal{Z} = \emptyset$, $k' = 0$, $\hat{\omega} = 0$ and $r_0 = g$ **while** $\|r_{k'}\|_2 > \epsilon \|g\|_2$ **do**

  $\mathcal{Z} \leftarrow \mathcal{Z} \cup \max \text{index} \left(J^T_{[1:N_G]} r_{k'} \right)$

  $\hat{\omega} \leftarrow \underset{\hat{\omega}' > 0}{\arg\min} \|g - J_{\mathcal{Z}} \hat{\omega}'\|_2^2$

  $r_{k'+1} \leftarrow g - J_{\mathcal{Z}} \hat{\omega}$

  $k' \leftarrow k' + 1$

end

$d \leftarrow k'$

$\hat{x}_k := x_{\mathcal{Z}_k}$, $1 \leq k \leq d$

For the primal quantity displacement u, we can identify the solution of the reduced problem $\hat{\mathbf{u}}^k_\mu \in \mathbb{R}^n$ with the reconstruction on the complete domain Ω: $\hat{u}^k_\mu = \sum_{i=1}^{n} \hat{u}^k_{\mu,i} \psi_i$. For the dual quantities, such identification does not exist. However, the behavior law has already been evaluated at the integration point of the reduced quadrature $\hat{x}_k$, $1 \leq k \leq d$. Since the evaluations are computed during the resolution of the reduced problem, we denote them by hats. For instance for the cumulated plasticity, $\hat{\boldsymbol{p}}_\mu \in \mathbb{R}^d$ is such that $\hat{p}_{\mu,k}$ is computed by the online evaluation of the behavior law solver at the reduced integration points $\hat{x}_k$, $1 \leq k \leq d$. To recover the cumulated plasticity on the complete structure Ω, a ROM-Gappy-POD procedure is used to reconstruct the fields on the complete domain, see Algorithms 3 and 4 and [41] for the original presentation of the Gappy-POD. In step 2 of Algorithm 3, EIM denotes the empirical interpolation method [42,43] and the set of integration point whose indices have been selected is still denoted $\{\hat{x}_k\}_{1 \leq k \leq m^p}$, where $n^p \leq m^p \leq n^p + d$. The dual quantities predicted by the reduced order model and reconstructed on the complete structure are denoted with tildes, for instance $\tilde{p}_\mu$ for the cumulated plasticity.

The ROM-Gappy-POD reconstruction is well-posed, since the linear system considered in the online stage of Algorithm 4 is invertible, see ([7], Proposition 1).

An interesting feature of our framework is the ability for it to be used sequentially or in parallel with distributed memory. Independently of the high-fidelity solver, the solutions can be partitioned between some subdomains and the reduced order framework can treat the data in parallel. The MPI communications are limited to the computation of the scalar products in line 1 of Algorithm 1 for the offline stage, and the scalar products in (6) and (7) in the online stage. Furthermore, these scalar products are well adapted to parallel processing: each process computes its independently contribution on its respective subdomain, and the interprocess communication is limited to an all-to-all transfer of a scalar. All the remaining operations in our framework are treated in parallel with no communication, in particular in the operator compression step, reduced quadrature formulae are constructed independently. A natural use for the parallel framework is in coherence with domain decomposition solvers (potentially from commercial codes), which conveniently produce solutions partitioned in subdomains. Actually in our framework, the three steps of the offline stage (data generation, data compression and operator compression), the online stage, the post-treatment and the visualization are all treated in parallel with distributed memory, see [7] for more details.

Algorithm 3: Dual quantity reconstruction of the cumulated plasticity p: offline stage of the reduced order model (ROM)-Gappy-POD.

Input: tolerance $\epsilon_{\text{Gappy-POD}}$, cumulated plasticity snapshots set $\{p_{\mu_i}\}_{1\leq i\leq N_c}$, indices of the integration points of the reduced quadrature formula
Output: indices for online material law computation, ROM-Gappy-POD matrix
1 Apply the snapshot POD (Algorithm 1) on the high-fidelity snapshots $\{p_{\mu_i}\}_{1\leq i\leq N_c}$ to obtain the vectors ψ_i^p, $1 \leq i \leq n^p$, orthonormal with respect to the $L^2(\Omega)$-inner product
2 Apply the EIM to the collection of vectors ψ_i^p, $1 \leq i \leq n^p$, to select n^p distinct indices and complete (without repeat) this set of indices by the indices of the integration points of the reduced quadrature formula
3 Construct the matrix $M \in \mathbb{N}^{n^p\times n^p}$ such that $M_{i,j} = \sum_{k=1}^{m^p} \psi_i^p(\hat{x}_k)\psi_j^p(\hat{x}_k)$ (Gappy scalar product of the POD modes)

Algorithm 4: Dual quantity reconstruction of the cumulated plasticity p: online stage of the ROM-Gappy-POD.

Input: online variability μ, indices for online material law computation, ROM-Gappy-POD matrix
Output: reconstructed value for p on the complete domain Ω
1 Construct $\boldsymbol{b}_\mu \in \mathbb{R}^{n^p}$, where $b_{\mu,i} = \sum_{k=1}^{m^p} \psi_i^p(\hat{x}_k)\hat{p}_{\mu,k}$, and $\hat{\boldsymbol{p}}_\mu \in \mathbb{R}^{m^p}$ is such that $\hat{p}_{\mu,k}$ is the online prediction of p at variability μ and integration point $\hat{x}_k$ (from the online evaluation of the behavior law solver)
2 Solve the (small) linear system: $M\boldsymbol{z}_\mu = \boldsymbol{b}_\mu$
3 Compute the reconstructed value for p on the complete subdomain Ω as $\tilde{p}_\mu := \sum_{i=1}^{n^p} z_{\mu,i}\psi_i^p$

4. A Heuristic Error Indicator

We look for an efficient error indicator in this context of general nonlinearities and nonparametrized variabilities. In model order reduction techniques, error estimation is an important feature, that becomes interesting under the condition that it can be computed in complexity independent of the number of degrees of freedom N of the high-fidelity model.

4.1. First Results on Errors and Residuals

We recall some notations introduced so far: bold symbols refer to vectors ($\boldsymbol{p}_\mu$ is the vector of components the value of the HF cumulated plasiticity field at reduced integration points), hats refer to quantities computed by the reduced order model ($\hat{u}_\mu$ is the reduced displacement and $\hat{\boldsymbol{p}}_\mu$ is the vector of components the value of the reduced cumulated plasticity at the reduced quadrature points), whereas tildes refer to dual quantities reconstructed by Gappy-POD (for instance $\tilde{p}$). Bold and tilde symbols, for instance $\tilde{\boldsymbol{p}}_\mu$, refer to the vectors of components the reconstructed dual quantities on the reduced integration points: $\tilde{p}_{\mu,k} = \tilde{p}_\mu(\hat{x}_k)$, $1 \leq k \leq m^p$. Notice that in the general case, $\tilde{\boldsymbol{p}}_\mu \neq \hat{\boldsymbol{p}}_\mu$: this discrepancy is at the base of our proposed error indicator. A table of notations is provided at the end of the document.

A quantification for the prediction relative error is defined as

$$E_\mu^p := \begin{cases} \frac{\|p_\mu-\tilde{p}_\mu\|_{L^2(\Omega)}}{\|p_\mu\|_{L^2(\Omega)}} & \text{if} \|p_\mu\|_{L^2(\Omega)} \neq 0 \\ \frac{\|p_\mu-\tilde{p}_\mu\|_{L^2(\Omega)}}{\max\limits_{\mu\in\mathcal{P}_{\text{off.}}} \|p_\mu\|_{L^2(\Omega)}} & \text{otherwise,} \end{cases} \quad (12)$$

where we recall that p_μ and $\tilde{p}_\mu$ are respectively the high-fidelity and reduced predictions for the cumulated plasticity field at the variability μ, and $\mathcal{P}_{\text{off.}}$ is the set of variabilities encountered during the offline stage.

Define the ROM-Gappy-POD residual as

$$\mathcal{E}^p_\mu := \begin{cases} \frac{\|\tilde{\boldsymbol{p}}_\mu - \hat{\boldsymbol{p}}_\mu\|_2}{\|\hat{\boldsymbol{p}}_\mu\|_2} & \text{if} \|\hat{\boldsymbol{p}}_\mu\|_2 \neq 0 \\ \frac{\|\tilde{\boldsymbol{p}}_\mu - \hat{\boldsymbol{p}}_\mu\|_2}{\max\limits_{\mu \in \mathcal{P}_{\text{off.}}} \|\hat{\boldsymbol{p}}_\mu\|_2} & \text{otherwise}, \end{cases}, \tag{13}$$

where $\|\cdot\|_2$ denotes the Euclidean norm. Notice that the relative error E^p_μ involves fields and L^2-norms whereas the ROM-Gappy-POD residual $\mathcal{E}^p_\mu$ involves vectors of dual quantities in the set of reduced integration points and Euclidean norms. In (13), $\|\tilde{\boldsymbol{p}}_\mu - \hat{\boldsymbol{p}}_\mu\|_2$ is the error between the online evaluation of the cumulated plasticity by the behavior law solver: $\hat{\boldsymbol{p}}_\mu$, and the reconstructed prediction at the reduced integration points $\hat{x}_k$: $\tilde{\boldsymbol{p}}_\mu$, $1 \leq k \leq m^p$. Let $B \in \mathbb{R}^{m^p \times n^p}$ such that $B_{k,i} = \psi^p_i(\hat{x}_k)$, $1 \leq k \leq m^p$, $1 \leq i \leq n^p$; by definition, $\tilde{p}_{\mu,k} = \sum_{i=1}^{n^p} z_{\mu,i}\psi^p_i(\hat{x}_k) = (B\boldsymbol{z}_\mu)_k$, $1 \leq k \leq m^p$. From Algorithm 3, $M = B^T B$ and from Algorithm 4, $\boldsymbol{b}_\mu = B^T \hat{\boldsymbol{p}}_\mu$, so that $\boldsymbol{z}_\mu = \left(B^T B\right)^{-1} B^T \hat{\boldsymbol{p}}_\mu$, which is the solution of the following unconstrained least-square optimization: $\boldsymbol{z}_\mu := \underset{\boldsymbol{z}' \in \mathbb{R}^n}{\arg\min} \|B\boldsymbol{z}' - \hat{\boldsymbol{p}}_\mu\|_2^2$. Hence, in (13), $\|\tilde{\boldsymbol{p}}_\mu - \hat{\boldsymbol{p}}_\mu\|_2$ is the norm of the residual of the considered least-square optimization.

Suppose $K := \{p_\mu$, for all possible variabilities $\mu\}$ is a compact subset of $L^2(\Omega)$ and define the Kolmogorov n-width by $d_n(K)_{L^2(\Omega)} := \inf\limits_{\dim(W)=n} d(K,W)_{L^2(\Omega)}$, where $d(K,W)_{L^2(\Omega)} := \sup\limits_{v \in K} \inf\limits_{w \in W} \|v - w\|_{L^2(\Omega)}$, with W a finite-dimensional subspace of $L^2(\Omega)$. The Kolmogorov n-width is an object from approximation theory; a presentation and discussion in a reduced order modeling context can be found in [44]. Denote also $\boldsymbol{\Pi}_\mu := \left(\left(p_\mu, \psi^p_i\right)_{L^2(\Omega)}\right)_{1 \leq i \leq n^p} \in \mathbb{R}^{n^p}$, where we recall that $\left\{\psi^p_i\right\}_{1 \leq i \leq n^p}$ are the Gappy-POD modes obtained by Algorithm 3 and where $(\cdot,\cdot)_{L^2(\Omega)}$ denotes the $L^2(\Omega)$ inner-product. All the dual quantities being computed by the high-fidelity solver at the N_G integration points, they have finite values at these points. Unlike the primal displacement field, the dual quantities are not directly expressed in a finite element basis, but through their values on the integration points. For pratical manipulations, we express the dual quantity fields as a constant on each polyhedron obtained as a Voronoi diagram in each element of the mesh, with seeds the integration points; the constants corresponding to the value of the dual quantity on the corresponding integration point.

We first control the numerator in the relative error E^p_μ with respect to the numerator in the ROM-Gappy-POD residual $\mathcal{E}^p_\mu$ in Proposition 1.

Proposition 1. *There exist two positive constants C_1 and C_2 independent of μ (but dependent on n^p) such that*

$$\|p_\mu - \tilde{p}_\mu\|^2_{L^2(\Omega)} \leq C_1\|B\boldsymbol{z}_\mu - \hat{\boldsymbol{p}}_\mu\|_2^2 + C_1\|\boldsymbol{p}_\mu - \hat{\boldsymbol{p}}_\mu\|_2^2 + C_2 d(K, \text{Span}\{\psi^p_i\}_{1 \leq i \leq n^p})^2_{L^2(\Omega)}. \tag{14}$$

Proof. There holds

$$\begin{aligned}
\|p_\mu - \tilde{p}_\mu\|^2_{L^2(\Omega)} &\le 2\left\|\sum_{i=1}^{n^p}\left((p_\mu,\psi_i^p)_{L^2(\Omega)} - z_{\mu,i}\right)\psi_i^p\right\|^2_{L^2(\Omega)} + 2\left\|p_\mu - \sum_{i=1}^{n^p}(p_\mu,\psi_i^p)_{L^2(\Omega)}\psi_i^p\right\|^2_{L^2(\Omega)} && \text{(15a)}\\
&= 2\sum_{i=1}^{n^p}\left((p_\mu,\psi_i^p)_{L^2(\Omega)} - z_{\mu,i}\right)^2 + 2\inf_{w\in\mathrm{Span}\{\psi_i^p\}_{1\le i\le n^p}}\|p_\mu - w\|^2_{L^2(\Omega)} && \text{(15b)}\\
&\le 2\sum_{i=1}^{n^p}\left(\Pi_{\mu,i} - z_{\mu,i}\right)^2 + 2\sup_{v\in K}\inf_{w\in\mathrm{Span}\{\psi_i^p\}_{1\le i\le n^p}}\|v - w\|^2_{L^2(\Omega)} && \text{(15c)}\\
&= 2\left\|M^{-1}M\left(\boldsymbol{\Pi}_\mu - \boldsymbol{z}_\mu\right)\right\|_2^2 + 2d(K,\mathrm{Span}\{\psi_i^p\}_{1\le i\le n^p})^2_{L^2(\Omega)} && \text{(15d)}\\
&= 2\left\|M^{-1}B^T\left(B\boldsymbol{\Pi}_\mu - \boldsymbol{p}_\mu + \boldsymbol{p}_\mu - \hat{\boldsymbol{p}}_\mu + \hat{\boldsymbol{p}}_\mu - B\boldsymbol{z}_\mu\right)\right\|_2^2 + 2d(K,\mathrm{Span}\{\psi_i^p\}_{1\le i\le n^p})^2_{L^2(\Omega)} && \text{(15e)}\\
&\le 6\left\|M^{-1}B^T\right\|_2^2\left(\|B\boldsymbol{\Pi}_\mu - \boldsymbol{p}_\mu\|_2^2 + \|\boldsymbol{p}_\mu - \hat{\boldsymbol{p}}_\mu\|_2^2 + \|B\boldsymbol{z}_\mu - \hat{\boldsymbol{p}}_\mu\|_2^2\right) + 2d(K,\mathrm{Span}\{\psi_i^p\}_{1\le i\le n^p})^2_{L^2(\Omega)} && \text{(15f)}\\
&\le C_1\|B\boldsymbol{z}_\mu - \hat{\boldsymbol{p}}_\mu\|_2^2 + C_1\|\boldsymbol{p}_\mu - \hat{\boldsymbol{p}}_\mu\|_2^2 + C_2 d(K,\mathrm{Span}\{\psi_i^p\}_{1\le i\le n^p})^2_{L^2(\Omega)}, && \text{(15g)}
\end{aligned}$$

where the triangular inequality and the Jensen inequality on the square function have been applied in (15a), and between (15e) and (15f). In (15g), the term $\|B\boldsymbol{\Pi}_\mu - \boldsymbol{p}_\mu\|_2^2$ has been incorporated in the term $C_2 d(K,\mathrm{Span}\{\psi_i^p\}_{1\le i\le n^p})^2_{L^2(\Omega)}$. This can be done since

$$\begin{aligned}
\|B\boldsymbol{\Pi}_\mu - \boldsymbol{p}_\mu\|_2^2 &= \sum_{k=1}^{m^p}\left(p_\mu(\hat{x}_k) - \sum_{i=1}^{n^p}\left(p_\mu,\psi_i^p\right)_{L^2(\Omega)}\psi_i^p(\hat{x}_k)\right)^2\\
&\le \frac{1}{\min_{1\le k'\le m^p}\nu_{k'}}\sum_{k=1}^{N_g}\nu_k\left(p_\mu(x_k) - \sum_{i=1}^{n^p}\left(p_\mu,\psi_i^p\right)_{L^2(\Omega)}\psi_i^p(x_k)\right)^2\\
&= \frac{1}{\min_{1\le k'\le m^p}\nu_{k'}}\int_\Omega\left(p_\mu(x) - \sum_{i=1}^{n^p}\left(p_\mu,\psi_i^p\right)_{L^2(\Omega)}\psi_i^p(x)\right)^2 dx\\
&\le \frac{1}{\min_{1\le k'\le m^p}\nu_{k'}} d(K,\mathrm{Span}\{\psi_i^p\}_{1\le i\le n^p})^2_{L^2(\Omega)},
\end{aligned} \tag{16}$$

where ν_k denotes the volume of the cell of the Voronoi diagram associated with integration point $\hat{x}_k$. □

We now control the numerator in the ROM-Gappy-POD residual $\mathcal{E}_\mu^p$ with respect to the numerator in the relative error E_μ^p in Proposition 1, leading to Corollary 1, which provides a sense a consistency: without any error in the reduced prediction, the ROM-Gappy-POD residual $\mathcal{E}_\mu^p$ is zero.

Proposition 2. *There exist two positive constants K_1 and K_2 independent of μ such that*

$$\|\tilde{\boldsymbol{p}}_\mu - \hat{\boldsymbol{p}}_\mu\|_2^2 \le K_1\left\|p_\mu - \tilde{p}_\mu\right\|^2_{L^2(\Omega)} + K_2\|\boldsymbol{p}_\mu - \hat{\boldsymbol{p}}_\mu\|_2^2. \tag{17}$$

Proof. There holds

$$\begin{aligned}
\|\tilde{\boldsymbol{p}}_\mu - \hat{\boldsymbol{p}}_\mu\|_2^2 &\le 2\|B\boldsymbol{z}_\mu - \boldsymbol{p}_\mu\|_2^2 + 2\|\boldsymbol{p}_\mu - \hat{\boldsymbol{p}}_\mu\|_2^2\\
&\le \frac{2}{\min_{1\le k'\le m^p}\nu_{k'}}\sum_{k=1}^{m^p}\nu_k\left(p_\mu(\hat{x}_k) - \sum_{i=1}^{n^p} z_{\mu,i}\psi_i^p(\hat{x}_k)\right)^2 + 2\|\boldsymbol{p}_\mu - \hat{\boldsymbol{p}}_\mu\|_2^2\\
&\le \frac{2}{\min_{1\le k'\le m^p}\nu_{k'}}\int_\Omega\left(p_\mu(x) - \sum_{i=1}^{n^p} z_{\mu,i}\psi_i^p(x)\right)^2 dx + 2\|\boldsymbol{p}_\mu - \hat{\boldsymbol{p}}_\mu\|_2^2\\
&= K_1\left\|p_\mu - \tilde{p}_\mu\right\|^2_{L^2(\Omega)} + K_2\|\boldsymbol{p}_\mu - \hat{\boldsymbol{p}}_\mu\|_2^2.
\end{aligned} \tag{18}$$

□

Corollary 1. *Suppose that the reduced solution is exact up to the considered time step at the online variability* μ*:* $p_\mu = \tilde{p}_\mu$ *in* $L^2(\Omega)$*. In particular, the behavior law solver has been evaluated with the exact strain tensor and state variables at the integration points* x_k*, leading to* $\hat{p}_\mu(\hat{x}_k) = p_\mu(\hat{x}_k)$*,* $1 \leq k \leq m^d$*. From Proposition 2,* $\|\tilde{p}_\mu - \hat{p}_\mu\|_2 = 0$*, and* $\mathcal{E}_\mu^p = 0$*.*

4.2. A Calibrated Error Indicator

As we will illustrate in Section 5, the evaluations of the ROM-Gappy-POD residual $\mathcal{E}_\mu^p$ (13) and the error E_μ^p (12) are very correlated in our numerical simulations. Our idea is to exploit this correlation by training a Gaussian process regressor for the function $\mathcal{E}_\mu^p \mapsto E_\mu^p$. At the end of the offline stage, we propose to compute reduced predictions at variability values $\{\mu_i\}_{1\leq i\leq N_c}$ encountered during the data generation step, and the corresponding couples $\left(E_{\mu_i}^p, \mathcal{E}_{\mu_i}^p\right)$, $1 \leq i \leq N_c$. A Gaussian process regressor is trained on these values and we define an approximation function

$$\mathcal{E}_\mu^p \mapsto \mathrm{Gpr}^p(\mathcal{E}_\mu^p) \tag{19}$$

for the error E_μ^p at variability μ as the mean plus three times the standard deviation of the predictive distribution at the query point $\mathcal{E}_\mu^p$. This is our proposed error indicator. If the dispersion around the learning data is small for certain values $\mathcal{E}_\mu^p$, then adding three times the standard deviation will not change very much the prediction, whereas for values with large dispersions of the learning data, this correction aims to provide an error indicator larger than the error. We used the GaussianProcessRegressor python class from scikit-learn [45]. Notice that although some operations in computational complexity dependent on N are carried-out, we are still in the offline stage, and they are much faster than the resolutions of the large systems of nonlinear Equations (2). If the offline stage is correctly carried-out and since $\mathcal{E}_\mu^p$ is highly correlated with the error, only small values for $\mathcal{E}_\mu^p$ are expected to be computed. Hence, in order to train the Gaussian process regressor correctly for larger values of the error, the reduced Newton algorithm (5) is solved with a large tolerance $\epsilon_{\mathrm{Newton}}^{\mathrm{ROM}} = 0.1$. We call these operations "calibration of the error indication", see Algorithm 5 for a description and Figure 3 for a presentation of the workflow featuring this calibration step.

Algorithm 5: Calibration of the error indicator.

Input: outputs of the data generation, data compression and operator compression steps of Section 3

Output: Approximation function $\mathcal{E}_\mu^p \mapsto \mathrm{Gpr}^p(\mathcal{E}_\mu^p)$ of the error E_μ^p

Initialization: $\mathcal{X} = \emptyset$

for $i \leftarrow 1$ **to** N_c **do**

  Construct and solve the reduced problem (5) with $\epsilon_{\mathrm{Newton}}^{\mathrm{ROM}} = 0.1$

  Compute the reconstructed plasticity $\tilde{p}_{\mu_i}$ using Algorithm 4 and $\mathcal{E}_{\mu_i}^p$ using (13)

  Compute the error $E_{\mu_i}^p$ using (12)

  $\mathcal{X} \leftarrow \mathcal{X} \cup \left(\mathcal{E}_{\mu_i}^p, E_{\mu_i}^p\right)$

end

Construct an approximation function $\mathcal{E}_\mu^p \mapsto \mathrm{Gpr}^p(\mathcal{E}_\mu^p)$ of the error E_μ^p using a Gaussian process regression and the data from $\mathcal{X}$

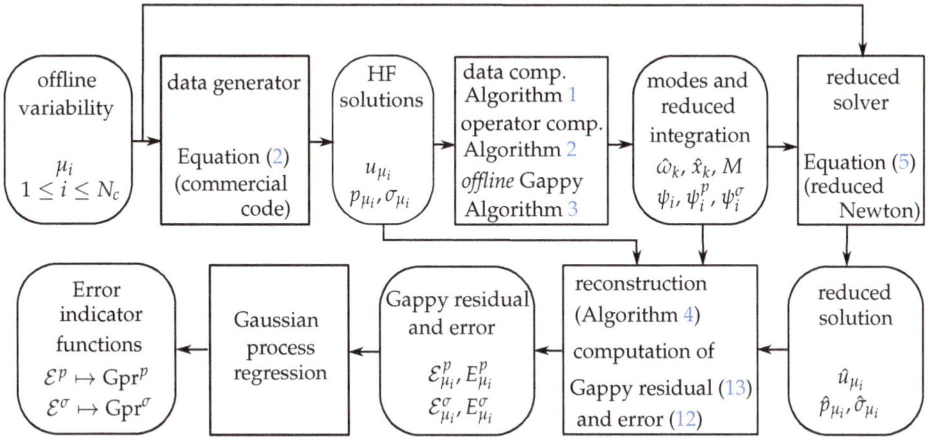

Figure 3. Workflow for the offline stage with error indicator calibration.

We recall that in model order reduction, the original hypothesis is the existence of a low-dimensional vector space where an acceptable approximation of the high-fidelity solution lies. The hypothesis is formalized under a rate of decrease for the Kolmogorov n-width with respect to the dimension of this vector space. The same hypothesis is made when using the Gappy-POD to reconstruct the dual quantities, which are expressed as a linear combination of constructed modes. For both the primal and dual quantities, the modes are computed by searching some low-rank structure of the high-fidelity data. The coefficients of the linear combination for reconstructing the primal quantities are given by the solution of the reduced Newton algorithm (5). After convergence, the residual is small, even in cases where the reduced order model exhibits large errors with respect to the high-fidelity reference: this residual gives no information on the distance between the reduced solution and the high-fidelty finite element space. However, in the online phase of the ROM-Gappy-POD reconstruction in Algorithm 4, the coefficients $\hat{p}_{\mu,k}$ contain information from the high-fidelity behavior law solver. Moreover, an overdetermined least-square is solved, which can provide a nonzero residual that implicitly contains this information from the high-fidelity behavior law solver. Namely the distance between the prediction from the behavior law and the vector space spanned by the Gappy-POD modes (restricted to the reduced integration points): this is the term $\|Bz_\mu - \hat{\boldsymbol{p}}_\mu\|_2$ in (14). Hence, the ability of the online variability to be expressed on the Gappy-POD modes is monitored through the behavior law solver on the reduced integration points. When the ROM is solved for an online variability not included in the offline variabilities, then the new physical solution cannot be correctly interpolated using the POD and Gappy-POD modes. Hence, the ROM-Gappy-residual becomes large. From Proposition 2, if $\|Bz_\mu - \hat{\boldsymbol{p}}_\mu\|_2 = \|\tilde{\boldsymbol{p}}_\mu - \hat{\boldsymbol{p}}_\mu\|_2$ is large, then the global error $\left\|p_\mu - \tilde{p}_\mu\right\|_{L^2(\Omega)}$ and/or the error at the reduced integration points $\hat{x}_k$ is large, which makes $\|Bz_\mu - \hat{\boldsymbol{p}}_\mu\|_2$ a good candidate for a enrichement criterion for the ROM. A limitation of the error indicator can occur if the online variability activates strong nonlinearities on areas containing no point from the reduced integration scheme, namely through the term $C_2 d(K, \mathrm{Span}\{\psi_i^p\}_{1\leq i\leq n^p})^2_{L^2(\Omega)}$ in (14).

We recall that the error indicator (19) is a regression of the function $\mathcal{E}^p_\mu \mapsto E^p_\mu$. In the online phase, we only need to evaluate $\mathcal{E}^p_\mu$ and do not require any estimation for the other terms and constants appearing in Propositions 1 and 2.

Equipped with an efficient error indicator, we are now able to assess the quality of the approximation made by the reduced order model in the online phase. If the error indicator is too large, an enrichment step occurs: the high-fidelity model is used to compute a new high-fidelity snapshot, which is used to update the POD and Gappy-POD basis, as well as the reduced integration schemes. Notice that for the enrichment steps to be computed, the displacement field and all the state variables of the previous time step need to be reconstructed on the complete mesh Ω to provide the high-fidelity solver with the correct material state. The workflow for the online stage with enrichment is presented in Figure 4.

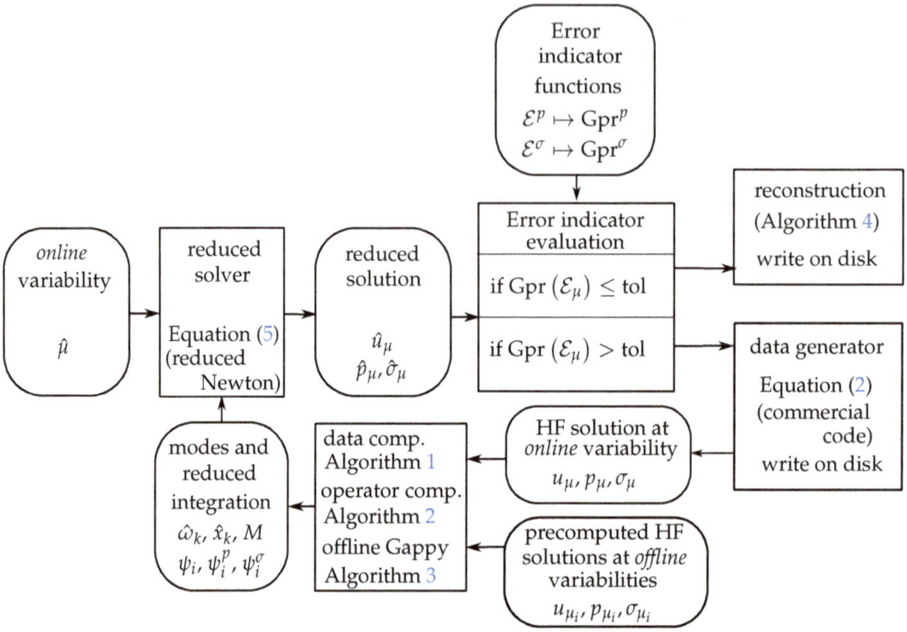

Figure 4. Workflow for the online stage with enrichment.

Remark 1 (Online efficiency). *The computation of the ROM-Gappy-POD residual (13) is efficient, since $\tilde{\boldsymbol{p}}_\mu$ and $\hat{\boldsymbol{p}}_\mu$ are already computed for the reconstruction, and m^p depending only on the approximation of $\sigma : \epsilon$ and p, it is independent of N. The evaluation of $\mathrm{Gpr}^p(\mathcal{E}_\mu^p)$ is also in computational complexity independent of N.*

If the enrichment is activated during the online phase, a high-fidelity solution is computed, which is a computationally demanding task. This is the price to add high-fidelity information in the exploitation phase. We will see in Section 5 that without this enrichment in our applications, the considered online variability on the temperature field strongly degrades the accuracy of the reduced order model prediction. The nonparametrized variability also induces online pretreatments in computational complexity depending on N, namely the precomputation of $\int_\Omega f_\mu \cdot \psi_i$ and $\int_{\partial\Omega_N} T_{N,\mu} \cdot \psi_i$ in (7), which is in practice much faster than other integrals that require behavior law resolutions.

Notice that the online stage can be further optimized by replacing the data compression and offline Gappy-POD steps by incremental variants, such as the incremental POD [46]. For the operator compression, the Nonnegative Orthogonal Matching Pursuit described in Algorithm 2 is not restarted from zero, but initialized by the current reduced quadrature scheme. Notice also that for the moment, the reduced order model is enriched using a complete precomputed reference high-fidelity computation, so that no speedup is obtained in practice. We still need to consider restart strategies to call the high-fidelity solver only at the time step of enrichment, from a complete mechanical state reconstructed from the prediction of the reduced order model at the previous time step, which will be the subject of future work.

When the framework is used in parallel, with subdomains, the calibration of the error indicator is local to each subdomain, so that the decision of enrichment in the full domain during the online stage can be triggered by a particular subdomain of interest.

5. Numerical Applications

We consider two behavior laws in the numerical applications:

(elas) Isotropic thermal expansion and temperature-dependent cubic elasticity: the behavior law is $\sigma = \mathcal{A} : \left(\epsilon - \epsilon^{\text{th}}\right)$, where $\epsilon^{\text{th}} = \alpha^{\text{th}}\left(T - T_0\right) I$, with I the second-order identity tensor and α^{th} the thermal expansion coefficient in MPa.K^{-1} depending on the temperature. The elastic stiffness tensor $\mathcal{A}$ does not depend on the solution u and is defined in Voigt notations by

$$\mathcal{A} = \begin{pmatrix} y_{1111} & y_{1122} & y_{1122} & 0 & 0 & 0 \\ y_{1122} & y_{1111} & y_{1122} & 0 & 0 & 0 \\ y_{1122} & y_{1122} & y_{1111} & 0 & 0 & 0 \\ 0 & 0 & 0 & y_{1212} & 0 & 0 \\ 0 & 0 & 0 & 0 & y_{1212} & 0 \\ 0 & 0 & 0 & 0 & 0 & y_{1212} \end{pmatrix}, \tag{20}$$

where the temperature T is given by the thermal loading, $T_0 = 20$ °C is a reference temperature and the coefficients y_{1111}, y_{1122} and y_{1212} (elastic coefficients in MPa) depend on the temperature. This law does not feature any internal variable to compute.

(evp) Norton flow with nonlinear kinematic hardening: the elastic part is given by $\sigma = \mathcal{A} : \left(\epsilon - \epsilon^{\text{th}} - \epsilon^P\right)$, where $\mathcal{A}$ and ϵ^{th} are the same as the (elas) law, ϵ^P is the plastic strain tensor. The viscoplastic part requires solving the system of ODEs:

$$\begin{cases} \dot{\epsilon}^P = \dot{p}\sqrt{\dfrac{3}{2}}\dfrac{s - \frac{2}{3}C\alpha}{\sqrt{\left(s - \frac{2}{3}C\alpha\right) : \left(s - \frac{2}{3}C\alpha\right)}}, \\ \dot{\alpha} = \dot{\epsilon}^P - \dot{p}D\alpha, \\ \dot{p} = \left\langle \dfrac{f_r}{K} \right\rangle^m, \end{cases} \tag{21}$$

where p is the cumulated plasticity, $f_r = \sqrt{\frac{3}{2}}\sqrt{\left(s - \frac{2}{3}C\alpha\right) : \left(s - \frac{2}{3}C\alpha\right)} - R_0$ defines the yield surface, α (dimensionless) is the internal variable associated to the back-stress tensor $X = \frac{2}{3}C\alpha$ representing the center of the elastic domain in the stress space, $s := \sigma - \frac{1}{3}\text{Tr}(\sigma)I$ (with Tr the trace operator) is the deviatoric component of the stress tensor, and $\langle\cdot\rangle$ denotes the positive part operator. The yield criterion is $f_r \leq 0$. The hardening material coefficients C (in MPa) and D (dimensionless), the Norton material coefficient K (in MPa.s$^{\frac{1}{m}}$), the Norton exponential material coefficient m (dimensionless), and the initial yield stress R_0 (in MPa) depend on the temperature. The internal variables considered here are ϵ^P, α and p, and the ODE's initial conditions are $\epsilon^P = 0$, $\alpha = 0$ and $p = 0$ at $t = 0$.

Two test cases are considered: an academic one in Section 5.1 and a high-pressure turbine blade setting of industrial complexity in Section 5.2.

5.1. Academic Example

We consider a simple geometry in the shape of a bow tie, to enforce plastic effects on the tightest area, see Figure 5. The structure is subjected to different variabilities of the loading (temperature, rotation, pressure), described in Figures 5–7. The axis of rotation is located on the left of the object along the x-axis, and the pressure field is represented in Figure 5. The rotation of the object is not computed: only the inertia effects are modeled in the volumic force term f in (1b). Four temperature fields are considered, two of them are represented in Figure 6 ("temperature_field_1" is a uniform 20 °C field, "temperature_field_2" is a 3D Gaussian with a maximum in the thin part of the object, close to an edge, "temperature_field_3" is proportional to "temperature_field_2", "temperature_field_4" obtained from "temperature_field_2" by random perturbation of 10% magnitude independently at each point). Notice that the irregularity of "temperature_field_4" will lead to small scaled structures in the cumulated plasticity and stress fields involving this variability. Notice also that the temperature field are not computed during the simulation: they are loading data for the mechanical computation. Figure 7 presents the three variabilities considered: computation 1 and computation 2 encountered in the offline phase, and new encountered in the online phase. The pressure loading is obtained by multiplying the pressure coefficient by the pressure field represented in Figure 5 (normals on the boundary are directed towards the exterior) and at each time step, the temperature field is obtained by linear interpolation between the previous and following fields in the temporal sequence. Notice that computation 1 and computation 2 are not defined on the same temporal range.

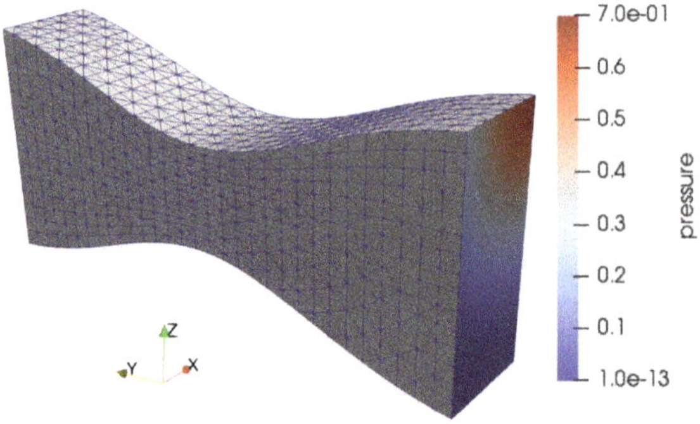

Figure 5. Academic test case: mesh and pressure field represented on its surface of application; the axis of rotation is located on the left of the object along the x-axis.

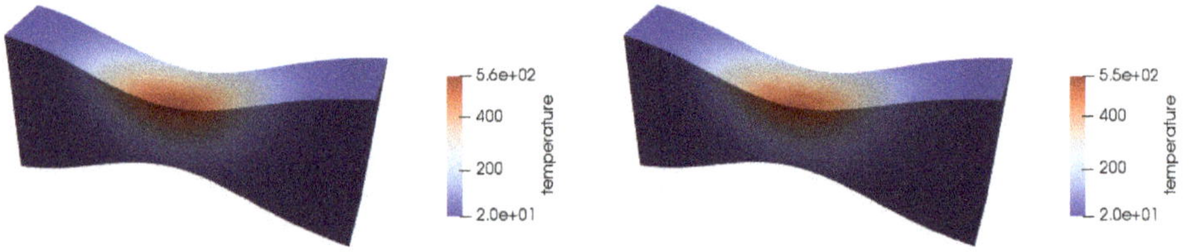

"temperature_field_2" "temperature_field_4"

Figure 6. Two different variabilities for the temperature loading (in °C) used in the academic test case.

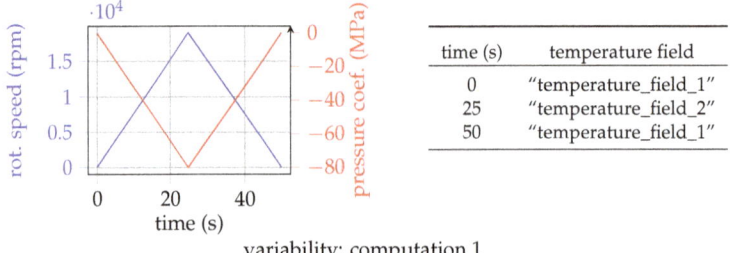

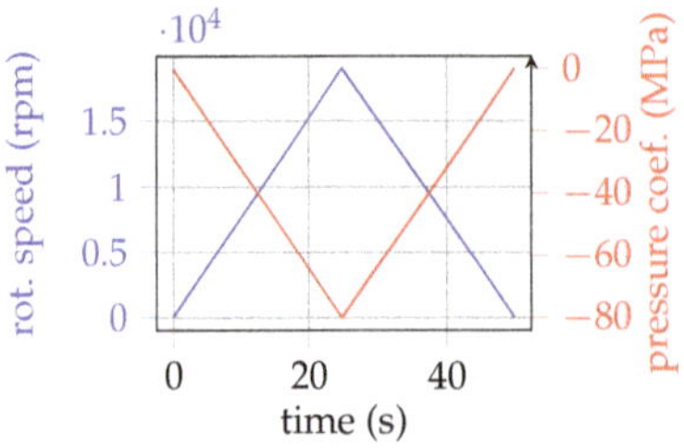

time (s)	temperature field
0	"temperature_field_1"
25	"temperature_field_2"
50	"temperature_field_1"

variability: computation 1

Figure 7. *Cont.*

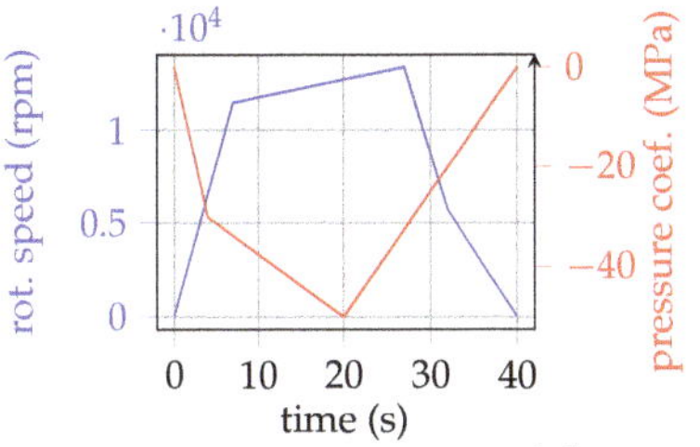

time (s)	temperature field
0	"temperature_field_1"
8	"temperature_field_2"
20	"temperature_field_3"
25	"temperature_field_2"
40	"temperature_field_1"

variability: computation 2

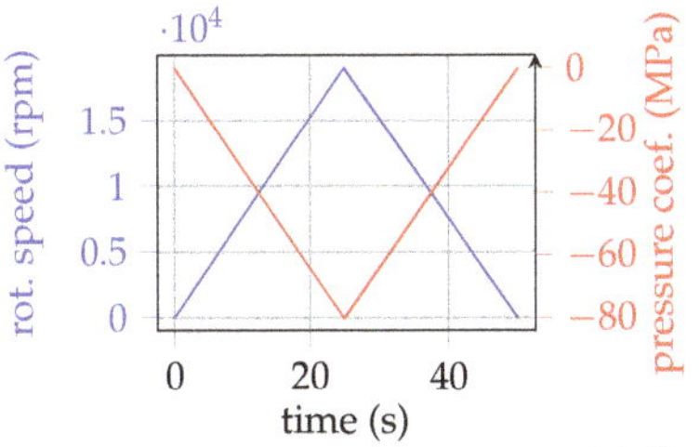

time (s)	temperature field
0	"temperature_field_1"
20	"temperature_field_2"
25	"temperature_field_4"
30	"temperature_field_2"
50	"temperature_field_1"

variability: new

Figure 7. Considered loading variabilities for the academic test case. (**left**) Rotation speed (—) and pressure coefficient (—) with respect to time. (**right**) Temporal sequence for the temperature field.

The characteristics for the academic test cases are given in Table 1.

Table 1. Characteristics for the academic test case.

number of dofs	78,120
number of (quadratic) tetrahedra	16,695
number of integration points	81,375
number of time steps	computation 1: 50, computation 2: 40, new: 50
behavior law	evp (Norton flow with nonlinear kinematic hardening)

The correlations between the ROM-Gappy-POD residual $\mathcal{E}$ (13) and the error E (12) on the dual quantities cumulated plasticity p and first component of the stress tensor σ_{11} are investigated in Table 2. The reduced solutions used for $\mathcal{E}$ correspond to the calibration step in the offline stage, in the second row of Figure 3, where we recall that the reduced Newton algorithm (5) is computed with a large tolerance $\epsilon_{\text{Newton}}^{\text{ROM}} = 0.1$ on the variabilities encountered in the data generation step. For the cumulated plasticity field, the values before the first plastic effects are neglected. A strong correlation appears in all the considered cases, although outliers are observed for the last time steps, where the building of residual stresses at low loadings are more difficult to predict with the ROM.

Table 2. Illustration of the correlation between the reduced order model (ROM)-Gappy-proper orthogonal decomposition (POD) residual $\mathcal{E}$ (13) and the error E (12) on the dual quantities cumulated plasticity p and first component of the stress tensor σ_{11}.

	p	σ_{11}
computation 1	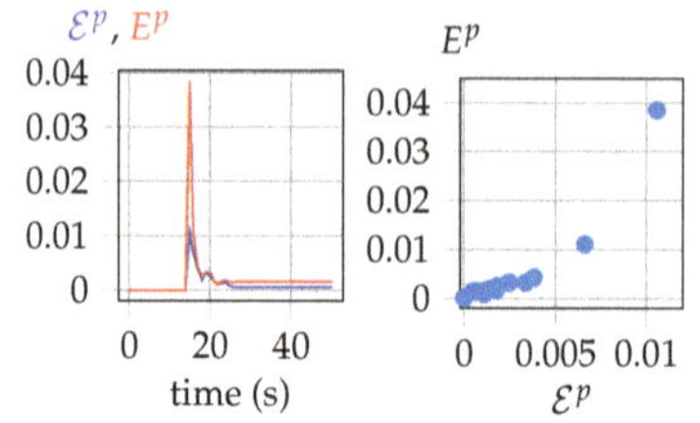	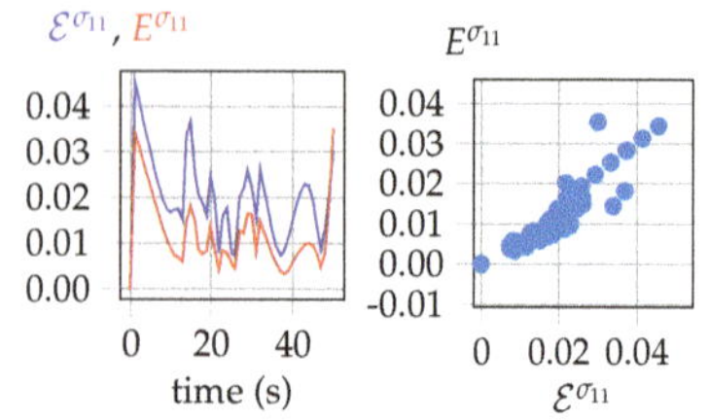
computation 2	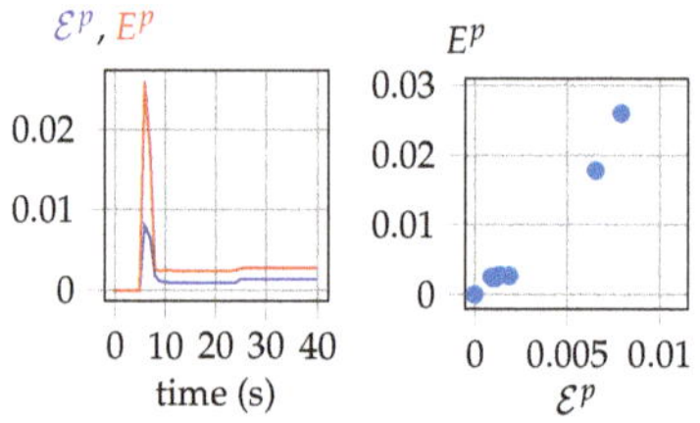	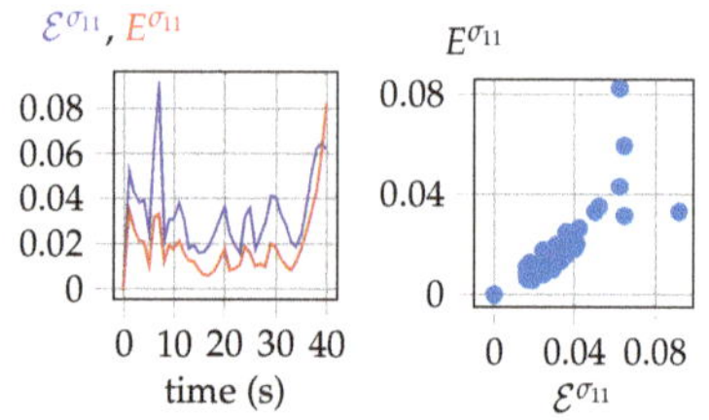

We now illustrate the quality of the error indicator (19), and its ability to increase the accuracy of the reduced order model when used in an enrichment strategy as described in the workflow illustrated in Figure 4. In Tables 3 and 4, we compare the error indicator (19) with the error (12) for various offline and online variabilities respectively without and with enrichment of the reduced order model. Although our error indicator is not a certified upper bound, we observe that thanks to the calibration process, its values are in the vast majority larger than the exact error, except in two regimes: (i) when the errors are very large (the calibration has been carried-out for mild errors, since we used the references from the offline variabilities and enforced reasonable errors in line 3 of Algorithm 5), and (ii) sometimes in the last time steps where the residual stresses build up and where we identified outliers in the Gaussian regressor process. In Table 3, we observe that without enrichment the errors are controlled whenever the online variability is contained in the offline variability. In the other cases, the error becomes very large, and the ROM prediction becomes useless. In Table 4, at the times when the ROM is enriched, both the error indicator and the error are set to zero, since the ROM prediction is replaced by a HF solution. The ROM is enriched when the $\mathrm{Gpr}^p(\mathcal{E}^p) > 0.2$ or $\mathrm{Gpr}^{\sigma_{11}}(\mathcal{E}^{\sigma_{11}}) > 0.2$. We observe that for cases where the online variability is included in the offline variability, the errors are still controlled and no enrichment occurs. In the other cases, the enrichment occurs a few times, so that the errors remain controlled below 0.2. For the online variability new, the ROM is enriched six times for an offline variability computation 1 and only three times for an online variability computation 1 and computation 2; in the latter case, the initial reduced order basis generates a larger base and needs less enrichment.

We now compare the reference HF prediction of the considered online variability with the ROM prediction without and with enrichment, in a case where this online variability is included in the offline variability (Figure 8) and in a case where it is not included (Figure 9). In Figures 8 and 9, dual quantities with index "ref." refers to the HF reference at the considered offline variability, "nores." to the ROM without enrichment and the absence of index to the ROM with enrichment. In the first case, the ROM predictions with and without enrichment are accurate (the magnitude of σ_{11} is small with respect to the ones of σ_{22}, so that the small differences observed in the second plot of Figure 8 are very small with respect to the magnitude of the tensor σ). In the second case, the ROM without enrichment leads to large errors, whereas the enrichment allows a good accuracy. We notice that due to the particular profile of the temperature loading "temperature_field_4" (c.f. Figure 6), the field σ_{11} is irregular. Even in such an unfavorable case, only three enrichment steps by HFM solutions allows a good accuracy for the ROM.

Table 3. Comparison of the error indicator (19) with the error (12) for various offline and online variabilities, without enrichment of the reduced order model. The category "offline" for the columns refers to the variabilities used in the data generation step of the offline stage, whereas the category "online" for the rows refers to the variability considered in the online stage.

Offline / Online	Computation 1		Computation 1 and Computation 2	
computation 1	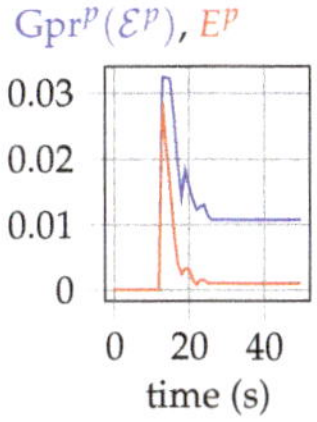	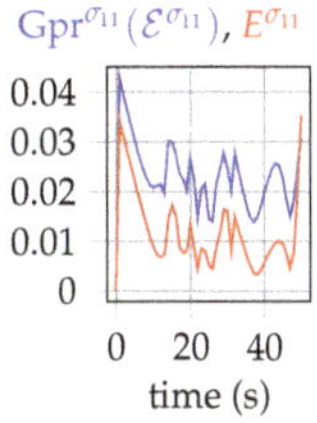	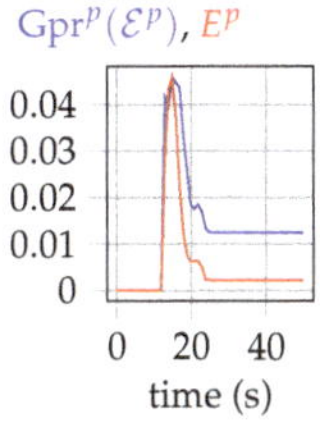	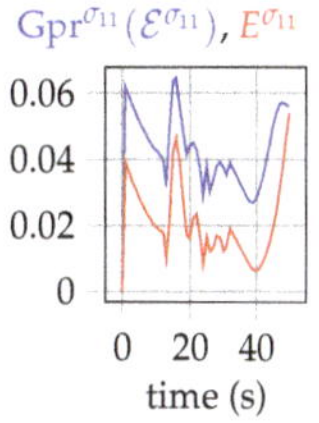
computation 2	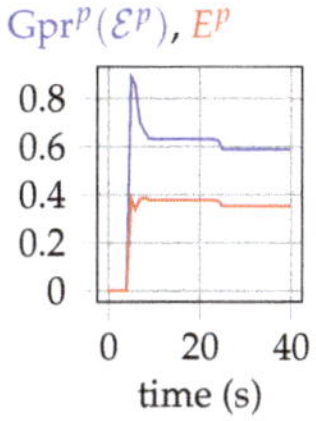	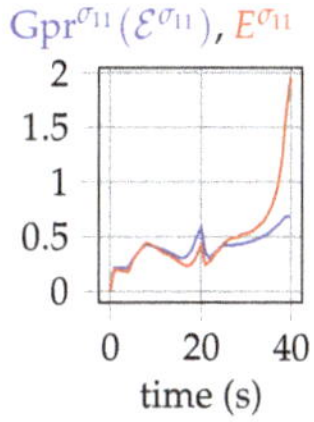	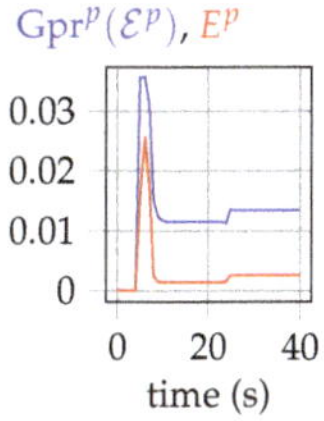	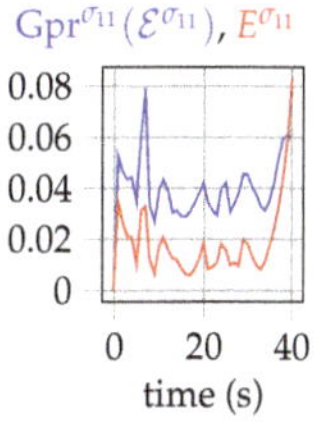
new	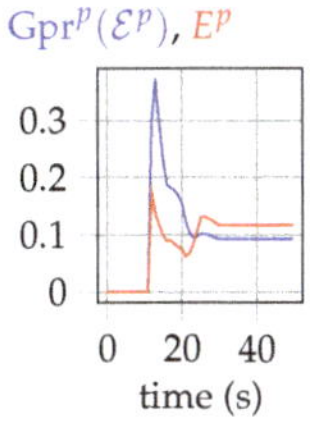	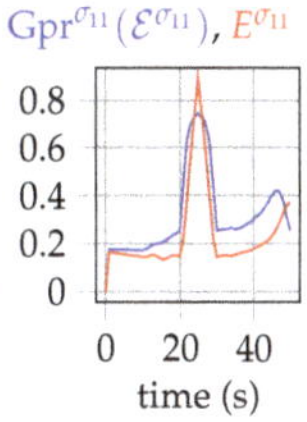	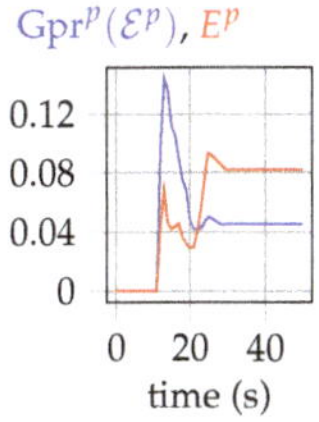	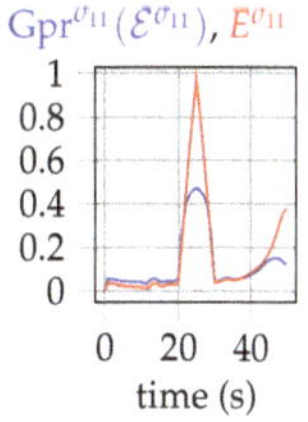

Table 4. Comparison of the error indicator (19) with the error (12) for various offline and online variabilities, with enrichment of the reduced order model.

Offline / Online	Computation 1		Computation 1 and Computation 2	
computation 1	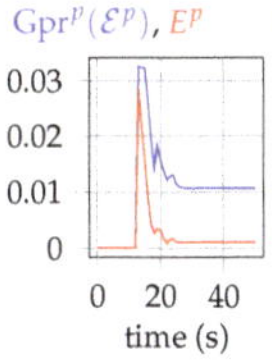	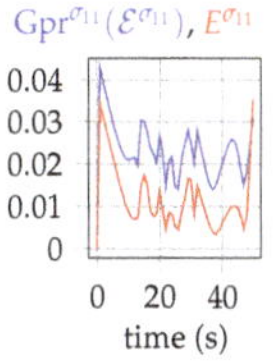	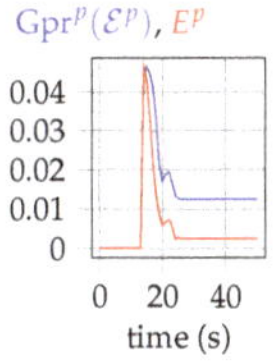	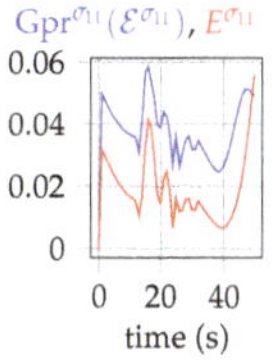

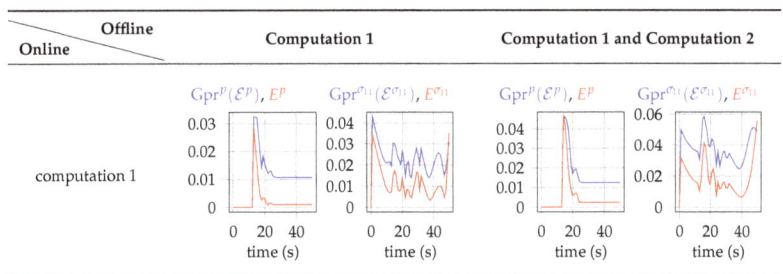

Table 4. *Cont.*

Online \ Offline	Computation 1		Computation 1 and Computation 2	
computation 2	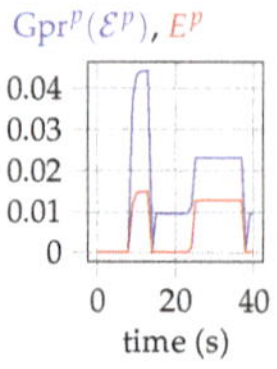	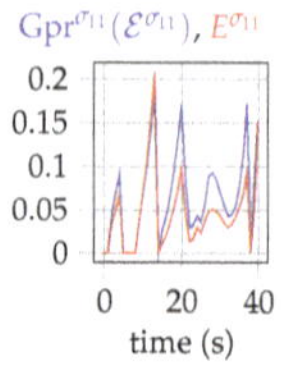	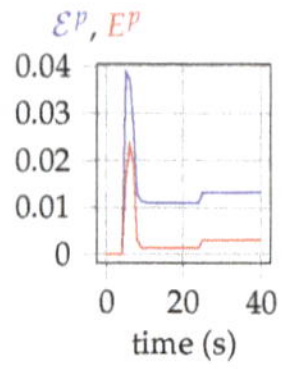	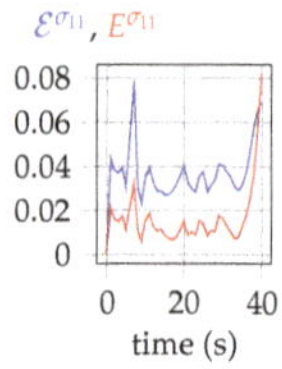
new	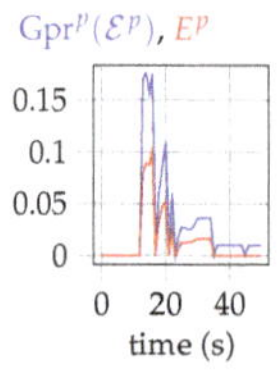	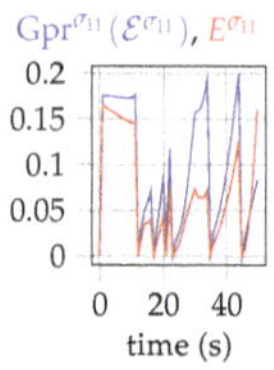	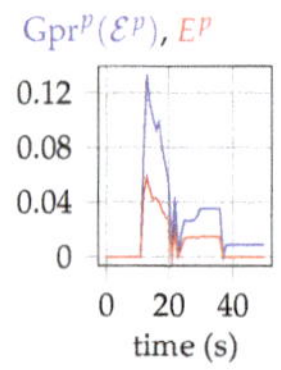	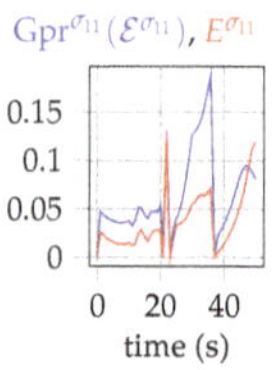

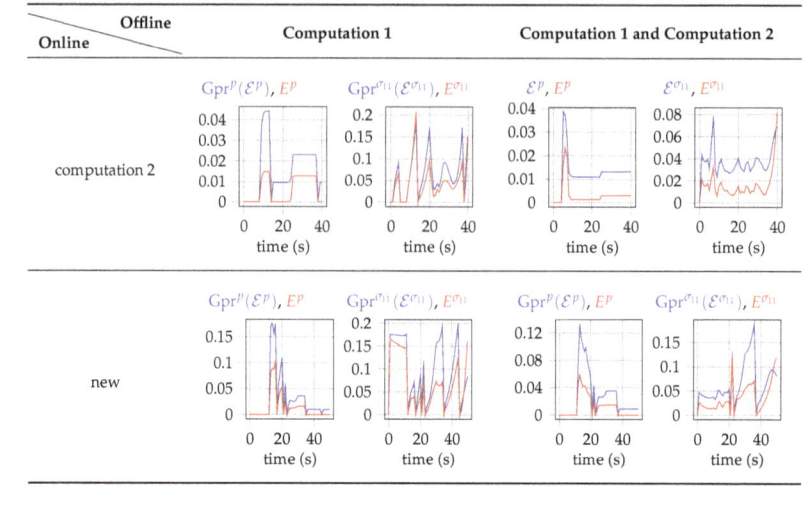

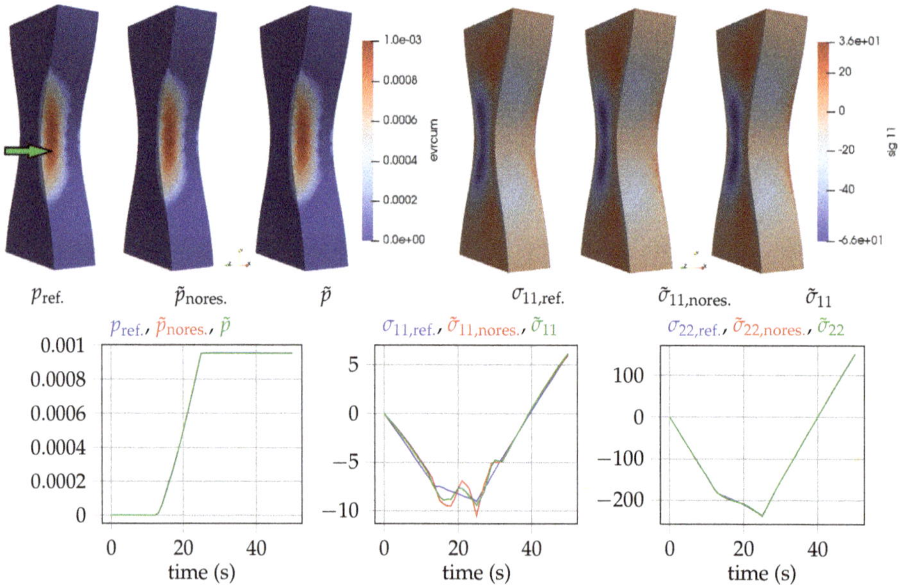

Figure 8. Offline variability: computation 1 and computation 2; online variability: computation 1. (**top**) Representation of dual fields for the reference high-fidelity (HF) prediction of the online variability, the reduced order model (ROM) without enrichment, and the ROM with enrichment ((**left**) p at $t = 50$ s and (**right**) σ_{11} at $t = 25$ s). (**bottom**) Comparison of p, σ_{11} and σ_{22} at the point identified by the green arrow on the top-left picture.

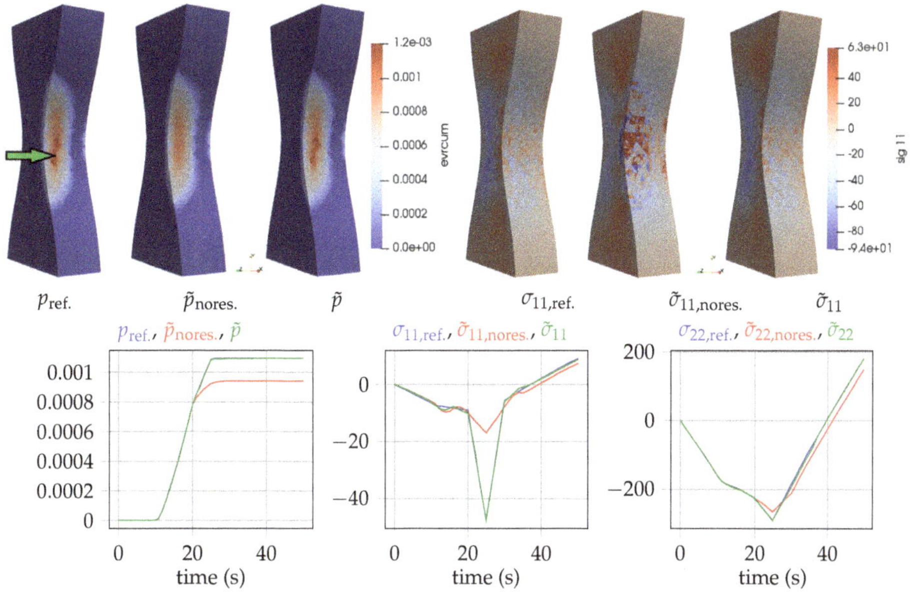

Figure 9. Offline variability: computation 1 and computation 2; online variability: new. (**top**) Representation of dual fields for the reference HF prediction of the online variability, ROM without enrichment, and ROM with enrichment ((**left**) p at $t = 50$ s and (**right**) σ_{11} at $t = 25$ s). (**bottom**) Comparison of p, σ_{11}, and σ_{22} at the point identified by the green arrow on the top-left picture.

5.2. High-Pressure Turbine Blade

We consider a simplified geometry of high-pressure turbine blade, featuring four internal cooling channels, introduced in [7]. The lower part of the blade, referred as the foot, is modeled by an elastic material (we are not interested in predicting the plastic effects in this zone since it does not affect the blade's lifetime) whereas the upper part is modeled by an elastoviscoplastic law. The HFM is computed in parallel using Z-set [33] with an Adaptive MultiPreconditioned FETI solver [47], see Figure 10.

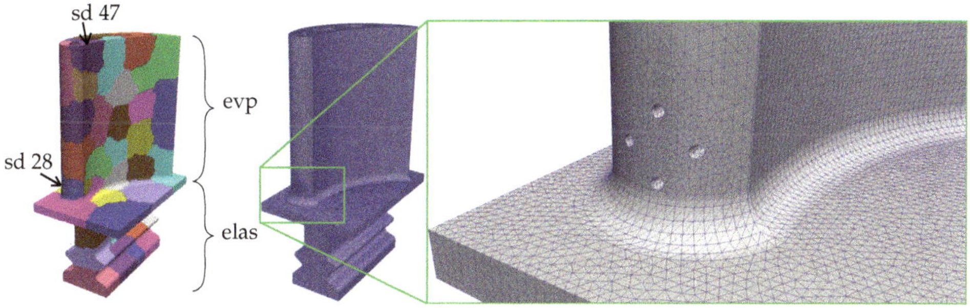

Figure 10. (**left**) Structure split in 48 subdomains—the top part of the blade's material is modeled by an elastoviscoplastic law and the foot's one by an elastic law; (**right**) mesh for the high-pressure turbine blade with a zoom around the cooling channels.

The loading is different from the application of [7] and is represented in Figure 11: 10 temperature fields were considered, the coolest were applied for the lowest rotation speeds, whereas the hottest were applied for the highest rotation speeds. The online variability differs from the offline variability during the three time steps located around the last three maxima of the rotation speed profile, where only the temperature fields changed as indicated by the two pictures at the right side of Figure 11. The maximum of the temperature is moved from the center to the front of the top part of the blade. As we will see, this local modification will lead to large errors for the ROM if no enrichment strategy is considered.

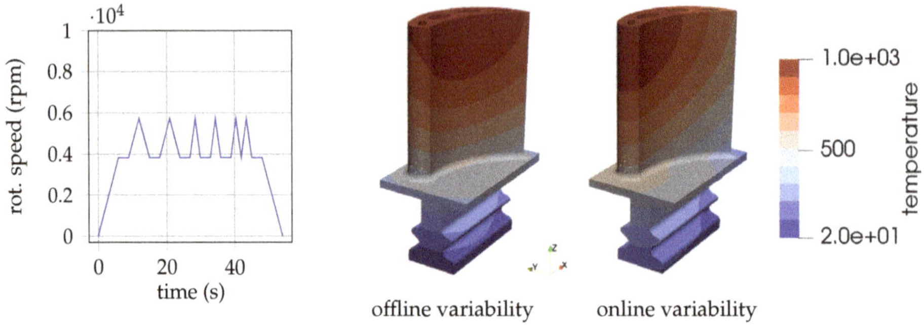

Figure 11. High-pressure turbine test case: (**left**) rotation speed with respect to time; (**right**) representation of maximum temperature fields used in the offline and online computations; the axis of rotation is located below the blade along the x-axis.

The characteristics for the high pressure turbine blade case are given in Table 5.

Table 5. Characteristics for the high-pressure turbine blade test case.

number of dofs	4,892,463
number of (quadratic) tetrahedra	1,136,732
number of integration points	5,683,660
number of time steps	50
behavior law for the foot	elas (temperature-dependent cubic elasticity and isotropic thermal expansion)
behavior law for the blade	evp (Norton flow with nonlinear kinematic hardening)

The computation procedure is presented in Table 6, all steps being computed in parallel with distributed memory, using MPI for the interprocess communications (48 processors within two nodes). The visualization is also parallel with distributed memory using a parallel version of Paraview [48,49].

Table 6. Description of the computational procedure.

Step	**Algorithm**
Data generation	AMPFETI solver in Z-set, $\epsilon^{\text{HFM}}_{\text{Newton}} = 10^{-5}$
Data compression	Distributed Snapshot POD, $\epsilon_{\text{POD}} = 10^{-5}$
Operator compression	Distributed NonNegative Orthogonal Matching Pursuit, $\epsilon_{\text{Op.comp.}} = 10^{-4}$
Reduced order model	$\epsilon^{\text{ROM}}_{\text{Newton}} = 10^{-4}$
Dual quantities reconstruction	Distributed Gappy-POD, $\epsilon_{\text{Gappy-POD}} = 10^{-5}$

The correlations between the ROM-Gappy-POD residual $\mathcal{E}$ (13) and the error E (12) on the dual quantities cumulated plasticity p and stress tensor σ are investigated in Table 7. This time, we carry-out the calibration process independently on each subdomain. The same conclusion as the academic test cases can be drawn for the correlations between the ROM-Gappy-POD residual $\mathcal{E}$ and the error E on the subdomains 28 and 47 (see Figure 10 for the localization of these subdomains).

Table 7. Illustration of the correlation between the ROM-Gappy-POD residual $\mathcal{E}$ (13) and the error E (12) on the dual quantities cumulated plasticity p and a component of the stress tensor σ.

	p	σ_{xx}
subdomain 28	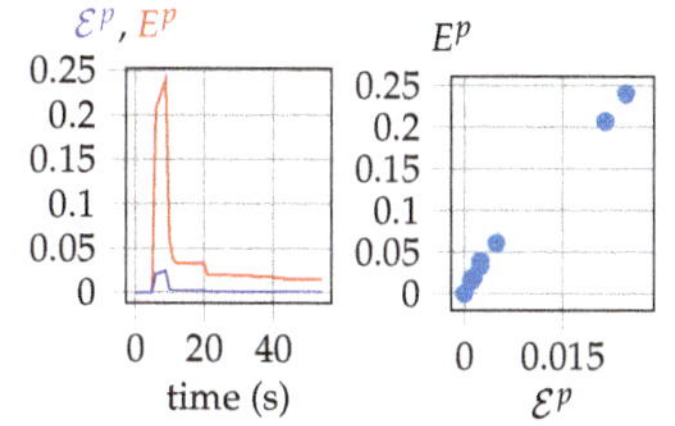	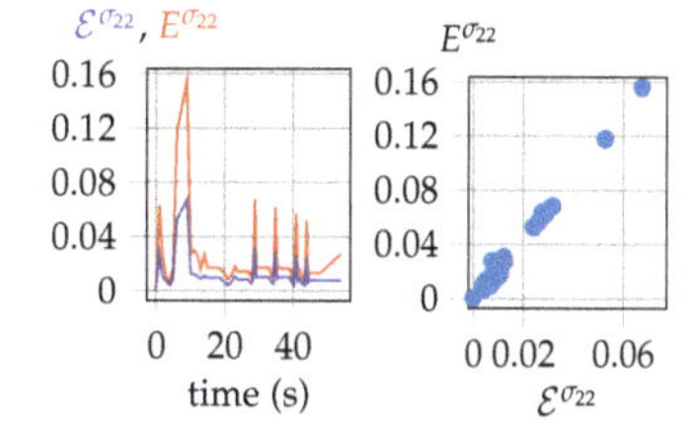
subdomain 47	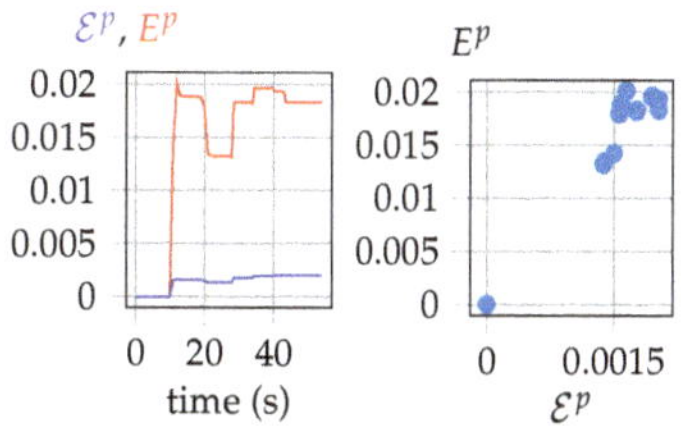	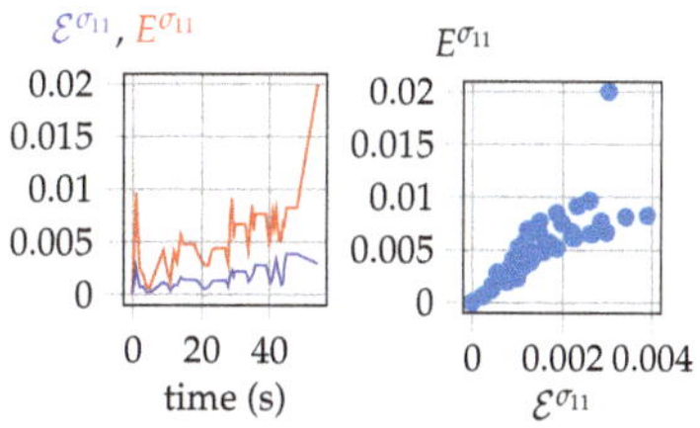

In Table 8, we compare the error indicator (19) with the error (12) for the considered offline and online variabilities. As for the academic test cases, the values of the error indicator are larger than the error except for very large errors (for which the ROM is useless), and sometimes in the last time steps, as residual forces build up. Without enrichment, the ROM makes very large error. We observe that the subdomain for which the enrichment criterion is used enables to control the error on the corresponding subdomain, whereas the error is larger in the other subdomain. This illustrates that local (in space) quantities of interest can be considered to prevent the enrichment steps to occur too often when it's not needed.

Table 8. Comparison of the error indicator (19) with the error (12) for the considered offline and online variabilities. The category "plot" for the columns refers to the subdomain for which the error indicator and the error are plotted, whereas the category "enrichment" for the rows refers to the subdomain of whom the indicator is used to decide the enrichment step.

Plot / Enrichment	Subdomain 28	Subdomain 47
no enrichment	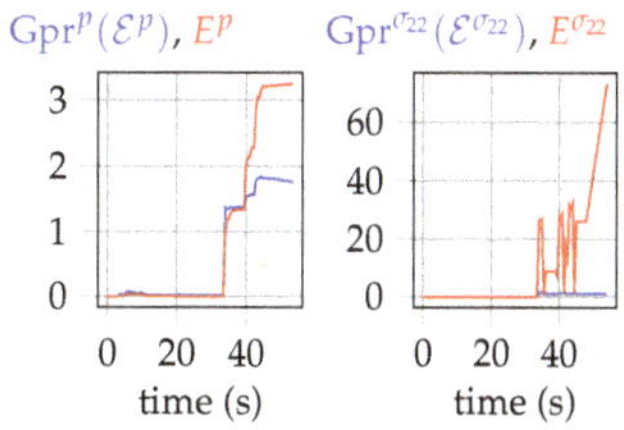	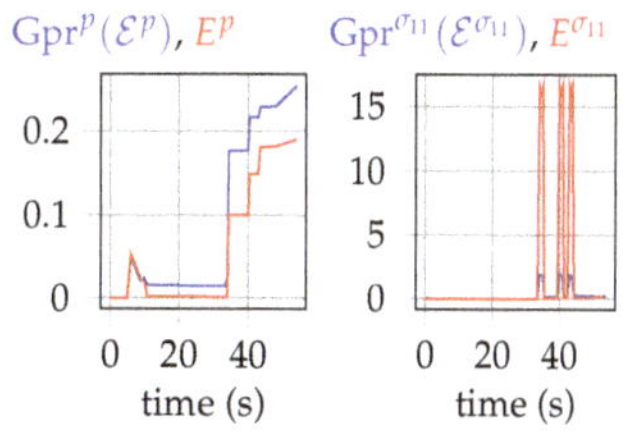

Table 8. *Cont.*

Enrichment \ Plot	Subdomain 28	Subdomain 47
monitoring subdomain 28	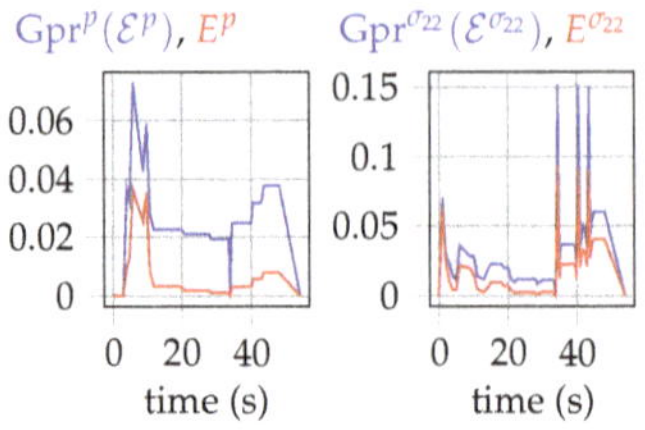	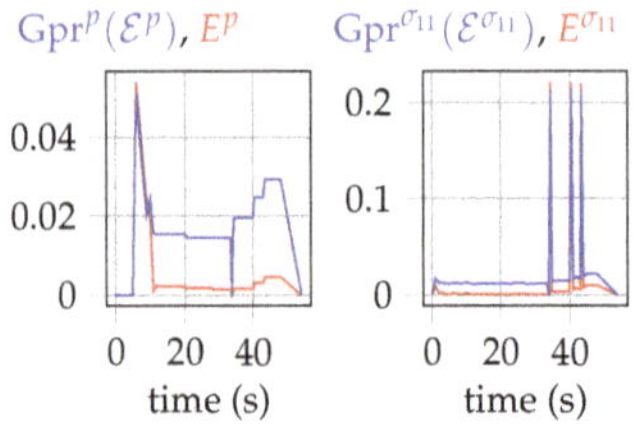
monitoring subdomain 47	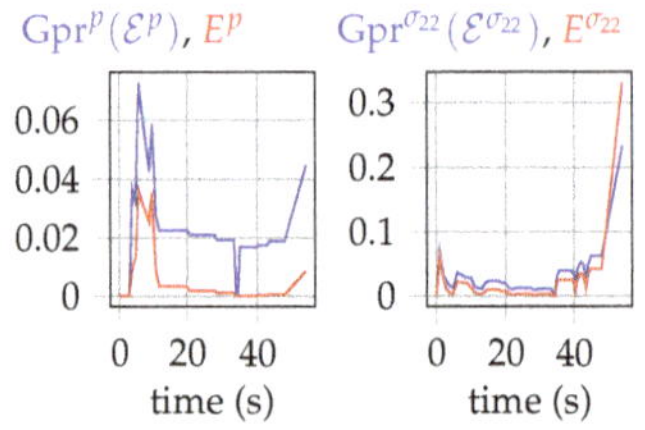	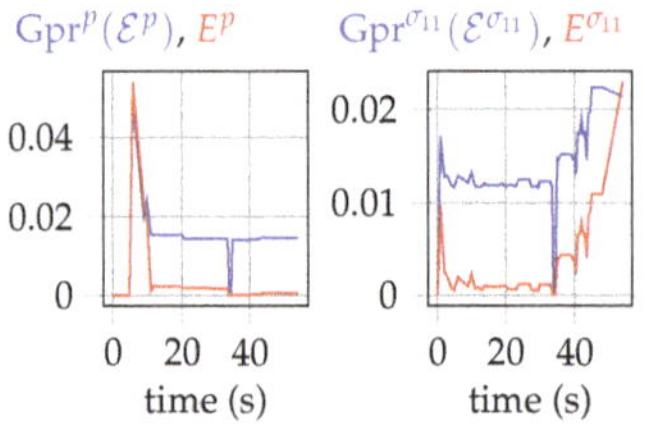

In Figures 12 and 13 are illustrated various predictions of dual quantities: the index "off." refers to the HF prediction for the offline variability, "ref." to the HF reference for the online variability, "nores." to the ROM without enrichment, "sd28" to the ROM with enrichment while monitoring the error indicator on subdomain 28, and "sd47" to the ROM with enrichment while monitoring the error indicator on subdomain 47. We observe that without enrichment, the ROM suffers from large errors. With enrichment, the monitored subdomain enjoys an accurate ROM prediction. Particularly in Figure 13, the conclusions hold when the HF reference for the online variability is visually different from the HF prediction for the offline variability.

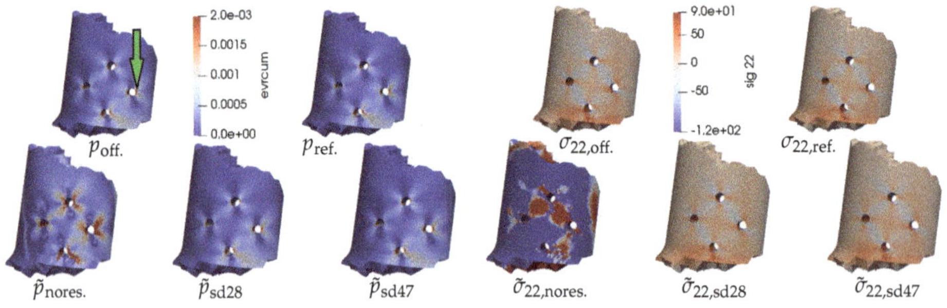

Figure 12. *Cont.*

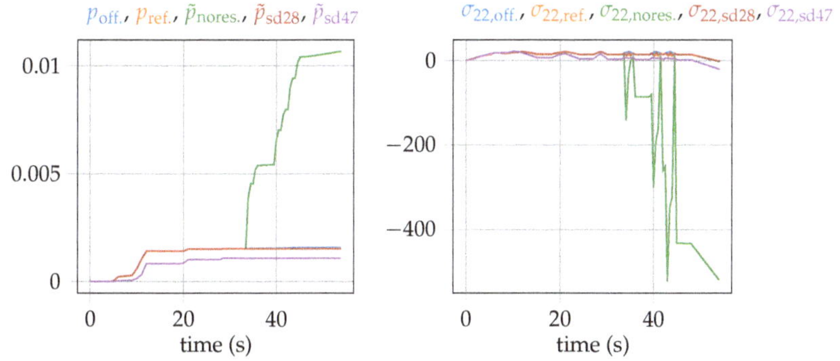

Figure 12. (**top**) Diverse HF and ROM dual quantity fields at t = 43.5 s for subdomain 28, (**left**) p, (**right**) σ_{22}; (**bottom**) comparison at the point identified by the green arrow on the top-left picture. The components of the stress tensor are in MPa.

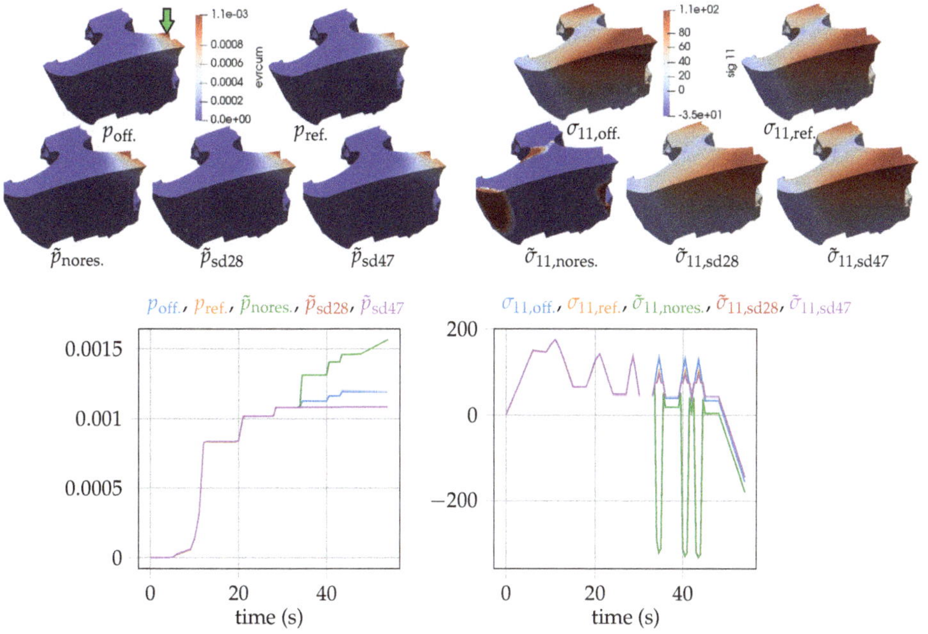

Figure 13. (**top**) Diverse HF and ROM dual quantity fields at t = 43.5 s for subdomain 47, (**left**) p, (**right**) σ_{11}; (**bottom**) comparison at the point identified by the green arrow on the top-left picture. The components of the stress tensor are in MPa.

Finally, we represent various predictions of dual quantities on the complete structure in Figure 14. The ROM without enrichment shows a cumulated plasticity with large errors around the cooling channel, whereas the stress prediction has large errors on the complete structure.

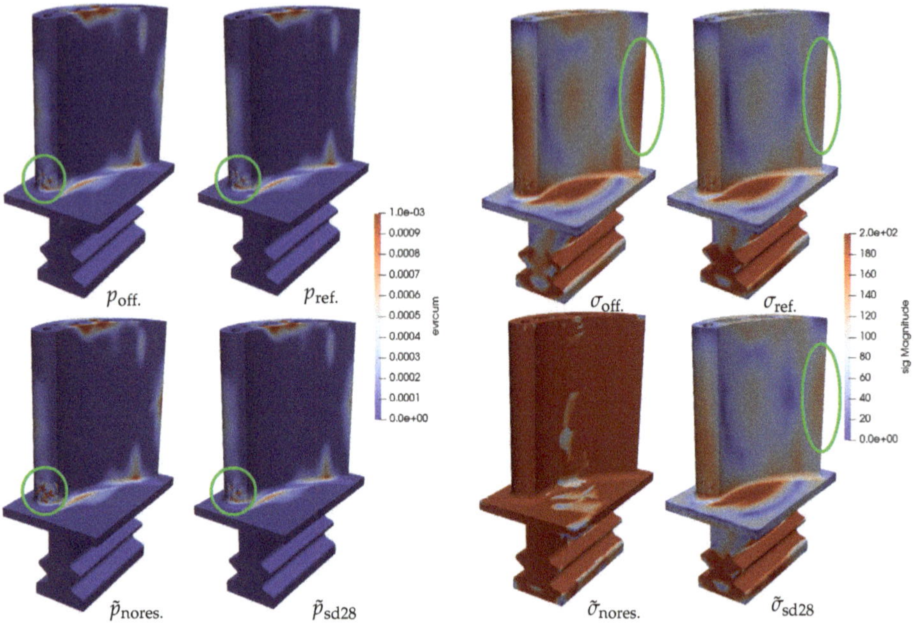

Figure 14. Complete ROM dual quantity fields at $t = 43.5$ s, with enrichment by monitoring subdomain 28. (**left**) Cumulated plasticity; (**right**) magnitude of the stress tensor.

The test cases presented in this section enable us to make two following observations:

[O1] in the a posteriori reduction of elastoviscoplastic computation, online variabilities of the temperature loading not encountered during the offline stage can lead to important errors,

[O2] the ROM-Gappy-POD residual (13) is highly correlated to the error (12), so that the proposed error indicator (19) can be used in the online stage as described in the workflow illustrated in Figure 4 to correct online variabilities of the temperature loading not encountered during the offline stage.

6. Conclusions and Outlook

In this work, we considered the model order reduction of structural mechanics with elastoviscoplastic behavior laws, with dual quantities such as cumulated plasticity and stress tensor as quantities of interest. We observed in our numerical experiments a strong correlation between the ROM-Gappy-POD residual of the reconstruction of these dual quantities and the global error. From this observation, we proposed an efficient error indicator by means of Gaussian process regression from the data acquired when solving the high-fidelity problem in the learning phase of the reduced order modeling. We illustrated the ability of the error indicator to enrich a reduced order model when the online variability cannot be predicted using the current reduced order basis, leading to an accurate reduced prediction.

For the moment, the reduced order model is enriched using a complete reference high-fidelity computation, and the POD and Gappy-POD are recomputed. In future work, we need to consider restart strategies to call the high-fidelity solver only at the time step of enrichment, from a complete mechanical state reconstructed from the prediction of the reduced order model at the previous time step, which can introduce additional errors. We also need to consider incremental strategies for the POD and Gappy-POD updates.

Author Contributions: Conceptualization, F.C.; Data curation, F.C.; Formal analysis, F.C.; Investigation, F.C.; Methodology, F.C. and N.A.; Software, F.C.; Validation, F.C. and N.A.; Visualization, F.C.; Writing—original draft, F.C.; Writing—review & editing, N.A.

Funding: This research was funded by the French Fonds Unique Interministériel (MOR_DICUS).

Conflicts of Interest: The authors declare no conflict of interest.

Abbreviations

The following abbreviations are used in this manuscript:

POD	Proper orthogonal decomposition
HF(M)	high-fidelity (model)
ROM	reduced order model

The following notations are used in this manuscript:

u	high-fidelity displacement field
$\hat{u}$	reduced displacement field
p	high-fidelity cumulated plasticity field
$\tilde{p}$	reduced cumulated plasticity field reconstructed by Gappy-POD
$\boldsymbol{p}$	vector of component the value of the high-fidelity cumulated plasticity field at the reduced integration points
$\hat{\boldsymbol{p}}$	vector of component the cumulated plasticity computed by the behavior law solver at the reduced integration points during the online phase. Notice that this vector is not obtained by taking the value of some field at the reduced integration points.
$\tilde{\boldsymbol{p}}$	vector of component the value of the reduced cumulated plasticity field reconstructed by Gappy-POD at the reduced integration points
E^p	relative error, defined in (12)
$\mathcal{E}^p$	ROM-Gappy-POD residual, defined in (13)
$\text{Gpr}^p\ (\mathcal{E}^p)$	proposed error indicator, defined in (19)
p_{off}	reference high-fidelity cumulated plasticity field at the considered offline variability
p_{ref}	reference high-fidelity cumulated plasticity field at the considered online variability
$\tilde{p}_{\text{nores}}$	reduced cumulated plasticity field reconstructed by Gappy-POD without enrichement (no restart)

The same notations as the ones on the cumulated plasticity are used for all the dual quantities.

References

1. Mazur, Z.; Luna-Ramírez, A.; Juárez-Islas, J.; Campos-Amezcua, A. Failure analysis of a gas turbine blade made of Inconel 738LC alloy. *Eng. Fail. Anal.* **2005**, *12*, 474–486. [CrossRef]
2. Cowles, B.A. High cycle fatigue in aircraft gas turbines—An industry perspective. *Int. J. Fract.* **1996**, *80*, 147–163. [CrossRef]
3. Schulz, U.; Leyens, C.; Fritscher, K.; Peters, M.; Saruhan, B.; Lavigne, O.; Dorvaux, J.M.; Poulain, M.; Mévrel, R.; Caliez, M. Some Recent Trends in Research and Technology of Advanced Thermal Barrier Coatings. *Aerosp. Sci. Technol.* **2003**, *7*, 73–80. [CrossRef]
4. Caron, P.; Lavigne, O. Recent studies at Onera on superalloys for single crystal turbine blades. *AerospaceLab* **2011**, *3*, 1–14.
5. Amaral, S.; Verstraete, T.; Van den Braembussche, R.; Arts, T. Design and Optimization of the Internal Cooling Channels of a High Pressure Turbine Blade—Part I: Methodology. *J. Turbomach.* **2010**, *132*, 021013. [CrossRef]
6. Verstraete, T.; Amaral, S.; Van den Braembussche, R.; Arts, T. Design and Optimization of the Internal Cooling Channels of a High Pressure Turbine Blade—Part II: Optimization. *J. Turbomach.* **2010**, *132*, 021014. [CrossRef]

7. Casenave, F.; Akkari, N.; Bordeu, F.; Rey, C.; Ryckelynck, D. A Nonintrusive Distributed Reduced Order Modeling Framework for Nonlinear Structural Mechanics—Application to Elastoviscoplastic Computations. *arXiv* **2018**, arXiv:1812.07228.
8. File:GaTurbineBlade.svg. Wikipedia, the Free Encyclopedia, Image under the Creative Commons Attribution-Share Alike 3.0 Unported license 2009. Available online: https://commons.wikimedia.org/wiki/File:GaTurbineBlade.svg (accessed on 18 April 2019).
9. Maday, Y.; Patera, A.T.; Turinici, G. A priori convergence theory for reduced-basis approximations of single-parameter elliptic partial differential equations. *J. Sci. Comput.* **2002**, *17*, 437–446. [CrossRef]
10. Machiels, L.; Maday, Y.; Oliveira, I.B.; Patera, A.T.; Rovas, D.V. Output bounds for reduced-basis approximations of symmetric positive definite eigenvalue problems. *C. R. Acad. Sci. Ser. I Math.* **2000**, *331*, 153–158. [CrossRef]
11. Maday, Y.; Patera, A.; Turinici, G. Global a priori convergence theory for reduced-basis approximations of single-parameter symmetric coercive elliptic partial differential equations. *C. R. Acad. Sci. Ser. I Math.* **2002**, *335*, 289–294. [CrossRef]
12. Yano, M. A Space-Time Petrov–Galerkin Certified Reduced Basis Method: Application to the Boussinesq Equations. *SIAM J. Sci. Comput.* **2014**, *36*, A232–A266. [CrossRef]
13. Ohlberger, M.; Rave, S. Nonlinear reduced basis approximation of parameterized evolution equations via the method of freezing. *C. R. Math.* **2013**, *351*, 901–906. [CrossRef]
14. Manzoni, A. An efficient computational framework for reduced basis approximation and a posteriori error estimation of parametrized Navier–Stokes flows. *ESAIM Math. Model. Numer. Anal.* **2014**, *48*, 1199–1226. [CrossRef]
15. Casenave, F. Accurate a posteriori error evaluation in the reduced basis method. *C. R. Math.* **2012**, *350*, 539–542. [CrossRef]
16. Casenave, F.; Ern, A.; Lelièvre, T. Accurate and online-efficient evaluation of the a posteriori error bound in the reduced basis method. *ESAIM Math. Model. Numer. Anal.* **2014**, *48*, 207–229. [CrossRef]
17. Buhr, A.; Engwer, C.; Ohlberger, M.; Rave, S. A numerically stable a posteriori error estimator for reduced basis approximations of elliptic equations. In Proceedings of the 11th World Congress on Computational Mechanics, WCCM 2014, 5th European Conference on Computational Mechanics, ECCM 2014 and 6th European Conference on Computational Fluid Dynamics, ECFD 2014, Barcelona, Spain, 20–25 July 2014; pp. 4094–4102.
18. Chen, Y.; Jiang, J.; Narayan, A. A robust error estimator and a residual-free error indicator for reduced basis methods. *Comput. Math. Appl.* **2019**, *77*, 1963–1979. [CrossRef]
19. Chinesta, F.; Leygue, A.; Bordeu, F.; Aguado, J.V.; Cueto, E.; González, D.; Alfaro, I.; Ammar, A.; Huerta, A. PGD-based computational vademecum for efficient design, optimization and control. *Arch. Comput. Methods Eng.* **2013**, *20*, 31–59. [CrossRef]
20. Ladevèze, P.; Chouaki, A. Application of a posteriori error estimation for structural model updating. *Inverse Probl.* **1999**, *15*, 49. [CrossRef]
21. Ladevèze, P.; Chamoin, L. Toward guaranteed PGD-reduced models. In *Bytes and Science*; CIMNE: Barcelona, Spain, 2013; pp. 143–154.
22. Chamoin, L.; Pled, F.; Allier, P.E.; Ladevèze, P. A posteriori error estimation and adaptive strategy for PGD model reduction applied to parametrized linear parabolic problems. *Comput. Methods Appl. Mech. Eng.* **2017**, *327*, 118–146. [CrossRef]
23. Chatterjee, A. An introduction to the proper orthogonal decomposition. *Curr. Sci.* **2000**, *78*, 808–817.
24. Sirovich, L. Turbulence and the dynamics of coherent structures, Parts I, II and III. *Q. Appl. Math.* **1987**, *XLV*, 561–590. [CrossRef]
25. Tröltzsch, F.; Volkwein, S. POD a-posteriori error estimates for linear-quadratic optimal control problems. *Comput. Optim. Appl.* **2009**, *44*, 83. [CrossRef]
26. Luo, Z.; Zhu, J.; Wang, R.; Navon, I.M. Proper orthogonal decomposition approach and error estimation of mixed finite element methods for the tropical Pacific Ocean reduced gravity model. *Comput. Methods Appl. Mech. Eng.* **2007**, *196*, 4184–4195. [CrossRef]
27. Kammann, E.; Tröltzsch, F.; Volkwein, S. A Method of a-Posteriori Error Estimation with Application to Proper Orthogonal Decomposition. Available online: https://pdfs.semanticscholar.org/7212/a310a9c0874d6e069e77b5f97aeb3f57f4df.pdf (accessed on 16 April 2019).

28. Henneron, T.; Mac, H.; Clenet, S. Error estimation of a proper orthogonal decomposition reduced model of a permanent magnet synchronous machine. In Proceedings of the 9th IET International Conference on Computation in Electromagnetics (CEM 2014), London, UK, 31 March–1 April 2014; pp. 1–6.
29. Wang, A.; Ma, Y. An error estimate of the proper orthogonal decomposition in model reduction and data compression. *Numer. Methods Part. Differ. Equ.* **2009**, *25*, 972–989. [CrossRef]
30. Ryckelynck, D.; Gallimard, L.; Jules, S. Estimation of the validity domain of hyper-reduction approximations in generalized standard elastoviscoplasticity. *Adv. Model. Simul. Eng. Sci.* **2015**, *2*, 6. [CrossRef]
31. Ryckelynck, D. Estimation d'erreur d'hyperréduction de problèmes élastoviscoplastiques. In Proceedings of the 21ème Congrès Français de Mécanique, 2013, Bordeaux, France, 26–30 August 2013.
32. Akkari, N.; Hamdouni, A.; Liberge, E.; Jazar, M. On the sensitivity of the POD technique for a parameterized quasi-nonlinear parabolic equation. *Adv. Model. Simul. Eng. Sci.* **2014**, *1*, 1–14. [CrossRef]
33. Mines ParisTech and ONERA the French Aerospace lab. Z-set: Nonlinear Material & Structure Analysis Suite. 1981–Present. Available online: http://www.zset-software.com (accessed on 16 April 2019).
34. Hernandez, J.A.; Caicedo, M.A.; Ferrer, A. Dimensional hyper-reduction of nonlinear finite element models via empirical cubature. *Comput. Methods Appl. Mech. Eng.* **2017**, *313*, 687–722. [CrossRef]
35. Farhat, C.; Avery, P.; Chapman, T.; Cortial, J. Dimensional reduction of nonlinear finite element dynamic models with finite rotations and energy-based mesh sampling and weighting for computational efficiency. *Int. J. Numer. Methods Eng.* **2014**, *98*, 625–662. [CrossRef]
36. Farhat, C.; Chapman, T.; Avery, P. Structure-preserving, stability, and accuracy properties of the energy-conserving sampling and weighting method for the hyper reduction of nonlinear finite element dynamic models. *Int. J. Numer. Methods Eng.* **2015**, *102*, 1077–1110. [CrossRef]
37. Paul-Dubois-Taine, A.; Amsallem, D. An adaptive and efficient greedy procedure for the optimal training of parametric reduced-order models. *Int. J. Numer. Methods Eng.* **2015**, *102*, 1262–1292. [CrossRef]
38. Amaldi, E.; Kann, V. On the approximability of minimizing nonzero variables or unsatisfied relations in linear systems. *Theor. Comput. Sci.* **1998**, *209*, 237–260. [CrossRef]
39. Yaghoobi, M.; Wu, D.; Davies, M.E. Fast non-negative orthogonal matching pursuit. *IEEE Signal Process. Lett.* **2015**, *22*, 1229–1233. [CrossRef]
40. Mallat, S.G.; Zhang, Z. Matching pursuits with time-frequency dictionaries. *IEEE Trans. Signal Process.* **1993**, *41*, 3397–3415. [CrossRef]
41. Everson, R.; Sirovich, L. Karhunen–Loève procedure for gappy data. *J. Opt. Soc. Am. A* **1995**, *12*, 1657–1664. [CrossRef]
42. Barrault, M.; Maday, Y.; Nguyen, N.C.; Patera, A.T. An 'empirical interpolation' method: Application to efficient reduced-basis discretization of partial differential equations. *C. R. Math.* **2004**, *339*, 667 – 672. [CrossRef]
43. Maday, Y.; Nguyen, N.C.; Patera, A.T.; Pau, S.H. A general multipurpose interpolation procedure: The magic points. *Commun. Pure Appl. Anal.* **2009**, *8*, 383–404. [CrossRef]
44. Maday, Y.; Mula, O.; Turinici, G. Convergence analysis of the Generalized Empirical Interpolation Method. *SIAM J. Numer. Anal.* **2016**, *54*, 1713–1731. [CrossRef]
45. Pedregosa, F.; Varoquaux, G.; Gramfort, A.; Michel, V.; Thirion, B.; Grisel, O.; Blondel, M.; Prettenhofer, P.; Weiss, R.; Dubourg, V.; et al. Scikit-learn: Machine Learning in Python. *J. Mach. Learn. Res.* **2011**, *12*, 2825–2830.
46. Ryckelynck, D.; Chinesta, F.; Cueto, E.; Ammar, A. On thea priori model reduction: Overview and recent developments. *Arch. Comput. Methods Eng.* **2006**, *13*, 91–128. [CrossRef]
47. Bovet, C.; Parret-Fréaud, A.; Spillane, N.; Gosselet, P. Adaptive multipreconditioned FETI: Scalability results and robustness assessment. *Comput. Struct.* **2017**, *193*, 1 – 20. [CrossRef]
48. Ahrens, J.; Geveci, B.; Law, C. *ParaView: An End-User Tool for Large Data Visualization, Visualization Handbook*; Elsevier: Amsterdam, The Netherlands, 2005.
49. Ayachit, U. *The ParaView Guide: A Parallel Visualization Application*; Kitware: Clifton Park, NY, USA, 2015.

MDPI
St. Alban-Anlage 66
4052 Basel
Switzerland
Tel. +41 61 683 77 34
Fax +41 61 302 89 18
www.mdpi.com

Mathematical and Computational Applications Editorial Office
E-mail: mca@mdpi.com
www.mdpi.com/journal/mca

www.ingramcontent.com/pod-product-compliance
Lightning Source LLC
LaVergne TN
LVHW071942080526
838202LV00064B/6650

* 9 7 8 3 0 3 9 2 1 4 0 9 9 *